# *Mass Spectrometry*

*Clinical and Biomedical Applications*
*Volume 2*

## MODERN ANALYTICAL CHEMISTRY

Series Editor: David M. Hercules
*University of Pittsburgh*

---

ADVANCES IN COAL SPECTROSCOPY
Edited by Henk L. C. Meuzelaar

APPLIED ATOMIC SPECTROSCOPY
Volumes 1 and 2
Edited by E. L. Grove

CHEMICAL DERIVATIZATION IN ANALYTICAL CHEMISTRY
Edited by R. W. Frei and J. F. Lawrence
Volume 1: Chromatography
Volume 2: Separation and Continuous Flow Techniques

COMPUTER-ENHANCED ANALYTICAL SPECTROSCOPY
Volume 1: Edited by Henk L. C. Meuzelaar and Thomas L. Isenhour
Volume 2: Edited by Henk L. C. Meuzelaar
Volume 3: Edited by Peter C. Jurs
Volume 4: Edited by Charles L. Wilkins

ION CHROMATOGRAPHY
Hamish Small

ION-SELECTIVE ELECTRODES IN ANALYTICAL CHEMISTRY
Volumes 1 and 2
Edited by Henry Freiser

LIQUID CHROMATOGRAPHY/MASS SPECTROMETRY
Techniques and Applications
Alfred L. Yergey, Charles G. Edmonds, Ivor A. S. Lewis, and Marvin L. Vestal

MASS SPECTROMETRY
Clinical and Biomedical Applications
Volumes 1 and 2
Edited by Dominic M. Desiderio

MODERN FLUORESCENCE SPECTROSCOPY
Volumes 1–4
Edited by E. L. Wehry

PRINCIPLES OF CHEMICAL SENSORS
Jiří Janata

TRANSFORM TECHNIQUES IN CHEMISTRY
Edited by Peter R. Griffiths

---

# *Mass Spectrometry*

## *Clinical and Biomedical Applications*
## *Volume 2*

*Edited by*

***Dominic M. Desiderio***

***Departments of Neurology and Biochemistry, and***
***The Charles B. Stout Neuroscience Mass Spectrometry Laboratory***
***The University of Tennessee, Memphis***
***Memphis, Tennessee***

***Springer Science+Business Media, LLC***

Library of Congress Cataloging in Publication Data

(Revised for vol. 2)

Mass spectrometry.

(Modern analytical chemistry) (The Language of science)
Includes bibliographical references and index.
Contents: v. 1–2. Clinical and biomedical applications
1. Mass spectrometry. 2. Biomolecules—Analysis. I. Desiderio, Dominic M. II. Series.
[DNLM: 1. Chemistry, Clinical—instrumentation. 2. Chemistry, Clinical—Methods. 2. Spectrum Analysis, Mass. QC 454.M3 M41421]
QP519.9.M3M352 1992 616.07'56 92-49302
ISBN 978-1-4899-1750-8

ISBN 978-1-4899-1750-8 ISBN 978-1-4899-1748-5 (eBook)
DOI 10.1007/978-1-4899-1748-5

*To*

*Jay, Annette, Rich, Anna,*
*Dominic, Kimberly, and Kirsten*

# Contributors

*Per E. Andrén,* Department of Biochemistry and Molecular Biology, and Analytical Chemistry Center, University of Texas Medical School, Houston, Texas 77225

*Richard M. Caprioli,* Department of Biochemistry and Molecular Biology, and Analytical Chemistry Center, University of Texas Medical School, Houston, Texas 77225

*Chhabil Dass,* Charles B. Stout Neuroscience Mass Spectrometry Laboratory, and Department of Neurology, University of Tennessee–Memphis, Memphis, Tennessee 38163

*Douglas A. Gage,* MSU–NIH Mass Spectrometry Facility, Department of Biochemistry, Michigan State University, East Lansing, Michigan 48824

*Zhi-Heng Huang,* MSU–NIH Mass Spectroscopy Facility, Department of Biochemistry, Michigan State University, East Lansing, Michigan 48824

*Shen-Nan Lin,* Department of Biochemistry and Molecular Biology, and Analytical Chemistry Center, University of Texas Medical School, Houston, Texas 77225

*Thomas D. McClure,* Department of Pharmacology/Toxicology, College of Pharmacy, University of Arizona, Tucson, Arizona 85721

*Curt B. Norwood,* EPA/ERL, Narragansett, Rhode Island 02882

*Karl H. Schram,* Department of Pharmacology/Toxicology, College of Pharmacy, University of Arizona, Tucson, Arizona 85721

*Cedric H. L. Shackleton,* Children's Hospital Oakland Research Institute, Oakland, California 94609

*Charles C. Sweeley,* MSU–NIH Mass Spectroscopy Facility, Department of Biochemistry, Michigan State University, East Lansing, Michigan 48824

*H. Ewa Witkowska,* Children's Hospital Oakland Research Institute, Oakland, California 94609

*Paul Vouros,* Department of Chemistry and Barnett Institute, Northeastern University, Boston, Massachusetts 02115

# *Preface*

– –  •–  •••  •••

•••  •– –•  •  •–•–  –  •–•  – – –  – –  •  –  •–•  –•– –

*M-A-S-S S-P-E-C-T-R-O-M-E-T-R-Y in Morse code*

This volume collects descriptions of selected recent developments in state-of-the-art mass spectrometric methods and reflects the broad-based approaches that mass spectroscopists apply to a variety of important clinical and biomedical problems. One chapter reviews current mass-spectrometric instrumentation and techniques, and other chapters describe the use of mass-spectrometric methods for the analysis of diacylglycerylphospholipids; modifications to DNA molecules; the characterization of variant hemoglobins; and characterization of urinary nucleosides. The final chapter describes the new technique of combined microdialysis/mass spectrometry.

This volume represents the collected efforts of several highly productive researchers who have developed new methods and instrumentation and have applied them to current research problems, such as lipid storage diseases, cancer, hemoglobinopathies, and brain neurochemistry. The chapters in Volumes 1 and 2 define the outlines of clinical and biomedical mass spectrometry and attest to the flexibility and creativity of mass spectroscopists and their interaction with biologic and clinical scientists.

The authors in this volume are to be congratulated for their writing efforts, their scientific vigor and rigor, their intellectual contributions, and the experimental details that are described in these chapters.

I thank each author for collaborating with me on the production of this volume, and I hope these chapters will help the practitioners of, and the newcomers to, the field of mass spectrometry.

Dominic M. Desiderio

*Memphis, Tennessee*

# Contents

*Chapter 2*

*Characterization of Diacylglycerylphospholipids by Fast Atom Bombardment Tandem Mass Spectrometry*

*Douglas A. Gage, Zhi-Heng Huang, and Charles C. Sweeley*



*Chapter 3*

*DNA Modifications: Investigations by Mass Spectrometry*

*Curt B. Norwood and Paul Vouros*

# 1

# *Mass Spectrometry*
## *Instrumentation and Techniques*

*Chhabil Dass*

## *1. INTRODUCTION*

Mass spectrometry (MS) was first used approximately 80 years ago by Sir J. J. Thomson for analysis of positive rays (1). Since then, it has undergone extensive innovations. The last decade has seen an especially rapid pace of development in both instrumentation and ionization techniques. MS is now probably our most versatile and comprehensive analytical tool. It has found applications in several areas of physics, chemistry, biology, medicine, geology, nuclear science, and environmental science. It is used routinely to obtain relative molecular weights ($M_r$) and structural information; to quantify at trace levels; to study ion chemistry and ion–molecule reactions dynamics; to provide data on physical properties such as ionizing energy, appearance energy, the enthalpy of a reaction, and proton affinities; and to verify theoretical predictions based on molecular orbital calculations.

Recent advances—the development of unique soft ionization techniques suitable for production of ions from nonpolar compounds, compounds of large $M_r$, and compounds of biological origin; the introduction of high-field and superfast magnets; the resurgence of time-of-flight (TOF)-MS; the emergence

---

*Chhabil Dass* • Charles B. Stout Neuroscience Mass Spectrometry Laboratory, and Department of Neurology, University of Tennessee–Memphis, Memphis, Tennessee 38163.

*Mass Spectrometry: Clinical and Biomedical Applications, Volume 2*, edited by Dominic M. Desiderio. Plenum Press, New York, 1994.

of tandem MS (MS/MS); the interfacing of high-resolution chromatographic techniques with MS; and much-improved detection devices and fast data systems—have all elevated MS to a level where it has become an indispensable weapon in the arsenal of biomedical research. This distinction is the result of the capability of MS to provide unsurpassed molecular specificity, high detection sensitivity, and unparalleled versatility in determining structures of unknown compounds. Molecular specificity results from the ability of MS to provide $M_r$ and structurally diagnostic fragment ions. The advent of electrospray ionization (ESI) (2–4) and matrix-assisted laser desorption ionization (MALDI) (5,6) methods have extended the upper mass range amenable to MS. Currently, several biopolymers with molecular mass over 100 kDa (kilodaltons) are analyzed routinely. Detection sensitivity in the femtomole-to-attomole range has been reported for quantification of neuropeptides (7).

In the past, the exorbitant cost of state-of-the-art mass spectrometers restricted MS to the domain of a very few select laboratories; for example, high-performance four-sector tandem mass spectrometers are still out of the reach of most researchers. With the introduction of reasonably priced gas chromatography (GC)-MS, high-performance liquid chromatography (HPLC)-MS, ESI-MS, and MALDI-MS commercial instruments, MS has become a much more accessible and essential component of any contemporary research laboratory. Almost every chemical and biotechnical industrial establishment in the United States has some sort of MS facility.

To perform an MS analysis, the analyte must first be converted to a gas-phase ionic species because it is easier experimentally to manipulate the motion of and to detect an ion than a neutral species. The excess energy transferred to the molecule during ionization leads to fragmentation. A mass analyzer separates ions according to their mass-to-charge ratio ($m/z$). A mass spectrum is a display of the $m/z$ of the ions and their relative abundances. In order that the ions move freely in space without colliding or interacting with other species, each of these steps is carried out under high vacuum.

Thus, essential components (shown in Fig. 1.1) of a mass spectrometer are the inlet system, to transfer a sample to the ion source; the vacuum system, to maintain very low pressure in the mass spectrometer; the ion source, to ionize neutral sample molecules; the mass analyzer, to provide $m/z$ information; the detector, to measure relative abundances of the mass-resolved ions; the electronics, to control the operation of various units; and the data system, to record, process, and store the data. The overall analytical capability of an MS system depends on the combined performance of these individual components.

The objective of this chapter is to familiarize nonpractitioners of MS with its basic equipment and experimental techniques. An account of various ionization methods, mass analyzers, tandem mass spectrometry, ion detectors, and some specialized techniques is presented.

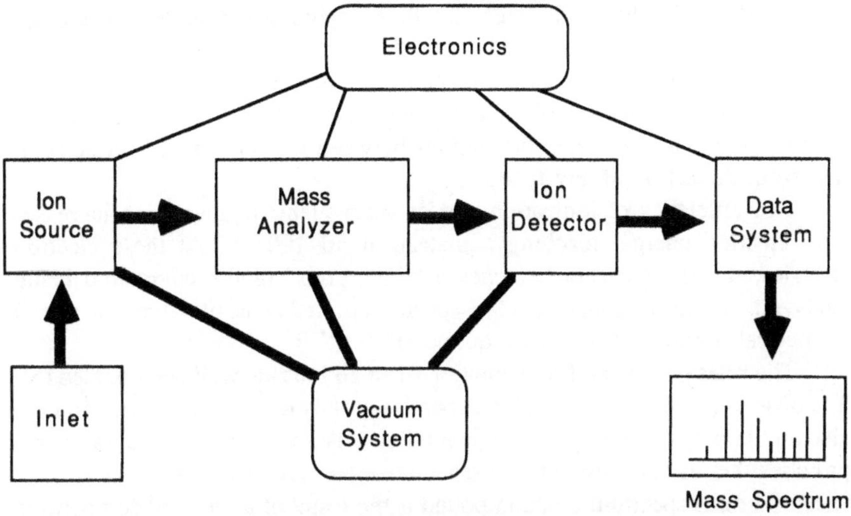


*Figure 1.1.* Schematic of a mass spectrometer.

## 2. *IONIZATION METHODS*

### *2.1. Electron Ionization*

Much of the success of an MS experiment depends on the way a neutral compound is transferred into the gas phase as an ionic species. Historically, electron ionization (EI) has been the most frequently used mode of ion formation. In this mode, a gaseous sample is bombarded with a beam of energetic electrons. This collision process displaces an electron from the molecule (M), converting it to a positive ion with an odd number of electrons, called a molecular ion ($M^{+\cdot}$) or a radical cation (Eq. 1.1):

$$M + e^{-} \rightarrow M^{+\cdot} + 2e^{-}. \quad (1.1)$$

In order for this process to occur, the energy of the bombarding electron must be greater than the ionization energy (IE) of the molecule. The energy in excess of IE can cause an ion to rearrange to its isomeric structure (e.g., M′ to M″ in Fig. 1.2) or allow it to fragment into structurally diagnostic fragments. Depending on the internal energy content $\varepsilon$, ions can be classified as stable (i.e., they have $\varepsilon$ below the isomerization barrier and retain their original structure), non decomposing (i.e., they have $\varepsilon$ below the threshold for first dissociation and reach the detector intact), and decomposing (i.e., they have $\varepsilon$ above the threshold for first dissociation and fragment before reaching the detector). Most decomposing ions have lifetimes less than 1 μsec and fragment in the ion source (see Fig. 1.2). Metastable ions are decomposing ions that fragment after leaving the ion source but before reaching the detector. For MS/MS applications, reactions of metastable and nondecomposing ions are studied in the field-free regions (FFR; i.e., the regions between the ion source and detector free from the action of any field).

The efficiency of ionization and of subsequent fragmentation increases with electron energy, reaching a plateau at 50–100 eV. At these electron energies, the EI spectrum becomes a "fingerprint" of the compound being analyzed. For this reason, most mass spectra reported in the literature and stored in spectral libraries have been acquired at 70 eV.

The rules governing fragmentation of small ions are well documented (8). The observed abundance of an ion depends not only on its rate of formation but also on the rates of subsequent fragmentations. The stability of those fragments provides the driving force for the reaction to proceed. Thus, the final appearance of a mass spectrum of a compound is the result of a series of competitive and consecutive unimolecular reactions, which have been described in terms of quasiequilibrium theory (9) and RRKM (Rice–Ramsperger–Kassel–Marcus) theory (10). The detailed discussion of these theories is beyond the scope of this chapter. However, in its simpler form, the rate constant $k$ of a fragmentation reaction is given by

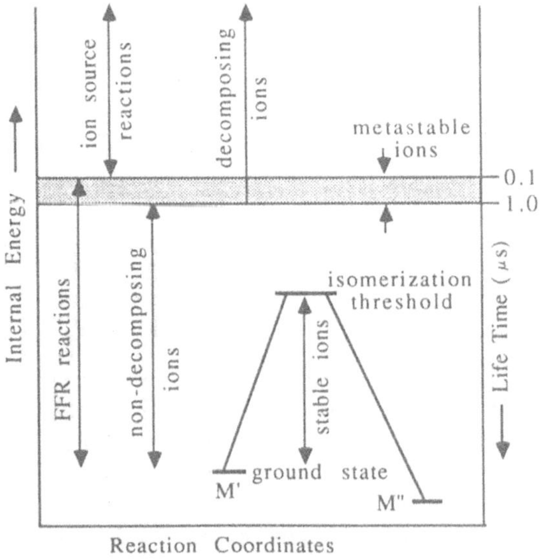


*Figure 1.2.* **Classification of ions on the basis of their energy content.**

$$k(\varepsilon) = \nu \left[ \frac{(\varepsilon - \varepsilon_0)}{\varepsilon} \right]^{s-1}, \tag{1.2}$$

where ε is the internal energy of the ion, $\varepsilon_0$ is the activation energy of the reaction, $s$ is an effective number of oscillators, and ν is a frequency factor measuring the entropy of activation (i.e., related to the geometry of the activated complex).

Figure 1.3 is a schematic of an EI source. Electrons are emitted from a heated filament F and accelerated toward a trap T at energies (usually 70 eV) determined by the difference in potential between the filament and the source chamber C. A portion of the electron current is used for ionization of the gaseous sample molecules, and the remainder is collected by the trap, which is held at a slightly positive potential with respect to the source block. The ionizing current $I_e$ is a measure of the number of electrons reaching the trap. A weak magnetic field **B**, applied parallel to the electron beam, collimates the beam, forcing it to follow a narrow and helical path, thus increasing the ionizing path length. The positive ions thus formed are pushed into the accelerating region by applying a positive potential to the repeller electrode R. Ions leave the source region at kinetic energies determined by the potential difference (in the kilovolt range for a magnetic sector mass spectrometer and a few volts for a quadrupole instrument) between the source block C and the exit slit S, which is held at ground potential.

The total positive ion current $I^+$ produced in the ion source is given by

$$I^+ = \beta Q_i L[N] I_e, \tag{1.3}$$

where β is the ion extraction efficiency, $Q_i$ is the total ionization cross-section. $L$ is the effective ionizing path length, and $[N]$ is the concentration of the sample molecules. Thus, keeping other parameters constant, $I^+$ increases in direct

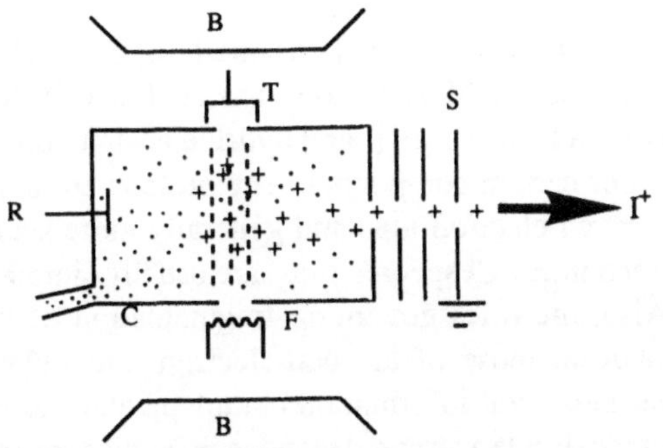


*Figure 1.3.* Schematic of an electron-impact ion source.

proportion to $I_e$. $Q_i$ depends on the chemical nature of the sample molecule and on the energy of the ionizing electron.

Although EI is an indispensable tool for the study of small molecules, it suffers from the disadvantage that many compounds are not stable under EI conditions. During EI, 2–8 eV of energy is transferred to the molecule, resulting in extensive fragmentation. As a consequence, the $M^{+\cdot}$ ion of certain compounds is either absent or of very low relative abundance. Because the $M^{+\cdot}$ ion is the most informative ion from the viewpoint of $M_r$ determination, its absence restricts the use of EI. A further drawback of EI is that many thermally labile and nonvolatile compounds are not amenable to EI. The volatility requirement limits the use of EI to compounds of lower molecular mass (<1000 Da).

## 2.2. *Chemical Ionization*

The need to produce molecular ion species of greater stability motivated the development of less energetic ionization methods, the foremost of which is chemical ionization (CI) (11,12). In this mode, a set of reagent ions is first produced by EI of a reagent gas at pressures between 0.1 and 1 torr, and the sample molecules are ionized by ion–molecule reactions with those stable reagent ions. The principle of CI is illustrated with methane as the reagent gas in the following equations:

$$CH_4 + e^- \rightarrow CH_4^{+\cdot} + 2e^- \quad \text{(then } CH_4^{+\cdot} \rightarrow CH_3^+ + \cdots \text{)} \tag{1.4}$$

$$CH_4 + CH_4^{+\cdot} \rightarrow CH_5^+ + CH_3^{\cdot} \tag{1.5}$$

$$CH_4 + CH_3^+ \rightarrow C_2H_5^+ + H_2 \tag{1.6}$$

$$M + CH_5^+ \rightarrow [M + H]^+ + CH_4. \tag{1.7}$$

The reagent ions formed in Eqs. (1.5) and (1.6) act as Brønsted acids and ionize the sample molecules M by proton transfer (Eq. 1.7). Addition reactions to form $[M + CH_5]^+$ and $[M + C_2H_5]^+$ and hydride abstraction to form $[M - H]^+$ are also observed for certain compounds. The protonated molecules $[M + H]^+$ formed by CI are even-electron ions and generally more stable than the odd-electron ions formed in EI. CI spectra, thus, are usually simple, with little or no fragmentation. Also, the rules governing fragmentation of the even-electron ions are different from those of the odd-electron ions (13); therefore, a CI spectrum provides structural information complementary to an EI spectrum.

A CI spectrum exhibits a strong dependence on the type of reagent gas and on the pressure in the ionization region. At higher pressures, stable ions are formed as a result of the cooling effect of thermalizing collisions with molecules of the neutral reagent gas. Some other commonly used reagent gases, along with the corresponding reagent ions and their proton affinities, are listed in Table 1.1. The use of several unusual positive ion CI reagents was reviewed recently (15).

The amount of energy transferred to the sample ion in CI is a function of the exothermicity of an ion–molecule reaction. For example, consider a general reaction in which a sample molecule M is ionized by proton transfer from $[B + H]^+$, a conjugated Brønsted acid of the reagent gas B:

$$M + [B + H]^+ \rightarrow [M + H]^+ + B. \tag{1.8}$$

This reaction occurs with high efficiency if the proton affinity (PA) of M is greater than that of B. (PA is defined as the heat liberated on protonation.) The extent of energy transferred during CI is given by Eq. (1.9), where $\Delta H^0$ is the heat of the reaction:

$$\Delta H^0 = -[PA(B) - PA(M)]. \tag{1.9}$$

Thus, for a given sample, the PA values included in Table 1.1 show that ionization with $H_3^+$ generates more energetic $[M + H]^+$ ions than those produced by CI with $CH_3NH_3^+$ ions. Furthermore, the high PA of ammonia and methylamine renders them ineffective as CI reagents for many organic compounds. This facet of CI is of advantage in mixture analysis, where a specific compound can be selectively ionized using a carefully chosen reagent gas. Therefore, it is possible to tailor a CI spectrum to suit a specific analytical application.

*Table 1.1*
*Positive Chemical Ionization Reagents*

| Reagent gas | Reagent ion | PA (kcal mol$^{-1}$)[a] |
|---|---|---|
| $H_2$ | $H_3^+$ | 101 |
| $CH_4$ | $CH_5^+$ | 132 |
| $H_2O$ | $H_3O^+$ | 167 |
| $CH_3OH$ | $CH_3OH_2^+$ | 182 |
| $C_2H_5OH$ | $C_2H_5OH_2^+$ | 186 |
| $i$-$C_4H_{10}$ | $i$-$C_4H_9^+$ | 196 |
| $NH_3$ | $NH_4^+$ | 204 |
| $CH_3NH_2$ | $CH_3NH_3^+$ | 211 |

[a]From Ref. 14.

Instrumentally, a CI ion source is similar to that used for EI but with slight modifications. The ionization chamber is made as gas-tight as possible by narrowing the electron beam's entry aperture and the ion exit slit. Also, the permanent magnet used for collimating the electron beam and the anode used for trapping electrons are no longer required because the electron beam does not penetrate completely through the gas chamber. In addition, a high-speed pumping system is required to maintain the source pressure to $<10^{-4}$ torr, and the analyzer region needs to be differentially pumped because of much increased source pressure.

### *2.2.1. Charge-Exchange Chemical Ionization (CE-CI)*

In this alternative mode of CI, odd-electron molecular ions are formed by exchange of an electron between the neutral analyte molecule and the reagent gas ions $G^{+\cdot}$, which are first formed by EI of the reagent gas:

$$G^{+\cdot} + M \rightarrow M^{+\cdot} + G. \tag{1.10}$$

The internal energy $\varepsilon$ of $M^{+\cdot}$ is given by

$$\varepsilon(M^{+\cdot}) = RE(G^{+\cdot}) - IE(M), \tag{1.11}$$

where $RE(G^{+\cdot})$ is the recombination energy of $G^{+\cdot}$. If $G^{+\cdot}$ is formed in the ground electronic state, then $RE(G^{+\cdot})$ is the negative of its vertical IE.

A variety of CE gases has been used for specific applications. These gases include (with increasing RE) toluene, benzene, NO, $CS_2$, COS, Xe, $CO_2$, CO, $N_2$, Ar, and He. The advantage of CE-CI is that the $M^{+\cdot}$ ions are formed within a well-defined and controlled energy distribution, which varies from near the threshold for ionization to the range where extensive fragmentation results. A CE-CI spectrum is similar to that obtained using EI. By using CE-CI with some of the above reagent gases, ions with energies below their threshold for isomerization have been generated (16,17).

### *2.2.2. Mixed Chemical Ionization*

When a spectrum having features of both EI and CI is required, a mixture of reagent gases such as $Ar/H_2O$ or $Ar/i\text{-}C_4H_{10}$ is employed. For example, CI with $C_4H_9^+$ ions (from isobutane) is usually only slightly exothermic and produces mainly the $[M + H]^+$ ions and a few fragments, whereas CE-CI with $Ar^{+\cdot}$ ions is highly exothermic and results in extensive fragmentation of the $M^{+\cdot}$ ions. Thus, a suitable reagent gas mixture provides a mixed spectrum in a single scan, from which the $M_r$ and structural information can easily be derived.

### 2.2.3. *Negative Chemical Ionization*

The production of negative ions in MS has been of interest in studying their ion chemistry and achieving increased detection sensitivity compared to EI and positive CI. In the gas phase, negative ions are formed mainly by resonance electron capture (REC) when electrons of near-thermal energies are available (18):

$$AB + e^{-}(\sim 0\ \text{eV}) \rightarrow AB^{-\cdot}. \tag{1.12}$$

Usually, an EI source generates electrons with energies >10 eV, but when a moderating gas such as $H_2$, $CH_4$, $i\text{-}C_4H_{10}$, $NH_3$, $N_2$, or Ar is employed, a high population of thermal-energy electrons is produced conveniently:

$$AB + e^{-}(0\text{–}15\ \text{eV}) \rightarrow A^{-} + B^{\cdot} \tag{1.13}$$

$$AB + e^{-}(>10\ \text{eV}) \rightarrow A^{-} + B^{-} + e. \tag{1.14}$$

With electrons of energies above thermal range, formation of negative ions is dominated mainly by dissociative electron capture (Eq. 1.13) and ion-pair formation processes (Eq. 1.14). Most $A^-$ ions formed by these two processes (e.g., $Cl^-$, $OH^-$, $CN^-$) are of low mass and therefore are of little structural significance.

The major advantage of REC-CI is increased detection sensitivity (100–1000×) compared to positive CI for compounds with high electron affinity. For example, compounds containing a nitro group, a halogen atom, or a conjugated π-electron system are all good substrates for REC-CI. The electron affinity of compounds devoid of these groups can be augmented by chemical derivatization with perfluoroacyl, perfluorobenzoyl, pentafluorobenzyl, or nitrobenzoyl groups. By using REC-CI, phenolic and primary amino compounds have been detected at the femtogram level (19).

Ion–molecule reactions with certain anions can also be used to generate negative ions. In most cases, these anionic reagents act as Brønsted bases $B^-$:

$$B^{-} + M \rightarrow [M - H]^{-} + BH. \tag{1.15}$$

This reaction is facilitated if the PA of the $B^-$ ion is greater than that of the $[M - H]^-$ ion. Some of the reagent ions that produce abundant negative molecular ions, with decreasing PA, include $NH_2^-$, $OH^-$, $O^{-\cdot}$, $CH_3O^-$, $F^-$, $O_2^{-\cdot}$, and $Cl^-$. Of these, $OH^-$, $O^{-\cdot}$, and $CH_3O^-$ have been employed more frequently and are generated in the CI source by EI of the respective reagent gas:

$$N_2O + e^- \rightarrow O^{-\cdot} + N_2$$

$N_2O/CH_4$ (or $i$-$C_4H_{10}$) mixture (1.16)

$$O^{-\cdot} + CH_4 \rightarrow OH^- + CH_3^{\cdot}$$

$N_2O/N_2$ mixture $$N_2O + e^- \rightarrow O^{-\cdot} + N_2 \quad (1.17)$$

$$CH_3NO_2 + e^- \rightarrow CH_3O^- + NO \quad (1.18)$$

## 2.3. *Desorption Ionization*

Although CI overcomes the problem of extensive fragmentation encountered in EI, the volatility limitation remains critical. This requirement restricts CI to compounds of molecular mass <1000 Da. Many compounds of biomedical relevance that contain polar functional groups are excluded from the protocol of CI analysis. Thermal energy supplied to break the intermolecular bonds in such polar compounds also causes their decomposition. Chemical derivatization can be used to enhance the volatility of such compounds. This process blocks those polar functional groups from participating in intermolecular hydrogen bonding. However, derivatization is not an attractive choice because of additional sample-handling steps and because of the uncertainty in determining the $M_r$ of the original compound when incomplete derivatization takes place and the number of derivatization sites is not known.

In the past few years, several desorption ionization techniques have been developed that can be used to transform involatile molecules into the gas phase. This development has helped to extend the upper mass range to several thousand daltons and has brought a variety of biomolecules into the realm of MS analysis.

### 2.3.1. *Field Desorption*

One of the earlier attempts to analyze nonvolatile samples was field desorption (FD) (20), in which ionization is assisted by a high electric field gradient (~$10^8$ V cm$^{-1}$). The actual mechanism of ion formation in FD has been the subject of much discussion; however, the broad consensus is that the strong electric field lowers the barrier for quantum electron tunneling from the molecule and facilitates desorption of ions by reducing the heat of desorption. For gaseous compounds, $M^{+\cdot}$ ions are formed by a process called field ionization. In FD, which is used for nonvolatile compounds, the sample is deposited on specially prepared needle-like structures grown on a thin wire filament. The emitter is heated to melt the sample and to facilitate the movement of dissolved salts such as NaCl and KCl. The freely moving protons and cations attach to the polar sample molecules to produce the $[M + H]^+$- and $[M + Na]^+$-type ions. The electric field desorbs those ions into the gas phase.

Because only a little excess internal energy is transferred to the molecule during FD, an FD spectrum contains mainly cationized molecular species and no fragment ions. Therefore, a major contribution of FD has been to determine the $M_r$ of several high-mass nonvolatile compounds. The lack of fragmentation in an FD spectrum could be a disadvantage for structure-determination studies. However, fragmentation can be induced to a certain extent by heating the filament to higher temperatures or collision activation in an FFR.

Despite this success, FD has not enjoyed wide popularity, mainly because operationally it is very difficult to make functional emitters; therefore, a successful analysis requires skilled manipulations. However, the recent introduction of the multichannel array detectors, coupled with the fact that background chemical noise in FD is virtually absent, may provide a renewed interest in FD.

### *2.3.2. Desorption Chemical Ionization*

Encouraged by the success of FD, an attempt was made to use direct chemical ionization (DCI) for nonvolatile samples (19,21). One of the premises of DCI is that the electric field generated by a plasma of CI ions may act as a powerful force to extract nonvolatile molecules into the gas phase. The sample is deposited on a thin wire filament and inserted directly into a CI source to enable the sample molecules to be bombarded by the plasma of CI ions. Heating the filament also assists desorption of the sample molecules.

As with conventional CI, DCI leads to the formation of $[M + H]^+$ ions. Fragment ions are formed by dissociation of these $[M + H]^+$ ions and by CI of the thermal degradation products of the neutral sample molecules. The actual mechanism of ion formation is not yet completely elucidated; nevertheless, DCI has been used to obtain mass spectra of many thermally labile polar compounds. However, DCI has not gained wide acceptance, perhaps again due to experimental uncertainties.

### *2.3.3. Plasma Desorption Ionization*

Californium-252 ($^{252}Cf$) plasma desorption (PD) ionization is one of a group of desorption ionization methods that use a high-energy particle beam for ionization and desorption of nonvolatile, polar, and thermally labile molecules (22,23). In general, these ionization techniques mix the sample with a suitable matrix and bombard the mixture with a high-energy particle beam to produce, just above the point of impact, a high concentration of the gas-phase neutral and ionic species. The underlying rationale in these sudden-energy methods is that, by depositing a large amount of energy into the molecule on a time scale that is fast compared with the vibrational time period, vaporization may occur before thermal decomposition begins. Conceptually, all of the desorption ionization methods described below are similar but differ in the way a primary beam is produced.

In PD, the primary beam consists of a plasma of high-energy (≈ 100 MeV range) ions generated by spontaneous fission of the radionuclide $^{252}Cf$. Each fission event releases simultaneously, in opposite directions, two fission fragments (e.g., $^{106}Tc^{22+}$ and $^{142}Ba^{18+}$, with kinetic energies of 100 and 80 MeV, respectively); one fission fragment is used to bombard the sample target, emitting secondary ions and neutrals, and the second fragment strikes the start detector. The secondary ions are accelerated to energies of 10–20 keV and subsequently mass-analyzed by a TOF mass spectrometer. Although a temperature of several thousand degrees is attained at the point of impact, the sample molecules do not decompose because energy transfer is very fast and mediated by the absorption, excitation, and relaxation processes that occur in the matrix.

The observed peaks in a PD spectrum of proteins and peptides are the molecular ion species with different charged states. The peaks are generally broad because they include contributions from metastable ions. In spite of the broadness of peaks, PD-MS has been used successfully to determine the $M_r$ of several proteins and peptides. To date, the highest-mass protein analyzed by this method is of 45,000 Da (24).

For a PD-MS analysis to succeed, great care must be exercised in preparing the sample. The presence of impurities has a debilitating effect on obtaining a high-quality spectrum; for example, alkali metal salts may reduce or quench entirely the molecular ion signal. It is recommended that, to remove such salts, the sample be adsorbed on a thin film of nitrocellulose and washed with ultrapure solvents (25).

The introduction of dedicated and relatively inexpensive commercial PD-MS (by BioIon Nordic AB, Uppsala, Sweden, and by PA Electron, Selmi, Ukraine) has ushered MS within the reach of many protein chemistry laboratories. Now it has become a routine matter to determine the $M_r$ of large biomolecules with a precision of ±0.03%, which exceeds the precision of the conventional electrophoresis and gel permeation chromatography techniques.

### *2.3.4. Secondary Ionization Mass Spectrometry*

Historically, secondary ionization mass spectrometry (SIMS) is the forerunner of the fast atom bombardment (FAB; discussed in Section 2.3.5.) ionization method. In SIMS, the sample is bombarded with a beam of keV-energy ions (26). The transfer of momentum from the beam to the area around the point of impact results in the sputtering of secondary ions related to the chemical environment of the surface. Static SIMS uses a primary beam of low ion-current densities (≈1 nA $cm^{-2}$), as compared to dynamic SIMS, where ion-current densities of up to 1 μA $cm^{-2}$ are used. Traditionally, SIMS has been used for determining the composition of surfaces and for their depth profiling.

For analysis of organic compounds, a sample is deposited as a monolayer on an etched metal (e.g., silver) surface, from which the impact of the ion beam emits cationized secondary ions. To prevent damage to the uppermost layer and to prolong emission of the secondary ion current, beams of low ion-current density must be employed. However, this arrangement results in poor yields of secondary ions, limiting the detection sensitivity available with SIMS.

### *2.3.5. Fast Atom Bombardment*

In SIMS, the secondary ion current is available only for a brief period. This disadvantage led to the development of FAB. In this technique, the sample is dissolved in a polar and relatively nonvolatile liquid matrix and bombarded with a beam of atoms of 6–10 keV energy (27). However, soon after the introduction of FAB, it was realized that the characteristics of the secondary ions do not change whether the sample is sputtered from a liquid matrix by bombardment with a beam of atoms or of ions (28). Because a matrix can absorb the impact of a high-flux atom or ion beam, damage to the sample layer is reduced correspondingly. Furthermore, the uppermost layer of the matrix is replenished continuously with sample molecules, resulting in prolonged sample ion currents. A constant and steady emission of secondary ions is essential for scanning instruments, high-resolution measurements, and MS/MS experiments.

The use of SIMS in conjunction with a liquid matrix has been given a new name, liquid-SIMS (LSIMS). Although conceptually LSIMS and FAB are identical, the former offers certain advantages over FAB. Because LSIMS employs ion beams of much higher energies (25–40 keV), high-mass ions are easily sputtered. Furthermore, it is easier to obtain a well-defined, electrically focused primary ion beam (not possible with a neutral atom beam), a process that results in enhanced secondary ion yields. Also, there is no gas load in an LSIMS ion source, and background noise is lower. These advantages have led to an improvement in detection sensitivity and high-mass capability. The current generation of mass spectrometers has begun to incorporate LSIMS guns in place of FAB guns. In most contemporary LSIMS ion guns, the ion beam consists of $Cs^+$ ions produced by thermal emission from a heated pellet of cesium aluminum silicate. Some LSIMS guns have also used liquid metals as the sources of the primary ion beam (29).

In contrast, FAB uses a beam of Ar or Xe atoms. The neutral gas atoms are first ionized by collisions with electrons moving in a saddle-field configuration. The ions formed are accelerated to the required potential (2–10 kV) and neutralized in a dense cloud of the excess neutral gas by REC to provide a beam of atoms of high translational energy.

FAB analysis can be performed in the positive- or negative-ion mode with equal ease. Although the mechanism of ion formation is not settled, it is generally accepted that the impact of an atom or ion beam sputters from the solution both neutrals and preformed ions, with additional ionization taking place in the region immediately above the matrix surface (selvedge region) by the gas-phase ion–molecule reactions (30). Most analytes readily accept or lose a proton to yield $[M + H]^+$ or $[M - H]^-$ ions, respectively. In a few cases, radical cations have also been observed (31). Unlike other desorption ionization techniques discussed in this chapter, the energy imparted to many samples during FAB ionization and desorption is sufficient to induce fragmentation. Thus, a FAB mass spectrum includes several fragments of structural significance. For example, the FAB spectrum of an octapeptide, dynorphin A fragment 1–8, shown in Fig. 1.4, readily provides its amino acid sequence.

The proper choice of a liquid matrix is very crucial in obtaining a high-quality FAB mass spectrum and in optimizing the sample ion current. The criteria for matrix selection have been discussed (32,33); the use of several

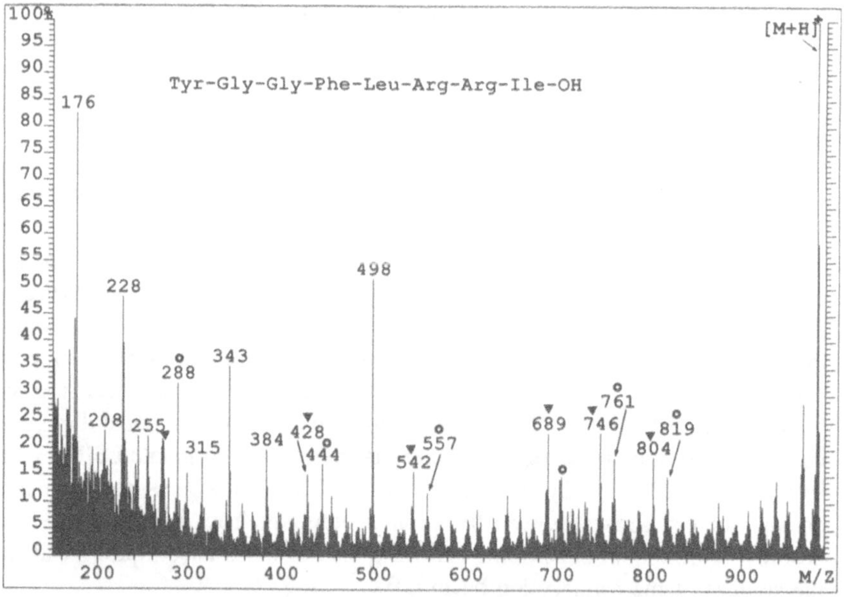


*Figure 1.4.* The positive-ion FAB mass spectrum of dynorphin A (1–8); the amino acid sequence of this peptide can be derived from two C-terminus ion-series labeled O and ▼.

matrices has been reviewed (32,34); and their important physicochemical properties have been tabulated (35). Glycerol is by far the most frequently used matrix. Some other commonly used matrices are α-thioglycerol, a mixture of dithiothreitol (DTT) and dithioerythritol (DTE; 5:1), thiodiglycol, 3-nitrobenzyl alcohol, tetraglyme, sulfolane, diethanolamine, and triethanolamine.

The matrix must be reasonably viscous, chemically inert, and nonvolatile, must possess good solvent and electrolytic properties, and must be mass-spectrally transparent. The acid–base and oxidizing–reducing nature of the matrix, the surfactancy of the solute relative to the matrix, and solute–solvent interactions all have a profound influence on the mass spectral appearance (36,37). For an intense $[M + H]^+$ ion signal, the matrix should be more acidic than the sample molecule, and for $[M - H]^-$ ion production, the matrix should be more basic. For example, glycerol, α-thioglycerol, and DTT/DTE are suitable acidic matrices. Also, α-thioglycerol, being more acidic than glycerol, provides improved $[M + H]^+$ ion yield. Similarly, for negative-ion analysis, the basic matrices such as diethanolamine and triethanolamine are preferred. The acidity and basicity of a matrix, however, can be augmented by addition of acids and bases, respectively.

Generally, ions in an FAB mass spectrum represent the surface composition of the matrix. For example, hydrophobic analytes tend to occupy the upper portion of a relatively hydrophilic matrix and are ionized preferentially relative to the hydrophilic compounds, which prefer to remain within the matrix (38). Thus, certain compounds remain undetected by FAB when present along with more hydrophobic compounds. Also, the presence of alkali salts tends to suppress ionization. Similarly, compounds with negative surface activity relative to the matrix may not produce any FAB signal. The surfactancy of a solute, however, can be manipulated by chemical derivatization with large alkyl groups or by adding surface-active agents to the matrix/analyte mixture (39).

Thus the detection sensitivity in FAB analysis is a function of the chemical composition of the solvent–solute mixture and of the presence of other unwanted analytes. The detection sensitivity is further compromised by the fact that the liquid matrix produces a high level of chemical background: ions at virtually every mass number are formed.

*Continuous-Flow Fast Atom Bombardment.* Continuous-flow (CF)-FAB diminishes significantly the problems associated with the use of a liquid matrix (40). In this kind of device, a stream of a very dilute aqueous solution of the matrix (1–10% by volume), usually glycerol, flows continuously through a specially designed probe. The sample is injected into this stream and bombarded by a high-energy beam in the conventional way. With this arrangement, the matrix now forms a very thin layer at the probe tip. As a result, the analyte-to-matrix ratio is increased and the matrix background is reduced significantly; both of these factors combine to yield a several-fold increase in detection sensitivity.

An additional advantage of this probe is that on-line direct analysis of biochemical processes can be accomplished (41). Also, this probe serves as an ideal interface for coupling of HPLC (42), capillary-zone electrophoresis (CZE) (43–45), and *in vivo* microdialysis (46) to a mass spectrometer. The last configuration especially permits *in situ* analysis of dialyzable compounds (e.g., neurotransmitters) from body tissues. For additional information, readers are referred to a recent monograph on this subject (47) and to Chapter 6 of this volume.

Several CF-FAB probes of different designs are in use; a schematic of the probe used for the VG Autospec mass spectrometer is shown in Fig. 1.5. The matrix solution is held in a pressurized glass vial and driven to flow with the help of a positive pressure of nitrogen through a silica capillary enclosed within a stainless steel probe shaft. The end of the capillary is in contact with a stainless steel screen, which acts as the probe tip. The probe shaft is inserted into the conventional FAB (or LSIMS) source. For stable operation, the liquid should not be allowed to accumulate at the probe tip. In this design, the probe tip is surrounded by a piece of absorbent filter paper to absorb the excess liquid. The sample is injected by means of a loop injector or mixed with the matrix–solvent mixture.

### *2.3.6. Laser Desorption Ionization*

Irradiation by an intense laser beam is another suitable mode to deposit a large amount of energy in sample molecules for their desorption into the gas phase (48). Two modes of laser desorption ionization (LDI) are commonly practiced. In the direct LDI mode, the sample is irradiated directly by a laser

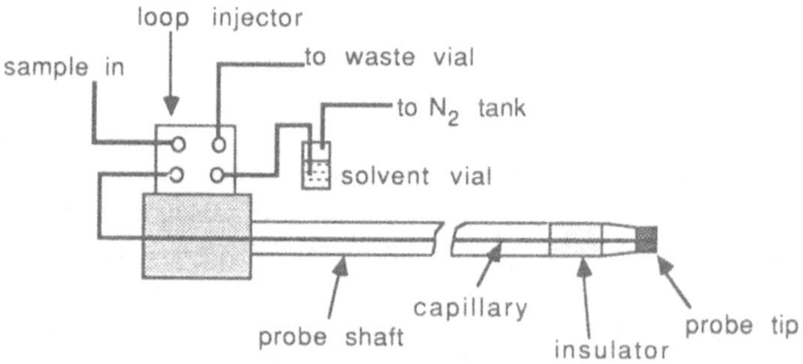


*Figure 1.5.* Schematic representation of a continuous-flow FAB probe.

beam, usually operating in the IR region [e.g., the Nd/YAG (neodymium/yttrium-aluminum-garnet; 1.06 μm) laser or the pulsed $CO_2$ laser (10.6 μm)]. A few examples have also used the frequency-quadrupled Nd/YAG laser emitting light in the UV range (at 266 nm).

It is believed that irradiation by a laser beam leads to desorption of the intact sample molecules. The ions characteristic of the sample are formed by the gas-phase attachment of codesorbed alkali-metal ions. For desorption of the sample molecules into the gas phase, the amount of energy deposited during irradiation is critical. The factors that determine the energy deposition are the laser intensity, the pulse width, and the absorption coefficient of the sample for the laser light. In a direct LDI mode, the extent of energy transfer is often difficult to control and may lead to excessive thermal degradation. Furthermore, not all compounds absorb radiation at the laser wavelength. Therefore, analysis of thermally labile molecules by the direct LDI mode is limited to compounds of molecular mass up to 1000 Da.

In the second mode, called matrix-assisted LDI (MALDI), the sample is mixed with a suitable matrix chosen according to the sole criterion that it absorb energy at the wavelength of the laser radiation (5,6). This matrix–wavelength combination permits a large amount of energy to be absorbed efficiently by the matrix and transferred efficaciously to the sample in a controlled manner. This process helps to desorb very large molecules into the gas phase without inducing their thermal degradation. The matrix also serves as a solvent for the analyte, so that the intermolecular forces are reduced and aggregation of the analyte molecules is held to a minimum. With this novel provision, very large ions, of up to $m/z$ = 300,000, have been detected (see Fig. 1.6) (48).

Several matrix–laser combinations have been tested successfully. Hillenkamp and Karas have found nicotinic acid to be a useful matrix with the frequency-quadrupled Nd/YAG laser at 266 nm (5). Vanillic acid, 2-pyra-

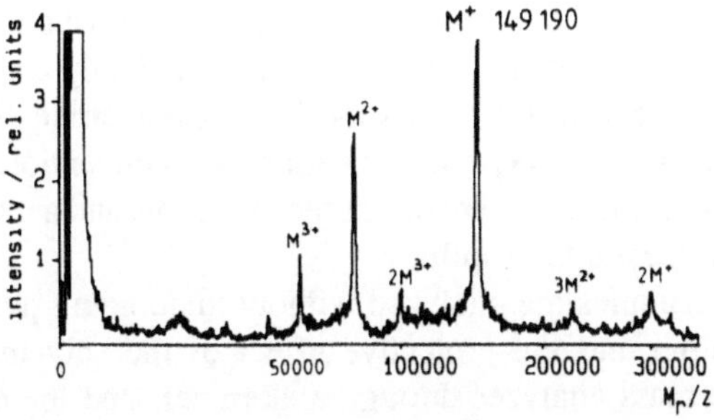


*Figure 1.6.* The MALDI spectrum of monoclonal antibody (IgG of mouse against a specific human lymphokine). Reprinted with permission from Hillenkamp and Karas, 1990, in *Methods in Enzymology* (J. A. McCloskey, ed.), Vol. 193, Academic Press, New York, pp. 280–295.

cinecarboxylic acid, thymine, and thiourea are also appropriate at this wavelength (49). A combination of a slurry of finely divided Pt in glycerol as an absorbing matrix and the $N_2$ laser at 337 nm has also been used for high-mass proteins (50). As with the other desorption–ionization methods, preparation of the sample for MALDI analysis requires utmost care. The homogeneity of the sample–matrix mixture is critical in obtaining good sample ion yields.

Positive- and negative-ion analysis can both be performed using MALDI. The MALDI mass spectra of proteins and peptides typically contain signals due to the molecular ions and their oligomers. Multiply charged ions of low abundance are also formed. With increasing mass of the analytes, multiply charged ions increase in abundance. Fragment ions are generally not observed. Therefore, MALDI has been used primarily to determine the molecular mass of proteins, although the precision is generally poor (0.1 to 0.5%).

### 2.4. *Spray Techniques*

The last few years have witnessed some important developments in ionization techniques capable of analyzing LC effluents directly. These techniques differ from the desorption methods discussed above in that the ions are generated in an excess of an ambient-bath gas rather than in vacuum.

#### 2.4.1. *Thermospray*

Thermospray is one of the most popular LC/MS interfaces (51). As the name implies, thermospray entails production of a jet of fine liquid droplets by heating. This device is used to transfer the sample to the ion source efficiently by controlled and partial evaporation of the LC effluent. It can handle flow rates in excess of 1 ml $min^{-1}$ and a wide range of samples. The LC effluent is passed through an electrically heated capillary mounted on a probe, which can be directly introduced into the ion source. The resulting adiabatic expansion and cooling of the liquid produces a supersonic jet of randomly charged fine liquid particles. Further vaporization occurs as the droplets strike the walls of the heated ion source. The source, a modification of a conventional CI source, is fitted with an EI gun, a low-current discharge electrode, and a mechanical pump located directly opposite the capillary.

Nonvolatile samples are analyzed without undergoing pyrolysis because of the rapid heating and the protective effect of the solvent. The ions are sampled into the mass analyzer through a skimmer, and the excess vapor is pumped away by the source auxiliary pump. Ionization of the analyte molecules occurs by means of one of three processes, namely, direct evaporation of the preformed ions from the charged droplets, CI initiated by the ionic components of the buffer used in the mobile phase, and CI with the reagent ions formed by EI or low-current discharge ionization of the mobile phase.

Direct ion evaporation is the preferred mode of ionization with ionic, polar, and nonvolatile samples and with aqueous mobile-phase systems. Ammonium acetate is the most widely used buffer because of its volatility. It is also a good source of $NH_4^+$ and $CH_3COO^-$ ions, both of which act as CI reagent ions to form positive and negative ions, respectively, by proton-transfer reactions. The basicity or acidity of the analyte relative to the mobile phase determines whether positive or negative ions result.

For volatile samples and a mobile phase containing a large fraction of an organic solvent, it is recommended that a process known as "filament-on operation" be used for initiating CI. If one switches on the EI filament, a plasma of ions from the organic mobile phase is formed. Alternatively, with a high percentage of the organic solvent, a discharge electrode is used to generate CI reagent ions.

Thermospray is a gentle mode of ionization and produces mostly intact molecular ions. Fragmentation is relatively sparse, and thus little structural information is available. Multiply charged ions of some analytes are also formed.

In order to obtain a true EI spectrum of samples from the LC effluent, a *particle beam interface*—also known as monodisperse aerosol generation interface for chromatography (MAGIC)—was designed (52). In the monodisperse aerosol generator, the LC effluent is passed through a small orifice to form a fine liquid jet, which breaks into a stream of uniform droplets under the influence of Rayleigh forces. A high-velocity stream of a gas is introduced at a right angle to the flow of liquid to disperse the particles. Vaporization of solvent occurs at atmospheric pressure in the desolvation chamber. The solvent vapors, along with the sample molecules, pass through a two-stage momentum separator to produce a particle beam of the sample. This beam is conveniently analyzed using a conventional EI or CI ion source.

### 2.4.2. *Electrospray Ionization*

Electrospray ionization (ESI) represents a significant advance in the MS analysis of high-mass proteins and peptides (2–4). With ESI-MS, the molecular masses of biopolymers of mass over 100 kDa can be determined with an accuracy of better than 0.01%. The basis of high-mass analysis is that an ESI mass spectrum is characterized by a cluster of peaks consisting of multiply charged ions of the general nature $[M + nH]^{n+}$; each peak differs from the neighboring peak by one charge (see Fig. 1.7). Because a mass spectrometer analyzes ions based on their $m/z$ ratios rather than their masses, the effect of

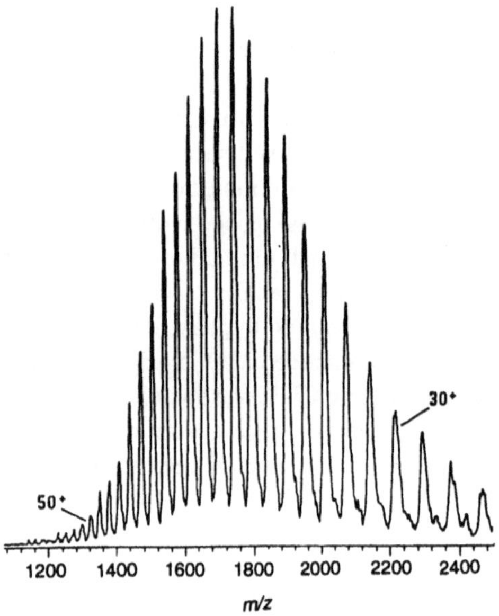


*Figure 1.7.* The positive-ion electrospray mass spectrum of bovine serum albumin ($M_r$=66,300). Each peak in this envelope differs from its neighboring peak by unit charge. Ions containing more than 50 positive charges can be recognized.

multiple charging is to reduce the *m/z* ratio, typically within the mass range of many quadrupole and low-cost magnetic sector mass spectrometers.

The origin of ESI can be traced to the early experiments of Dole and co-workers, who used ESI to produce molecular beams of very-high-mass macroions of polystyrene (53,54). Recent developments are credited to Fenn and co-workers (2,3,55,56) and to Aleksandrov *et al.* (57). Because of the high voltages involved in ESI, the combination of an ESI ion source and a quadrupole mass analyzer was the first interface that was easy to fabricate. Consequently, many quadrupoles have been retrofitted with an ESI source. However, ESI ion sources are now available commercially that can be used with magnetic sector instruments. Figure 1.8 is a schematic diagram of the ESI ion source used with the VG Autospec magnetic sector mass spectrometer.

In ESI, a continuous stream (a few μl min$^{-1}$) of a sample in a suitable solvent, usually 1:1 (v:v) $CH_3OH$—$H_2O$ containing 1% $CH_3COOH$, flows through a stainless steel capillary tube held at a few kV relative to the walls of the chamber and the counter electrode. The effluent from the needle tip emerges into a chamber maintained at nearly atmospheric pressure as a fine spray of charged droplets. A flow of hot-bath gas, usually nitrogen, assists in evaporation of the solvent from those charged droplets. As the droplets shrink, the

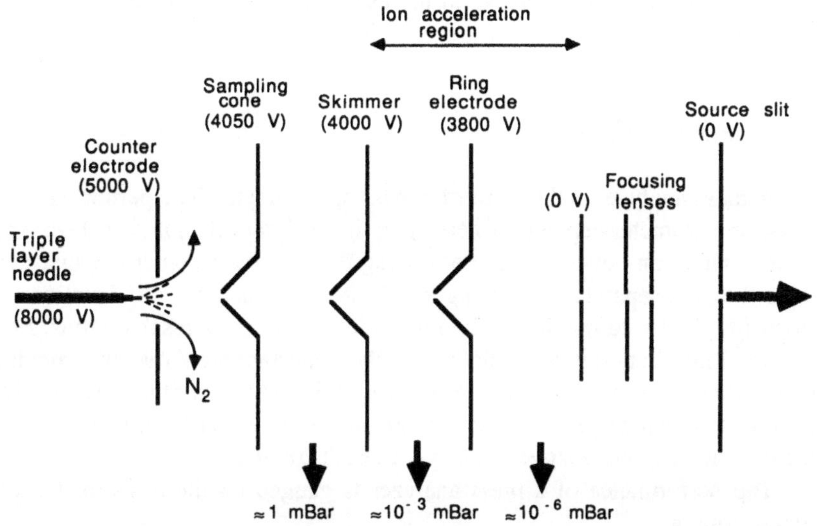


*Figure 1.8.* Schematic of an electrospray ion source, depicting typical voltages for 4-keV ions.

charge density on their surface increases until the Rayleigh instability limit is reached. At this point, repulsive Coulombic forces are of the order of droplet cohesive forces and cause fission of the droplets into smaller charged droplets. This fission process is repeated several times until the drop size is reduced to a limit where the electric field due to the surface charge density is sufficient to assist direct ion evaporation. Before entering the MS ion source, these ions pass through additional chambers maintained at progressively reduced pressures, where most of the neutrals are removed by auxiliary pumping.

Some researchers have used a high-velocity annular flow of gas to enhance nebulization of the ES liquid (58,59). This pneumatically assisted ES process has been termed "ionspray" and has the advantage that flow rates of up to ~100 μl $min^{-1}$ can provide a stable spray. These types of ES sources are ideally suited for analysis of effluents from microbore LC column.

In the short time since its inception, ESI-MS has been successfully used to determine the $M_r$ of large proteins and polypeptides (4), to distinguish posttranslational modifications in biomolecules (4), to identify aberrant states of proteins and peptides (60), to quantify neuropeptides (61), and as a device to couple HPLC (59) and CZE (62–64) separation techniques to MS instruments. Furthermore, because the efficiency of collision-induced dissociation (CID) is significantly enhanced for multiply charged ions compared to the case of singly charged ions, ESI in conjunction with MS/MS can be a highly potent analytical tool (65). An increase in the use of ESI-MS to solve problems in biochemistry, immunology, and genetics is anticipated in future.

## 3. *MASS ANALYZERS*

A mass analyzer is the heart of a mass spectrometer. The performance of a mass spectrometer depends largely upon the design of the mass analyzer and the associated ion optics. Ions, after exiting the ion source, enter the analyzer, where they are separated according to $m/z$. A mass analyzer also functions as a focusing device to maximize the intensities of the mass-resolved ions. Momentum, kinetic energy, and velocity are the important properties of a moving charge particle. A mass analyzer makes use of these properties to mass-resolve various ions. All analyzers function on one of three basic principles, viz., electrostatic analysis, magnetic analysis, and TOF analysis.

The performance of a mass analyzer is gauged on the basis of the following criteria:

1. mass range, the maximum allowance mass that can be analyzed;
2. resolution, the separation of two adjacent peaks;
3. scan speed, because a high scan speed is required for rapidly changing events (e.g., the eluant from capillary GC, or a pulsed event) and a low speed is obligatory for accurate mass measurement; and
4. detection sensitivity, the smallest amount of an analyte that can be detected.

Another important criterion in choosing a mass analyzer is the ease with which certain ionization and other ancillary equipment, such as multichannel array detectors and chromatographs, can be outfitted.

High resolution or resolving power ($R$) is necessary for accurate mass measurement, for accuracy in quantification, and for unambiguous precursor-ion selection in MS/MS experiments. Accuracy in mass measurement is needed to determine the structure of an analyte and to ascertain the fragmentation mechanism of an ion. The resolving power of a mass spectrometer, its ability to separate two neighboring ions, is expressed as $R = m/\Delta m$ with reference to Fig. 1.9, where two equal-height peaks of mass $m_1$ and $m_2$ are considered resolved when the valley between the two is 10% of the height of either one i.e., each contributes 5% to the valley), $m$ is the average mass of those two peaks, and $\Delta m$ is their mass difference. In practice, $R$ is determined instrumentally by electrical switching methods from a single symmetric peak of mass $m$ (e.g., as shown for the right-hand peak in Fig. 1.9) and width $\Delta m$ (at 5% height).

As the mass of the ions increases, the need for higher resolution becomes

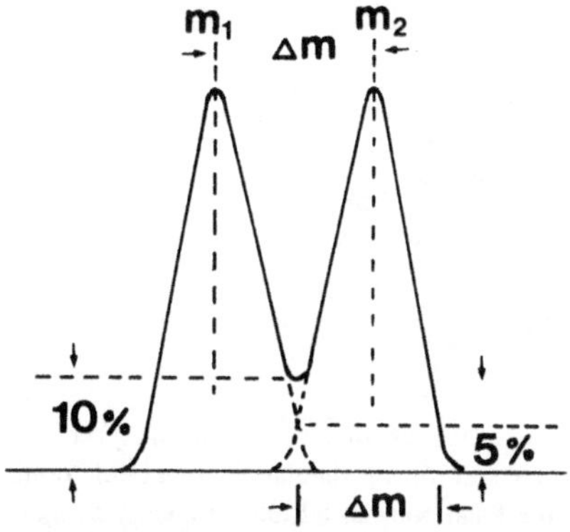


*Figure 1.9.* Mass resolution (10% valley definition) of two ions of mass $m_1$ and $m_2$.

correspondingly more critical. For example, to mass-resolve ethylene$^{+\cdot}$ ($m$ = 28.031300) and $CO^{+\cdot}$ ($m$ = 27.994915), a resolution of 770 is needed. Now consider the two ions $C_{13}H_{16}$—$C_2H_4^+$ and $C_{13}H_{16}$—$CO^+$. Although the mass difference is still 0.036385, the latter doublet requires a resolution of 5500 for separation of the two signals. Another experimentally complicating fact is that, with an increase in mass, the number of possible elemental compositions increases exponentially. Thus, for unequivocal assignment of chemical composition to an ion, the accuracy of mass measurement must be extremely high.

### *3.1. Electrostatic Analyzer*

When ions move parallel to an electric field, they experience an acceleration, as in an ion source. The kinetic energy KE of the accelerated ion is given by

$$\text{KE} = zV = \tfrac{1}{2}\, mv^2, \tag{1.19}$$

where $m$, in kilograms (1 kg = $6.023 \times 10^{26}$ Da), is the mass of the ion; $v$, in meters per second, is the velocity of the ion; $z$ is the charge of the ion ($z = ne$, where $n$ is the integral number of units of charge and $e$ is the fundamental unit of charge = $1.602 \times 10^{-19}$ C); and $V$, in volts, is the potential difference to which the ion is subjected. Thus, from Eq. (1.19), it is obvious that when two ions of different masses but the same charge are accelerated through an identical potential difference, their kinetic energy upon leaving the accelerator is the same.

An electrostatic analyzer (ESA) consists of two parallel plates (Fig. 1.10), one of which is held at a positive potential and the other at a negative potential

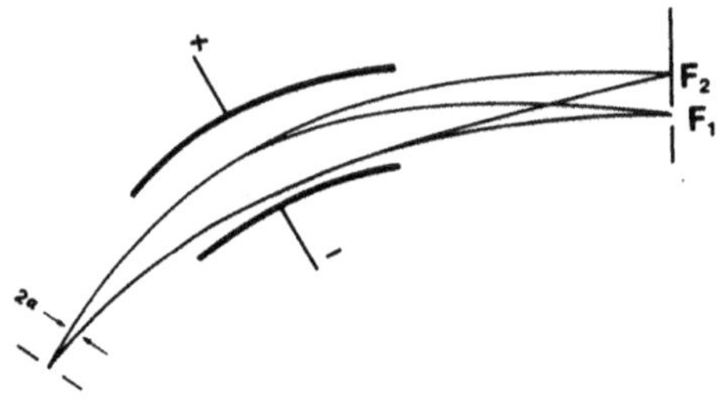


*Figure 1.10.* The focusing action of an electrostatic analyzer. All ions of the same energy emanating from a point source and with an angular divergence of 2α are brought to focus at a single point. For example, $F_1$ is the focal point of all ions of energy $E_1$ and mass $m_1$ and $m_2$, and $F_2$ of ions of energy $E_2$ and mass $m_1$ and $m_2$.

of equal magnitude (66,67). When ions pass through this radial electric field, they move in a circular trajectory and are dispersed according to their kinetic energy. In other words, ions of a given energy only will be focused at a point. Thus, the ESA is used as an energy filter to produce an ion beam of nearly homogeneous energy. The ESA is also used for direction focusing. All ions of identical energy and emanating from the same point with small angular divergence are brought to focus (e.g., at points $F_1$ and $F_2$ in Fig. 1.10). The motion of an ion in this analyzer is governed by

$$\frac{mv^2}{r} = zE \quad \text{and} \quad r = \frac{2V}{E}, \tag{1.20}$$

where $r$, in meters, is the radius of curvature; $E$, in volts per meter, is the electric field strength; and $m$, $v$, $z$, and $V$ are as defined above.

### 3.2. Magnetic Analyzer

When a charged particle enters a magnetic field (Fig. 1.11), it travels a circular path in a plane perpendicular to the direction of the field, according to

$$\frac{mv^2}{r} = zvB \quad \text{and} \quad \frac{mv}{z} = Br, \tag{1.21}$$

where $B$, in tesla (T), is the magnetic field strength (66,67). Thus, the magnetic sector acts as a momentum analyzer and disperses the ions according to their momentum-to-charge ratios.

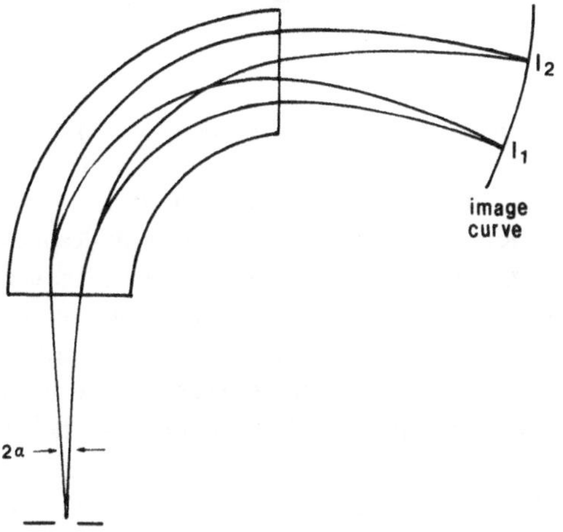


*Figure 1.11.* Mass separation by a magnetic analyzer. All ions of mass $m_1$ and $m_2$ emanating from a point source with an angular divergence of $2\alpha$ are brought to focus on the image curve at points $I_1$ and $I_2$, respectively.

Rearranging Eq. (1.21) and substituting the value of $v$ from Eq. (19), one obtains

$$\frac{m}{z} = \frac{B^2r^2}{2V}. \tag{1.22}$$

Thus, a magnetic sector also separates ions according to their $m/z$ ratios, provided all ions are of equal kinetic energy. If $V$ is constant and $B$ is scanned, then all ions follow the same trajectory and are collected at a single focal point, resulting in a mass spectrum. Alternatively, keeping $B$ constant, all ions that follow paths of different radii are collected simultaneously in the same plane, as is done in the double-focusing mass spectrometers of the Mattauch–Herzog geometry. Rearranging Eq. (1.22) in terms of radius $r$, one obtains

$$r = \frac{1}{B}\left(\frac{mV}{z}\right)^{1/2}. \tag{1.23}$$

By logarithmic differentiation of Eq. (1.23), it becomes clear that, to obtain the best separation for different masses, an ion beam must be homogeneous in energy (i.e., $\Delta V$ must be as small as possible):

$$\frac{\Delta r}{r} = \frac{1}{2}\frac{\Delta m}{m} + \frac{1}{2}\frac{\Delta V}{V}. \tag{1.24}$$

### 3.3. Double-Focusing Magnetic Sector Mass Spectrometer

From the above discussion, we learn that an ESA produces an ion beam homogeneous in energy and a magnetic sector provides a mass-resolved beam. When a proper combination of these two analyzers is used, it is possible to correct simultaneously for both direction and velocity inhomogeneities (66,67). These so-called double-focusing mass spectrometers are capable of achieving very high mass resolution, provided aberrations from the second-order terms caused by field inhomogeneities and fringing fields are held to a minimum. This condition is generally achieved by use of additional electrostatic lenses.

A variety of different arrangements of electrostatic and magnetic analyzers is in use. When, as in Fig. 1.12, an ESA (E) precedes a magnetic analyzer (B), the combination results in a forward-geometry (EB) double-focusing mass spectrometer. In a reverse-geometry instrument (Fig. 1.13), a magnetic analyzer precedes an ESA (i.e., the geometry is BE). In an instrument of Nier–Johnson geometry, the two fields deflect the ions in the same direction (i.e., in a C shape), and the point of double-focusing is obtained when the velocity- and direction-focusing curves are made to intersect. A detector is placed at this point, and the mass spectrum is acquired by changing the magnetic field strength. Nier–Johnson instruments are limited in sensitivity because, at any one time, only one ion is detected. In the Mattauch–Herzog design, the ions are deflected in opposite directions (i.e., in an S shape), and the double-focusing of all masses is achieved in the same plane (Fig. 1.14). All mass-resolved ions are detected simultaneously by placing a focal plane detector (e.g., a photographic plate or a multichannel array detector) at the exit boundary of the magnet. These instruments provide increased sensitivity and accuracy in mass measurement. However, they did not gain wider acceptance because of the cumbersome photographic plate detector. The recent introduction of multichannel array

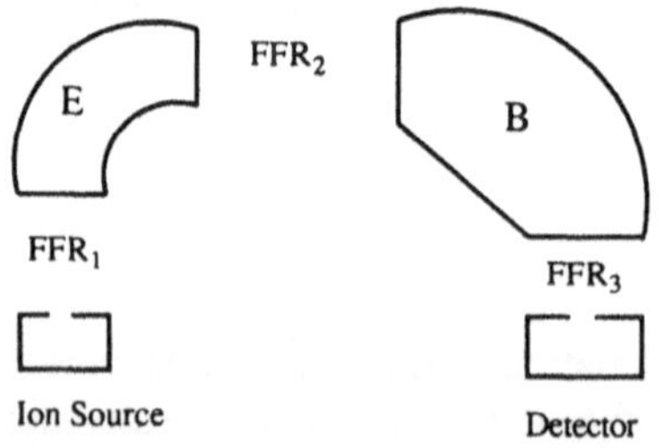


*Figure 1.12.* A forward-geometry (EB) Nier–Johnson magnetic sector mass spectrometer.

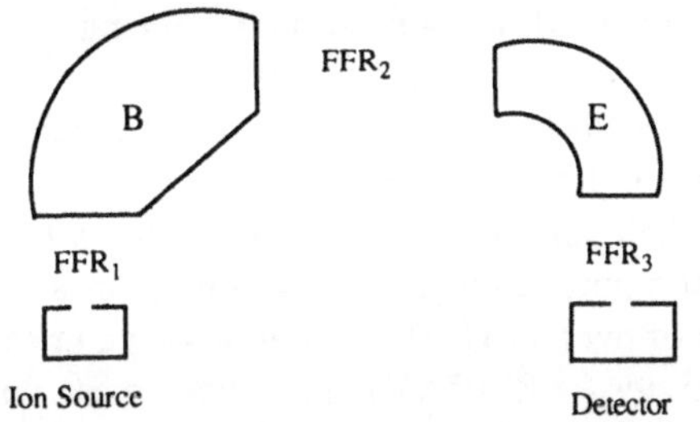


*Figure 1.13.* A reverse-geometry (BE) Nier–Johnson magnetic sector mass spectrometer.

detectors (68) has rekindled interest in Mattauch–Herzog geometry mass spectrometers (69). A multichannel array detector can also be used with Nier–Johnson mass spectrometers to take advantage of their limited focal plane width (5–10% of the central mass).

The major advantages of double-focusing mass spectrometers are their high mass range, high resolution, accuracy of mass determination, reasonable scan speed, and high dynamic range. Furthermore, magnetic sector instruments are ideal for studying reactions in a FFR, including high-energy CID, charge permutation reactions, reactions of metastable ions, and kinetic energy release (KER) experiments.

To produce an ion beam of lower energy spread [i.e., to minimize the $\Delta V/V$ term in Eq. (1.24)] and to improve ion extraction efficiency, the ion source is operated at high voltage (6–10 kV). This requirement, however, creates arcing

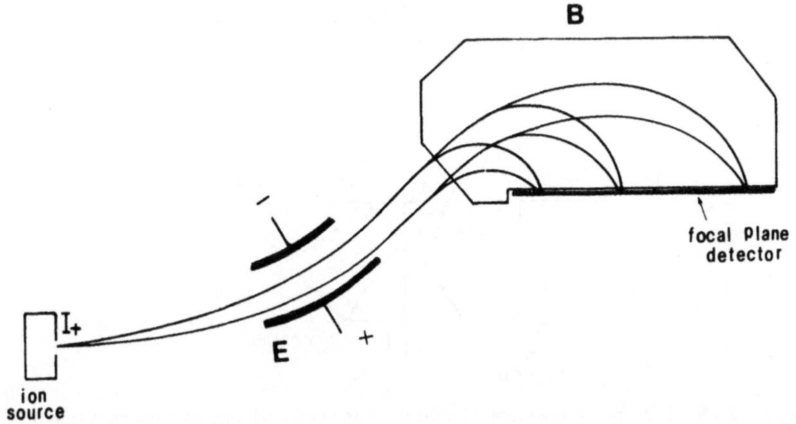


*Figure 1.14.* Principle of ion detection by a forward-geometry Mattauch–Herzog magnetic sector mass spectrometer.

problems with certain modes of ionization that operate at atmospheric pressure (e.g., thermospray and electrospray) or when effluents from LC columns are to be analyzed. Further disadvantages of these instruments are their exorbitant cost and very low ion transmission.

Aside from their high cost, these state-of-the-art magnetic sector mass spectrometers with improved magnets and ion optics are capable of achieving a mass resolution of over 100,000 and a mass range in excess of 15,000 Da at the highest sensitivity (at 8–10 kV). The upper mass range can be extended further by reducing the accelerating potential (i.e., up to 120,000 Da at 1 kV), but at the cost of reduced sensitivity.

### *3.4. Quadrupole Mass Spectrometer*

A quadrupole is a path-stability device in which mass separation is accomplished solely by using an electric field. A stable trajectory is created by applying radio frequency (rf) and dc potentials to four rods of hyperbolic cross-section arranged in a square array (Fig. 1.15); in practice, cylindrical rods are also used effectively. Opposite rods are electrically connected. One pair receives a positive dc potential $U$ and an rf potential $V_0 \cos(\omega t)$, where $\omega$ is the angular frequency and $t$ is the time; the other pair is at $-U$ dc and an rf voltage also of magnitude $V_0 \cos(\omega t)$, but out of phase by 180°. This arrangement creates an oscillating field such that the potential $\Phi$ at any point within the rods is given by

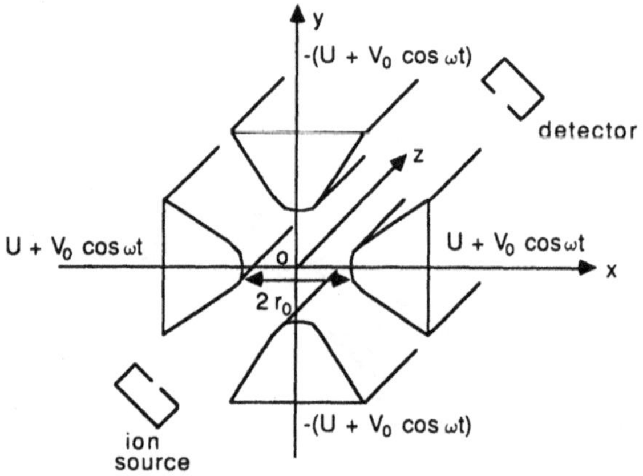


*Figure 1.15.* Principle of operation of a quadrupole mass analyzer. At a certain value of $U$ and $V_0$, ions of specific $m/z$ undergo stable oscillations in the $x$ and $y$ directions and travel in the $z$ direction; all other ions follow unstable trajectories and are lost.

$$\Phi = [U + V_0 \cos(\omega t)] \frac{x^2 - y^2}{r_0}, \tag{1.25}$$

where $x$ and $y$ are distances from the center of the field and $2r_0$ is the spacing between opposite rods. With a value of $U$ and $V_0$ such that $U/V_0 = 0.168$, the ions of a specific $m/z$ ratio adopt stable oscillations in the $x$ and $y$ directions and are transmitted through the parallel array of the rods (i.e., in the $z$ direction); all other ions have unstable trajectories because they tend toward infinite displacement from the center and are lost. At a fixed frequency, mass scanning is accomplished by varying $U$ and $V_0$, keeping their ratio constant.

For effective mass separation, an ion must spend enough time in the quadrupole field, because the greater the number of oscillations, the greater the mass resolution. Therefore, the velocities of the ions entering the quadrupole field must be kept low. This condition is achieved by producing the ions at low accelerating potentials (typically <100 V); however, a lower accelerating potential means poor transmission of ions from the source. Nevertheless, the overall sensitivity of quadrupole mass analyzers is higher than that of magnetic sector instruments because no slits are used, resulting in higher transmission along the rods. The mass resolution of the quadrupole system increases with mass because heavier ions spent more time within the oscillating field, but the overall resolution is limited to unit mass only.

The upper mass range of most quadrupoles is of the order of 1000 Da. In modern instruments, this range has been extended to 4000 Da by the use of accurately machined hyperbolic rods. The other attractive features of a quadrupole mass filter are low cost, compactness, high scan speeds, and linear mass scale. Also, performance is independent of the energy distribution of ions. All these features make it attractive for use with ion sources that require high pressures and high voltages for ion production and that produce ions with high energy spread. Also, quadrupoles are easy to interface with various inlet systems. However, compared to the magnetic sector instruments, the quadrupoles are inferior in mass range, mass resolution, and accuracy of mass measurement.

### *3.5. Time-of-Flight Mass Spectrometer*

A time-of-flight (TOF) mass analyzer uses the simplest concept in the analysis of charged particles (70). After initial acceleration through the source, ions are allowed to drift in a field-free region. As discussed above [Eq. (1.19)], all ions accelerated to a fixed potential leave the ion source with the same kinetic energy. Therefore, rearranging Eq. (1.19), the velocity with which they travel is different; lighter ions move faster than heavier ones:

$$v = \left(\frac{2zV}{m}\right)^{1/2}. \tag{1.26}$$

The $v$, $z$, $V$, and $m$ are as defined in Section 3.1. The time of arrival of ions at the detector placed at a distance $d$ from the source is given by

$$t = d/v = d\left(\frac{m}{2zV}\right)^{1/2}. \tag{1.27}$$

Thus, the arrival time of two different ions is proportional to the square root of their mass:

$$\frac{t_1}{t_2} = \frac{v_2}{v_1} = \left(\frac{m_1}{m_2}\right)^{1/2} \tag{1.28}$$

All ions of the same mass and kinetic energy arrive together at the end of the flight tube in a group. The flight time of each ion is used to establish the mass scale. For example, ions of masses 2000 and 2500 Da, when accelerated through a potential of 6000 V, take 41.56 and 46.47 μsec, respectively, to travel 1 m. The short time between the arrival of neighboring masses demands a high level of accuracy in measurement of the arrival times. This requirement limits the resolution attainable by TOF analyzers. Most commercial instruments have a resolution of only a few hundred. Resolution can be improved by increasing the path length, by reducing the energy spread of the ion beam, and by forming all ions in the same plane (e.g., as in SIMS and PD ionization sources).

In modern instruments, energy spread is reduced by incorporating a reflecting electrostatic lens assembly, called a reflectron, which consists of a series of rings with increasing voltage (71). It is placed at the end of the flight tube. Ions penetrate the reflectron until they reach zero energy and are then reflected back with energies equivalent to their original value, but with opposite velocities. Ions of the same mass, but at a higher energy, travel a longer path in the reflectron than ions of lower energy. The resolution of such reflection instruments is in the region of 5000 to 10,000.

One basic requirement for a TOF mass analysis is that all ions must leave the ion source at exactly the same time. When an EI source is fitted to a TOF, this requirement is fulfilled by switching on the filament for a short duration or by leaving the filament on all the time and pulsing the accelerating potential. $^{252}$Cf-PD and LDI techniques are ideal as the source of ions for a TOF analyzer because they function in the pulse mode.

No upper mass limit exists for a TOF analyzer. This welcome feature, coupled with the fact that the $^{252}$Cf-PD and MALDI ionization techniques can transform high-mass ions into the gas phase, has revived an interest in TOF mass analyzers. Dedicated commercial instruments are available outfitted with $^{252}$Cf-PD or MALDI sources.

Other attractive features of the TOF mass filters are high scanning speed (10,000 spectra per second), high detection sensitivity, simplicity of operation, and low cost. Ion transmission through the flight tube is very high because all ions that exit the ion source are potentially transmitted and detected.

### *3.6. Fourier Transform Ion Cyclotron Resonance Mass Spectrometer*

A Fourier transform mass spectrometer (FTMS) is based on the principle of ion cyclotron resonance (ICR), in which ionization, mass analysis, and detection take place in the same region, generally in a 1-in cubic cell positioned in a strong magnetic field (72). Ions are formed in the pulsed mode, and trapped within the cell by applying an electric potential to the front and rear plates. The containment of ions within a magnetic field forces the ions to move in circular orbits, described by Eq. (1.21). The time required to complete a single revolution is given by

$$t = \frac{2\pi r}{v} \tag{1.29}$$

$$t = 2\pi\left(\frac{m}{zB}\right) \tag{1.30}$$

$$\omega_c = \frac{zB}{2\pi m}, \tag{1.31}$$

where $r$, $v$, $z$, $m$, and $B$ are as defined earlier. The cyclotron frequency $\omega_c$, the number of revolutions per second, is calculated according to Eq. (1.31). Thus, the cyclotron frequency of an ion is related to the magnetic field strength and to the $m/z$ ratio of the ion.

In an FTMS analysis, a broad-band excitation pulse is applied to the side plates of the cell. The ions whose cyclotron frequency matches the applied frequency absorb energy from that pulsed signal and are promoted to orbits of a larger radius. When these coherently moving ions are in the proximity of the receiving plates, they transmit a complex rf signal (i.e., the image current) that contains frequency components characteristic of each ion. By applying a Fourier transform, this time-domain signal is converted to a frequency-domain signal to provide a mass spectrum.

The important features of an FTMS are high detection efficiency, high resolving power, high mass range, high scan speed, multichannel analysis, the possibility of higher-order MS/MS analysis, and low dynamic range. However, to obtain good resolution and sensitivity, the pressure inside the cell must be low to avoid ion–molecule reactions and scattering of the ions.

### *3.7. Ion-Trap Mass Spectrometer*

An ion trap (IT) is a three-dimensional analogue of a quadrupole mass filter (73). It consists of a doughnut-shaped central ring electrode and two end-cap electrodes, as shown in Fig. 1.16. For ideal field geometry, the electrodes must be of hyperbolic cross-section. The field is generated by applying an rf voltage to the central ring electrode and maintaining the end-cap electrodes at ground potential.

Like an FTMS, an IT also uses temporal sequences for mass analysis. A pulse of an electron beam enters through one of the end-cap electrodes to ionize gaseous molecules present in the ion trap. At certain rf amplitude, all ions above a specific *m/z* follow trajectories that keep them confined within the volume described by the electrodes. Increasing the rf voltage destabilizes the ion trajectories and ejects the ions sequentially from the ion trap region through perforations in the other end-cap electrode. The ejected ions are detected by an external detector to provide a mass spectrum.

The performance characteristics of an IT are similar to those of a quadrupole mass filter, viz., high scan speed, high ion transmission, low cost, and simplicity of operation. Because of these attributes, ITMS instruments are

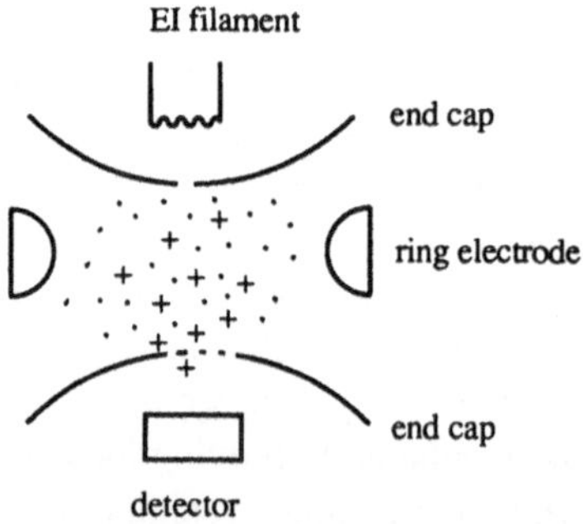


*Figure 1.16.* Schematic representation of an ion-trap mass spectrometer.

becoming increasingly popular both as analytical devices and as research tools.

## 4. TANDEM MASS SPECTROMETRY

The popularity of tandem mass spectrometry (MS/MS) has grown very rapidly (74,75). Its usefulness is unsurpassed in structure determination and identification of compounds in complex mixtures. MS/MS takes advantage of fragmentation reactions that occur in the FFRs of multisector instruments (see Fig. 1.2). As the name implies, MS/MS uses two stages of mass analysis. The first stage performs mass selection of a specified ion from a mixture of ions formed in the ion source. This precursor ion undergoes fragmentation in the intermediate region. The second stage mass-analyzes the product ions formed in the intermediate step. This operation is akin to GC/MS in that the first stage of MS/MS separates a mixture of ions into individual components (as GC resolves a mixture of compounds) and the second stage obtains mass spectra of each mass-resolved ion.

The major advantage of MS/MS is its molecular specificity, which is the result of an incontrovertible link established between the precursor ion and all of its product ions. Those product ions must derive exclusively from the preselected precursor. Since the introduction of desorption ionization techniques such as FAB and LSIMS for analysis of compounds of biomedical relevance, MS/MS has gained even more critical significance. A FAB spectrum is beset with chemical background noise caused by the matrix ions. Also, many isolates of biological samples are mixtures, even after the best possible chromatographic separation. MS/MS overcomes most of those difficulties.

The most common mode of MS/MS operation is the *product ion spectrum* (i.e., the spectrum of the product ions formed by dissociation of a mass-selected precursor). This spectrum is useful in the structural analysis of the preselected ion. The second type of spectrum obtained in MS/MS is the *precursor ion spectrum,* in which information about all precursor ions that fragment to a preselected product ion is obtained. In the *constant neutral loss spectrum,* by selecting a specific neutral loss, information about all precursors that undergo loss of that common neutral is obtained. These last two modes of operation are specially useful for mixture analysis. For example, in all opioid peptides, the sequence of the first four amino acids is Tyr-Gly-Gly-Phe. The FAB mass spectra of all these peptides contain an ion of $m/z$ 425, and they also exhibit the loss of a neutral moiety of 107 Da (76). The ion of $m/z$ 425 consists of the first four amino acids, and the neutral loss of 107 Da is the result of cleavage of the Tyr side chain. When a mixture of those neuropeptides is analyzed using the precursor ion and constant neutral loss scans, the resulting spectra contain information about all the neuropeptides that fragment to yield $m/z$ 425 and exhibit the loss of 107 Da, respectively.

### 4.1. Magnetic Sector MS/MS Instruments

As was discussed in Section 3.3, most high-performance magnetic sector instruments consist of an electrostatic analyzer and a magnetic analyzer. The FFRs before both of these analyzers are useful for MS/MS studies. When an ion of mass $m_1$ fragments in the FFR before an electric sector, the product ion ($m_2$) does not have the same kinetic energy as the precursor ion because the mass has changed and the velocity has remained the same. To pass this newly formed ion through the electric sector, the electric field $E_1$ must be changed to a new value $E_2$ such that

$$E_2 = \frac{m_2}{m_1} E_1 . \tag{1.32}$$

Thus, by scanning the electric sector, the energy spectrum, which gives information about all product ions, is obtained.

When an ion fragments in the FFR before a magnetic analyzer, the product ion is observed not at mass $m_2$ but at

$$m^* = \frac{m_2^2}{m_1} . \tag{1.33}$$

#### 4.1.1. Mass-Analyzed Kinetic Energy Spectrometry

To obtain an MS/MS spectrum with a two-sector mass spectrometer, it is advantageous that the magnetic analyzer precede the electrostatic analyzer (i.e., it uses the BE geometry) (77). With this type of instrument, the precursor ion is mass-selected unambiguously by adjusting the magnetic field and is allowed to fragment in the region ($FFR_2$) between the two analyzers (Fig. 1.13). The mass analysis of those fragment ions is obtained by scanning the electric sector [Eq. (1.32)]. This technique is known as mass-analyzed ion kinetic energy spectrometry (MIKES) (77,78). As mentioned above, a fragmentation in a FFR is accompanied by a release of kinetic energy, a process that causes broadening of peaks in the MIKES spectrum and results in a very poor product ion resolution ($<300$). In spite of this drawback, MIKES has been used extensively for several years to solve a number of ion chemistry, analytical, and bioanalytical problems (79).

*Table 1.2*
*Linked-Field Scans for MS/MS*

| Scan type | Reaction region | Information obtained | Collision energy | Instrument |
|---|---|---|---|---|
| V | $FFR_1$ | Precursor ion spectrum | high | EB |
| E | $FFR_2$ | Product ion (MIKES) spectrum | high | BE |
| | $FFR_3$ | | | EBE |
| B/E | $FFR_1$ | Product ion spectrum | high | EB,BE,EBE,BEB,EBQQ, BEQQ |
| | $FFR_2$ | | | EBE,BEB |
| $B^2E$ | $FFR_1$ | Precursor ion spectrum | high | EB,BE,EBE,BEB,EBQQ, BEQQ |
| | $FFR_2$ | | | EBE,BEB |
| $B^2E$ | $FFR_2$ | Precursor ion spectrum | high | BE,BEQ |
| | $FFR_3$ | | | EBE,EBEQ |
| $(B/E')^2(1\text{-}E')$ | $FFR_1$ | Neutral loss spectrum | high | EB,BE,EBE,BEB,EBQQ, BEQQ |
| $Q_3$ | $Q_2$ | Product ion spectrum | low | $Q_1Q_2Q_3$(TSQ) |
| $Q_2$ | $Q_1$ | | | EBQQ,BEQQ |
| $Q_1$ | $Q_2$ | Precursor ion spectrum | low | $Q_1Q_2Q_3$ |
| B | $Q_1$ | | | EBQQ,BEQQ |
| $Q_1$–$Q_3$ | $Q_2$ | Neutral loss spectrum | low | $Q_1Q_2Q_3$ |
| $B^2$–$Q_2$ | $Q_1$ | | | EBQQ,BEQQ |

### *4.1.2. Linked-Field Scans*

Because a vast majority of double-focusing instruments is of the forward geometry design, a strong need was felt to develop MS/MS techniques for use with these instruments. This objective was accomplished using linked-field scans, in which two of the fields (*V*, *E*, or *B*) are scanned simultaneously, keeping their ratio constant (80). A list of commonly used linked-field scans is provided in Table 1.2. The mass selection and fragmentation in the EB instruments occur in $FFR_1$ (see Fig. 1.12). For example, the product ion spectrum is obtained by the ***linked-field scan at constant B/E***. Because the double-focusing principle still applies to the product ion analysis, the mass resolution is much higher compared to the MIKES spectrum but somewhat dependent on the kinetic energy release. The major drawback of this type of scan, however, is that mass selection of the precursor is at lower resolution (typically 300 to 400), so that the specificity of the analysis is somewhat compromised. In spite of this limitation, our laboratory has used the linked-field scan at constant *B/E* for a number of years for qualitative and quantitative analysis of several biologically active neuropeptides (81).

The *linked-field scan at a constant* $B^2/E$ provides the precursor ion spectrum, and the neutral loss spectrum is obtained when $(B/E')^2(1 - E')$, where $B$ is the magnetic field required to transmit $m_2$ and $E' = E_2/E_1$, is held constant throughout the scan.

These linked-field scans are also applicable to reverse-geometry instruments, with the difference that for the precursor ion scan, the product $B^2E$ is held constant and the mass selection and fragmentation occur in the $FFR_2$ (Fig. 1.13).

### *4.1.3. Three- and Four-Sector Tandem Mass Spectrometers*

In an effort to overcome the resolution problem inherent with MIKES and linked-field scans, three- and four-sectors instruments are used to obtain high quality MS/MS spectra (69). The most common designs are of geometry EBE (Kratos MS-50TA, Manchester, UK), BEB (VG ZAB-3F, Manchester, UK), EBEB (JEOL HX110/HX110, Tokyo, Japan, and VG 70-SE-4F, Manchester, UK), BEEB (VG ZAB-SE-4F, Manchester, UK), and BEBE (Kratos Concept IIHH, Manchester, UK).

In the current four-sector instruments, it is possible to incorporate an optional intermediate-region ion source for mass calibration of MS-2 and also with the intention that the MS-2 be able to function as a stand-alone mass spectrometer. The latest of the four-sector tandem mass spectrometers is the VG ZAB-T, which uses a reverse-geometry instrument for MS-1 and the reverse-geometry Mattauch–Herzog design for MS-2 (69). The special features of this instrument are the use of an inhomogeneous electric field produced by a series of shim electrodes for the second ESA and a multichannel array detector for the second stage mass analysis.

The three-sector instruments (EBE and BEB) permit high-resolution (over 100,000) selection of the precursor ion; however, product ion analysis is limited to low to modest resolution. The BEB design is superior to the EBE geometry in that the first magnet mass-selects the precursor ion and the remaining EB portion provides a high-resolution mass analysis of the product ions. Of course, the four-sector instruments have the advantage of retaining the high resolving power in MS-1 while providing improved mass resolution in MS-2. However, it must be remembered that an upper limit (<10,000) of resolution exists in MS-2 because of the energy spread of the product ions. In addition to the intermediate-region reactions, these multisector instruments also have access to reactions in other FFRs. A unique feature of these multisector instruments is the possibility of higher-order ($MS^n$) MS/MS experiments (82).

The four-sector mass spectrometers have been used for obtaining the amino acid sequences of peptides and proteins and for determining posttranslational modifications (83). Those experiments have established that the upper mass limit for CID of peptides is around 3000 Da. Coupled with enzymolysis to obtain smaller fragments, it is possible to use these instruments to derive the amino acid sequence of high-mass proteins. However, the exorbitant cost of the four-sector instruments limits their use to a few selected laboratories.

### *4.2. Triple-Sector Quadrupole*

To benefit from the unique advantages of quadrupoles for MS/MS analysis, the triple-sector quadrupole (TSQ) was designed (84). In this instrument, three quadrupoles are arranged in sequence (Fig. 1.17); the first ($Q_1$) and third ($Q_3$) quadrupoles function in the normal way (i.e., as mass filter devices), whereas the center quadrupole ($Q_2$) operates in the rf mode only and serves as an ion containment region where the low-energy CID experiments are performed.

To obtain a product ion spectrum, $Q_1$ is adjusted to allow mass selection of the precursor ion and $Q_3$ is scanned. This procedure is reversed when the precursor ion spectrum is required (i.e., $Q_3$ is used to mass-select the desired product ion and $Q_1$ is scanned). For the neutral loss scan, the fields of $Q_1$ and $Q_3$ are offset by a factor related to the mass of the neutral and then scanned in tandem to maintain that mass difference.

The important advantages of a TSQ are low cost, operational simplicity, straightforward scan laws, linear mass scale, and unit mass resolution. However, the resolution of the precursor ion selection and product ion mass analysis and the mass range (<4,000) of TSQ are much lower than in the four-sector instruments discussed above. Also, an inability to perform high-energy CID limits their use for certain types of reactions (e.g., charge permutation). Nevertheless, the TSQ is an effective alternative to the high cost of the four-sector mass spectrometers and to the low-resolution mass analysis limitation of the two-sector instruments.

### *4.3. Hybrid Tandem Mass Spectrometers*

Hybrid tandem mass spectrometers are constructed using two different mass analysis principles. Thus, the advantages of the best performance features of both are utilized (75). The most common design uses a magnetic sector MS

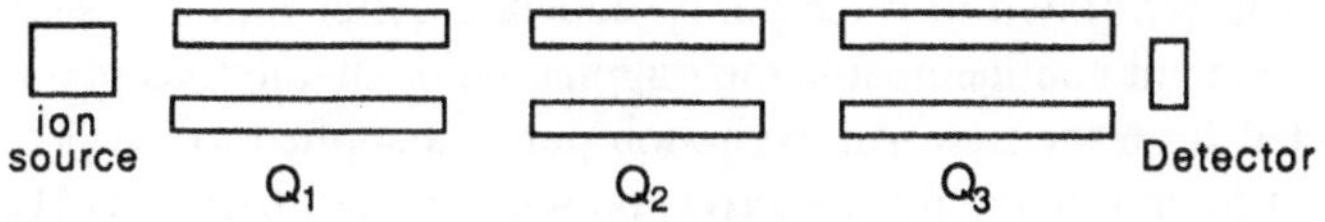


*Figure 1.17.* Diagram of a triple-sector quadrupole tandem mass spectrometer.

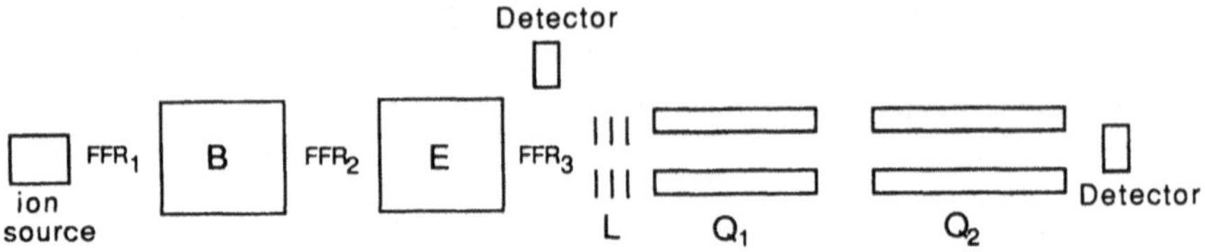


*Figure 1.18.* Diagram of a hybrid tandem mass spectrometer that employs a reverse-geometry magnetic sector for MS-1 and a quadrupole for MS-2.

for MS-1 and a quadruple for MS-2 (Fig. 1.18). Now a single instrument can be used to access both low- and high-energy regimes for collision activation. Forward- and reverse-geometry double-focusing magnetic sector instruments have both been utilized for this purpose. Because of the mismatch in translational energies of the ions in these two analyzers, the intermediate region must incorporate a deceleration lens assembly L to convert the keV to eV ions. The addition of an off-axis detector before this lens assembly permits the first stage to be used as a stand-alone high-performance (high resolution, accurate mass determination, and high mass range) double-focusing mass spectrometer.

In MS/MS applications using these hybrid geometries, mass selection of the precursor is at high resolution (using the double-focusing principle of the magnetic sector) and the product ion analysis at unit mass resolution (using a quadrupole as MS-2). The first quadrupole ($Q_1$) functions as an ion containment region and as a low-energy collision cell, and the second quadrupole ($Q_2$) as a scanning device to obtain MS/MS data. The upper mass range of the current high-performance quadrupoles is 4000 Da. This mass limit is more than adequate for MS/MS studies of most biological compounds because, as described above, it is not possible to fragment singly charged ions of mass above 3000 Da by collision activation. The added advantage of these instruments is low cost compared to the four-sector tandem instruments.

## 4.4. *Tandem Ion Trap Mass Spectrometry*

Whereas in the instruments discussed above, different stages of MS/MS analysis are separated in space, the FTMS (85) and ITMS (86) act as tandem-in-time mass spectrometers. Although these two instruments differ in design, MS/MS operation is conceptually similar. Ion formation, excitation, and detection are all performed in a temporal sequence. After initial steps of ion formation and their confinement in the trapping field, all ions except the precursor are ejected from the trap. An excitation pulse is applied to the precursor ion, causing it to undergo collision activation with an inert bath gas. The resulting product ions are mass-analyzed as usual. Because all steps are performed within the same space, the possibility of interference due to side reactions exists. These instruments are also appropriate for higher-order MS/MS experiments.

### *4.5. Metastable Ions*

As discussed in Section 2.1, metastable ions fragment after leaving the ion source but before reaching the detector. These metastable ions are characterized by a narrow range of energies just above the threshold for the lowest dissociation channel (shaded portion in Figure 1.2), and fragment at rates of $10^5$–$10^6$ $sec^{-1}$. In the conventional magnetic scan spectrum, they appear as diffuse peaks at the nonintegral mass given by Eq. (1.33). The broadness of the peak is due to the partitioning of the internal energy of the fragmenting ion into translational energy of the products. The width of the peak and the magnitude of the kinetic energy release are characteristic parameters of the structure of the fragmenting ion (77). These pieces of information, along with the ion abundances in the metastable spectrum, have been used to distinguish ionic structures (82).

### *4.6. Collision-Induced Dissociation*

In softer ionization techniques (CI, FD, FAB, LSIMS, ES, and LDI), the fragmentation pattern is relatively weak and yields little structural information. Furthermore, the population of the spontaneously fragmenting (i.e., metastable) ions sampled by MS/MS is usually low. Therefore, to induce fragmentation of the energetically cool mass-selected ions (nondecomposing ions of Fig. 1.2), the MS/MS instruments are usually fitted with some means of additional fragmentation. The most common of these methods is collision activation, in which the fast-moving mass-selected ion collides with a neutral target, often a helium atom. During this excitation process, a part of the initial translational energy of the incident ion is converted into excitation energy, leading to its fragmentation:

$$m^{+\cdot} + N \rightarrow [m^{+\cdot}]^* \rightarrow (m^{+\cdot}, m_1^+, m_2^+, m_3^+, \text{etc.}) + N \qquad (1.34)$$

where $m^{+\cdot}$ is the mass-selected ion, $m_1^+$, $m_2^+$, etc. are its fragments, and N is the neutral target. This process is called collisionally activated dissociation (CAD) or collision-induced dissociation (CID). In high-energy collisions (e.g., for ions that have translational energies in the keV region), the initial collision leads to a vertical electronic excitation of the incident ion. The excitation energy is rapidly converted statistically into various vibrational modes, ultimately leading to fragmentation, a process reminiscent of EI. However, during this unimolecular dissociation of the excited ions, a part of the excess energy is released as translational energy, a process that causes broadening of the peaks in the CID spectrum.

In low-energy collisions (for ions of energies <100 eV, which are accessible in quadrupoles and FTMS), excitation occurs mainly to higher vibrational states followed by unimolecular dissociation of the excited ion. The low-energy ions spend more time in the reaction region and undergo a greater number of collisions. Because of the relatively long life of the collision complex, the low-energy ions have a higher probability to participate in ion–molecule reactions. Also, the target gas has a profound effect on the nature of the MS/MS spectrum. Thus, significant differences are expected in the low- and high-energy CID spectra.

Collision activation also yields charge permutation reactions, some of which are endothermic and observed only in the high-energy regimes.

Charge exchange $$m^{+\cdot} + N \rightarrow m + N^{+\cdot} \quad (1.35)$$

Charge inversion $$m^{-\cdot} + N \rightarrow m^{+\cdot} + N + 2e^- \quad (1.36)$$

Charge stripping $$m^{+\cdot} + N \rightarrow m^{++\cdot} + N + e^- \quad (1.37)$$

Charge ionization $$m + N \rightarrow m^{+\cdot} + N + e^- \quad (1.38)$$

These reactions are diagnostic of the structural differences between the ions. Reaction (1.38) has been used in conjunction with a novel technique called *neutralization–reionization MS,* in which source-formed ions are first neutralized in an FFR and then ionized using collision activation (87).

Although not yet popular, alternative modes of ion activation have been developed by some researchers. These modes are surface-induced dissociation (in which the incident ion is impacted onto a metal surface (88), electron impact activation (a phenomenon similar to EI) (89), and photodissociation (using high-energy photons, e.g., from a laser beam) (90). Space does not permit a detailed discussion of these newer specialized techniques; for further reading, the references cited here may be consulted.

## 5. *ION DETECTORS*

The purpose of a detector is to measure the current due to the mass-analyzed ion beam. Sensitivity, accuracy, resolution, and response time are the most important characteristics of any ion detector. One of the simplest devices is a *Faraday cup detector,* which consists of a conical metal cup connected to an amplifier. The charge transferred to the cup by the incoming ion beam is measured directly by the voltage drop across a large feedback resistance ($10^{11}$ Ω). This detector is very useful for high-precision measurements (e.g., in isotope ratio measurements). However, the response of a Faraday cup detector is slow, making it unsuitable for scanning mass spectrometers.

The most widely used detector is an *electron multiplier* (EM), in which a series of dynodes, usually made of Be/Cu, are connected together through a chain of resistors of equal value (Fig. 1.19). A high voltage (up to −3,000 V) applied across the first dynode (called the conversion dynode) and the final dynode (anode) is divided into equal steps between the dynodes. The ion beam strikes the conversion dynode, causing it to emit secondary electrons in direct proportion to the number of incident ions. The secondary electrons are accelerated toward and made to strike the second dynode, which emits more secondary electrons. Thus, amplification of secondary electrons occurs at each successive stage. An EM exhibits fast response time, high sensitivity, and high gain (which depends upon the number of conversion stages used, generally 16 or 20).

A *channel electron multiplier* (CEM) also functions on the same principle but uses a continuous arrangement of electrodes. The detector is a horn-shaped glass tube; either it is doped with Pb or its inner surface is coated with Be. A high voltage applied across the two ends of the tube is distributed uniformly throughout the length of the tube. A CEM is more compact and less expensive than a discrete EM. It is more commonly used in quadrupole mass filters.

The gain of a CEM is a function of the length-to-diameter ratio of the tube. Therefore, microchannel plates (MCP) using fiber optic cables of very small diameter (~10 μm) can serve as very efficient CEM. In this arrangement, two MCPs made of millions of individual channels are arranged in a chevron configuration. The ion beam strikes the front of the MCPs and secondary electrons, after several-fold multiplication, emerge from the rear face of the MCPs. The MCP-EM is used in TOF instruments and in multichannel array detectors.

Some new generations of mass spectrometers use a *photomultiplier de-*

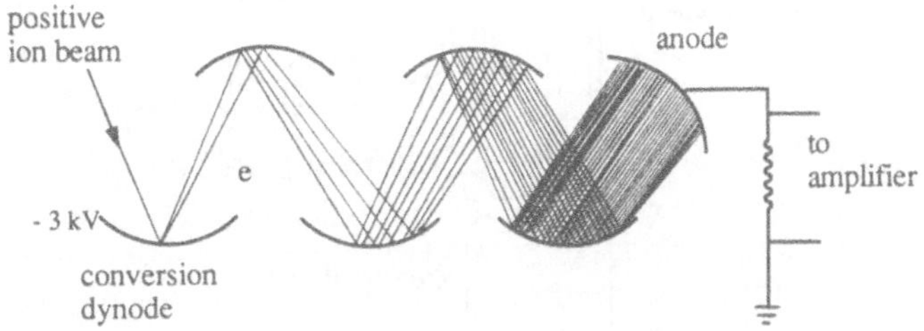


*Figure 1.19.* Principle of operation of an electron multiplier.

*tector* in place of an EM. In this type of detector, the ions first strike a scintillation material, and the emitted photons are detected by a conventional photomultiplier.

One drawback of an EM is that the yield of secondary electrons depends upon the velocity of the striking ions. Because all ions reaching the detector have the same kinetic energy, their velocity is an inverse function of their mass [Eq. (1.26)]. Therefore, the detection efficiency of high-mass ions decreases commensurately. In order to augment the velocity of the incident ions, post-acceleration is used in modern mass spectrometers. For positive ion detection, the beam is first deflected toward an off-axis *post acceleration detector* (PAD), which is held at a potential of –5 to –30 kV. The secondary electrons emitted from this detector are then accelerated toward an EM.

The detection of negative ions using an EM is less efficient because those ions are slowed down as they approach the first dynode (held at a large negative potential), resulting in reduced yields of secondary electrons. Postacceleration is, thus, a necessary step in negative ion detection. The PAD is held at a positive potential to permit acceleration of the incoming negative ions. The impact of negative ions on the target of the PAD yields secondary positive ions, which are then detected by an EM in the usual way.

Bateman and Bott (91) have designed an ion detector that incorporates two conversion dynodes—one for positive-ion and another for negative-ion detection, a phosphor, and a photomultiplier (Fig. 1.20). For positive ion detection, the conversion dynode is maintained at –10 to –20 kV and the secondary electrons are accelerated toward a phosphor, the voltage of which is varied between 10 and 20 kV. For negative-ion detection, the incoming ions are deflected first toward a cylindrical conversion dynode, which is held at half the phosphor voltage. The electrons emitted from this electrode travel toward the

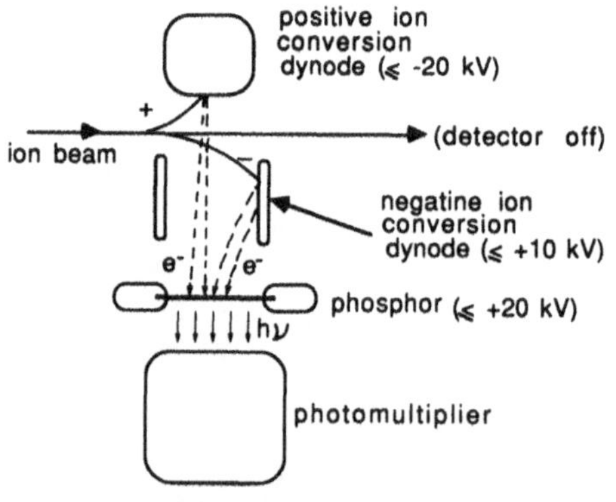


*Figure 1.20.* Schematic of a combined positive- and negative-ion postacceleration detector.

phosphor under the influence of the penetrating electric field of the positively charged phosphor.

As discussed above, a serious disadvantage of the focal point mass spectrometers is that, at any one instant, only an ion of one particular $m/z$ is detected, while the ion current due to other ions is lost. An ideal arrangement would provide simultaneous detection of all mass-resolved ions. A *photoplate detector* used with a Mattauch–Herzog instrument fulfills this requirement. The ions impinging on a photoplate produce darkening in proportion to the number of incident ions. A microdensitometer measures the relative position and optical density of the lines to indicate the mass and intensity of the ions, respectively. Although simultaneous detection of all ions by a photoplate is very advantageous, using the photoplate detector is very tedious.

An alternative to photoplate detection is a *multichannel array detector*. The detector, developed by Cottrell and Evans (68) uses a CEM made of a pair of MCPs arranged in a chevron configuration, a phosphor screen, a fiber optic coupling, and a 1024-channel photodiode array (Fig. 1.21). The mass-resolved ion beam strikes the entrance face of the MCP assembly; at the phosphor screen, the secondary electrons emerging from the MCP assembly are converted to photons, which travel through the fiber optic into a charge- or plasma-coupled device (CCD or PCD), where they are converted to electric charge. This detector was developed for use with a Nier–Johnson mass spectrometer, which has a useful focal plane width of 5–10% of the central mass. A simultaneous detection of a mass window >4% is possible with this 25-mm array detector.

The VG ZAB-T four-sector tandem MS is fitted with a 152-mm array consisting of 6144 channels interfaced to a 4096-channel photodiode array

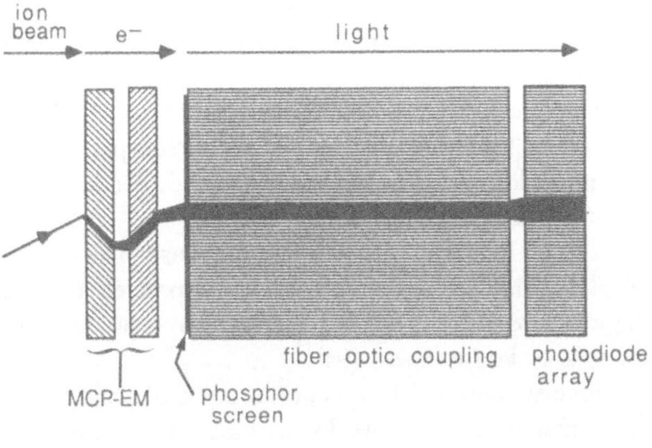


*Figure 1.21.* Principle of operation of a multichannel array detector.

(69). A mass range of 2:1 is accessible when an intermediate ion source is used. Biemann and co-workers (92) have used a novel arrangement to extend the mass range of an array detector without increasing its size. By their addition of two quadrupole lenses after the second-stage magnet of a four-sector EBEB instrument, the mass range of a 50-mm array was extended from 6.6% to 30%.

The array detector used in Finnigan MAT instruments is based on the principle of position- and time-resolved ion counting (PATRIC) (93). The ion beam is converted in the usual manner into electrons by using a pair of MCPs, but the electron cloud produced is detected by a multianode assembly instead of a photodiode array. The anode assembly consists of a large number of conductive strips connected by identical capacitors. Charge-sensitive amplifiers located at each end of the anode assembly provide position and time information for every ion. The detectors cover a mass range of 8%. Because ions are detected individually, the array can also be used to obtain a mass spectrum by scanning over the complete mass range.

One of the major advantages of array detectors is a significant increase in detection sensitivity (100-fold or more). However, the entire impact of this advantage is not felt when FAB or other desorption modes of ionization are used because the chemical background noise inherent with these desorption methods is also amplified concurrent with the analyte signal. For this reason, the full potential of multichannel array detectors can be realized when they are used in MS-2 of a tandem mass spectrometer.

## 6. *ACCURATE MASS MEASUREMENT*

Accurate mass data permit assignment of elemental composition to an ion. This information is essential in structure determination and in interpretation of mass spectra of an unknown or of a mixture of compounds. Ideally, accurate mass measurement is accomplished using the double-focusing magnetic sector mass spectrometers.

One of the simplest ways to obtain the accurate mass of a compound is to record its spectrum at high resolution (ca. 10,000) in the presence of a reference compound. From the known mass reference ions (see Section 7), the masses of the sample ions are determined to within <5 ppm.

Alternatively, all ions are recorded simultaneously on a photographic plate using a focal plane Mattauch–Herzog mass spectrometer. Using a microdensitometer, the exact position of each mass is then accurately measured with respect to the ions of known mass (94).

The most precise (within 0.1–0.3 ppm) method of accurate mass measurement is the peak-matching technique. In this method, the magnetic field is held constant and the accelerating voltage is changed in such a manner that a reference ion of a known exact mass and the unknown ion alternatively follow the same trajectory. In practice, the lower of the two masses is displayed on an oscilloscope screen and the accelerating voltage is varied by a precision (within 1 ppm) voltage divider until the two ions are superimposed. The unknown mass is measured from this ratio, using Eq. (1.39). To achieve higher accuracy in mass measurement, the mass of the calibrant ion must be within 2% of the unknown mass. In the current generation of mass spectrometers, peak-matching is executed under software control. From Eq. (1.22), we have $m \propto 1/V$; therefore

$$m_2 = m_1 \frac{V_2}{V_1}. \tag{1.39}$$

Accurate mass data can easily be obtained for samples that are accessible to EI-MS analysis. In contrast, such measurements are difficult with FAB because the alkali salts used as reference materials tend to suppress the sample ion current. Also, the sample current is not very steady, and the reference ions are often not sufficiently close. For accurate mass measurement in FAB, it is advisable to use the matrix ions as the reference ions, or to use some internal standards such as polyethylene or polypropylene glycols. A compound closely related in mass and chemical characteristics to the analyte can also be used as a reference compound.

## 7. *MASS CALIBRATION STANDARDS*

As discussed in Section 6, a reference compound is needed for accurate mass measurement. A reference compound is also required for calibration of the data system and for tuning and performance checks of the instrument. A computer establishes a mass scale from the known masses of reference ions to assign mass values to the sample ions.

The reference compound should provide a sufficient number of ions of high relative intensities spaced regularly throughout the scan range, should be readily available, should be chemically inert, and should be sufficiently volatile that it can be introduced into and readily pumped away from the ion source. Compounds commonly used as standards for various ionization modes are listed in Table 1.3.

Perfluorokerosene (PFK) is the most widely used mass calibrant in EI-MS. The low-boiling mixture (PFK-L) is used up to 600 Da and the high-boiling (PFK-H) up to 900 Da. Perfluorotributylamine [PFTBA; $(C_4F_9)_3N$; also known as heptacosa] is another frequently used EI calibrant, especially for quadrupole

*Table 1.3*
*Mass Reference Standards*

| Compound | Structure | Ionization mode | Mass range (Da) |
|---|---|---|---|
| Perfluorotributylamine | $(C_4F_9)_3N$ | EI,CI | 0–600 |
| Perfluorokerosene | $CF_3(CF_2)nCF_3$ | EI,CI | 0–900 |
| Ultramark F series (Fomblin) | $CF_3O(CFCF_3CF_2O)_m$-$(CF_2O)_nCF_3$ | EI,CI,FAB | up to 3500 |
| Ultramark 1621 | perfluoroalkoxycyclo-triphosphazenes | EI,CI | up to 3000 |
| Tris(perfluoroalkyl)-*s*-triazine | | EI | up to 1500 |
| Glycerol | $CH_2OHCHOHCH_2OH$ | FAB | 20–1200 |
| Cesium iodide | CsI | FAB | 130–30,000 |
| CsI/glycerol | | FAB | 90–3500 |
| CsI/NaI/RbI | | FAB | 20–1000 |
| LiI/NaI | | FAB | 0–1000 |
| Polyethylene glycol | $H(OCH_2CH_2)nOH$ | FAB/ES/TS | 50–2000 |
| Polypropylene glycol | $H[OCH(CH_3)CH_2]nOH$ | ES | up to 3000 |
| Myoglobin | | ES | 600–2500 |
| Lysozyme | | ES | 1000–2100 |
| Polypropylene glycol sulfate | | ES (−*ve*) | 300–1700 |

mass filters, where *m/z* of 69 to 502 is used as a measure of system performance. For high-mass (up to 3000 Da) analysis in EI-MS, triazines and Ultramark compounds are appropriate as reference mass markers.

In FAB or LSIMS, the glycerol matrix can also serve as a mass calibrant up to *m/z* ca. 1200. Intense cluster ions of *m/z* (92*n* + 1) and (92*n* − 1) (where *n* is an integer, 92 being the mass of glycerol) are seen in the positive- and negative-ion analyses, respectively. Alkali metal halides (e.g., CsI, LiI, NaI, and RbI) are the most appropriate reference compounds for FAB-MS. For example, CsI yields a series of ions beyond *m/z* 30,000 of compositions. $Cs(CsI)n^+$ and $I(CsI)n^-$ in the positive- and negative-ion modes, respectively. One difficulty with these standards is that the adjacent peaks are too far apart. However, a mixture of NaI, CsI, and RbI can be used to minimize those gaps. In our laboratory, a saturated solution of CsI in glycerol is successfully used as a mass calibrant up to *m/z* 3000.

## 8. *QUANTIFICATION*

A mass spectrometer is one of the most sensitive of analytical instruments. It has the capability of detecting a single ion. Therefore, one of the important applications of MS is quantification. However, to achieve the high accuracy of quantitative analysis, the use of an *internal standard* is mandatory. An internal standard accounts for any fluctuation in the MS response and the sample losses in various sample handling and chromatographic steps. An ideal internal standard is one whose physical and chemical properties match the analyte. A stable isotope-labeled analog of the analyte fulfills this criterion.

Quantification is accomplished using the *selected ion monitoring* (SIM) mode of ion detection, in which the ion current from only one or a few selected ions rather than a complete mass spectrum is recorded. During the scan of a complete spectrum, all ions other than the one arriving at the detector are ignored. On the other hand, in SIM, the detector is used exclusively to monitor only the mass-selected ion(s). In this way, most of the ion current at the selected mass value is monitored, thereby significantly improving the detection sensitivity compared to that of the full-scan spectrum. The molecular ion or any other intense peak along with the corresponding ion of the internal standard is chosen for the SIM quantification experiment.

The disadvantage of the SIM technique is the lack of molecular specificity. In SIM, it is presumed that observed ion current arises only from the analyte. In practice, it is likely that some other interfering ions, due the presence of other constituents of the sample, may contribute to the mass of the analyte ion. Many biological samples are mixtures, even after extensive chromatographic separation. Furthermore, the use of a matrix in FAB produces background noise at every mass.

To improve the molecular specificity of the measurement, a technique known as *selected reaction monitoring* (SRM), is used. This method is an MS/MS technique in which a preselected analyte ion is allowed to fragment by collision activation in a FFR, and one or more specific product ions are monitored. The molecular specificity is the result of the structural link maintained between this exclusively mass-selected precursor ion and its product ions.

## 9. *CONCLUSIONS*

A brief discussion of various ionization techniques, different types of mass analyzers, tandem mass spectrometry, and a few important specialized techniques was presented in this chapter. It is obvious that no one ionization technique is universal and no one mass analyzer can solve all the analytical problems. For volatile samples, EI and CI are the preferred ionization modes, whereas large and more complex biological compounds are analyzed using desorption ionization and spray techniques. A mixture of biological compounds can be handled by coupling high-resolution separation techniques such as HPLC and CZE and CF-FAB or electrospray.

The choice of the mass analyzer is dictated by the mass range, resolution, scan speed, and detection sensitivity. Often, cost of the instrument plays a very important role in selection of a mass spectrometer. A wide selection of instruments from low-cost ITMS to high-performance four-sector magnetic deflection instruments is available. When molecular specificity is the important criterion of analysis, MS/MS is the method of choice.

ACKNOWLEDGMENT. The author wishes to thank the National Institutes of Health (Grant NS 28025) for financial support.

## *REFERENCES*

1. Thomson, J. J., 1913, *Rays of Positive Electricity and Their Applications to Chemical Analysis,* Longmans, London.
2. Fenn, J. B., Mann, M., Meng, C. K., Wong, S. F., and Whitehouse, C. M., 1989, Electrospray ionization for mass spectrometry of large biomolecules, *Science* **246:**64–71.
3. Fenn, J. B., Mann, M., Meng, C. K., Wong, S. F., and Whitehouse, C. M., 1990, Electrospray ionization—principles and practice, *Mass Spectrom. Rev.* **9:**37–70.
4. Smith, R. D., Loo, J. A., Ogarzalek Loo, R.R., Busman, M., and Udseth, H. R., 1991, Principles and practice of electrospray ionization—mass spectrometry for large polypeptides and proteins, *Mass Spectrom. Rev.* **10:**359–451.
5. Karas, M., and Hillenkamp, F., 1988, Laser desorption ionization of proteins with molecular masses exceeding 10,000 daltons, *Anal. Chem.* **60:**2299–2301.
6. Karas, M., Bahr, U., and Giebmann, U., 1991, Matrix-assisted laser desorption ionization mass spectrometry, *Mass Spectrom. Rev.* **10:**335–357.
7. Tolun, E., Dass, C., and Desiderio, D. M., 1987, Trace level measurement of enkephalin peptides at the attomole/femtomole level by FAB-MS. Optimization of FAB matrix conditions, *Rapid Commun. Mass Spectrom.* **1:**77–89.
8. McLafferty, F. W., 1980, *Interpretation of Mass Spectra,* University Science Books, Mill Valley, CA.
9. Rosenstock, H. M., Wallenstein, M. B., Wahrhaftig, A. L., and Eyring, H., 1952, Absolute rate theory of isolated systems and the mass spectra of polyatomic molecules, *Proc. Natl. Acad. Sci. U.S. A.* **38:**667–678.
10. Marcus, R. A., 1952, Unimolecular dissociations and free radical recombination reactions, *J. Chem. Phys.* **20:**359–362.
11. Munson, M. S. B., and Field, F. H., 1966, Chemical ionization mass spectrometry, *J. Am. Chem. Soc.***88:**2621–2630.
12. Harrison, A. G., 1983, *Chemical Ionization Mass Spectrometry,* CRC Press, Boca Raton, FL.
13. McLafferty, F. W., 1980, Unimolecular decompositions of even-electron ions, *Org. Mass Spectrom.* **15:**114–121.
14. Lias, G. G., Bartmess, J. E., Liebman, J. F., Holmes, J. L., Levin, R. D., and Mallard, W. G., 1988, *J. Phys. Chem. Ref. Data* **17:**Suppl. 1 (1988).
15. Vairamani, M., Mirza, U. A., and Srinivas, R., 1990, Unusual positive ion reagents in chemical ionization mass spectrometry, *Mass Spectrom. Rev.* **9:**235–258.
16. Dass, C., and Gross, M. L., 1983, Electrocyclic ring opening of 1-phenylcyclobutene and 3-phenycyclobutene radical cations, *J. Am. Chem. Soc.* **105:**5724–5729.

17. Dass, C., and Gross, M. L., 1985, The question of cyclic versus acyclic ions: the structure of $[C_{10}H_{12}]^{+\cdot}$ gas-phase ions, *Org. Mass Spectrom.* **20:**34–40.
18. Hunt, D. F., Stafford, Jr., G. C., Crow, F. W., and Russell, J. W., 1976, Pulsed positive negative ion chemical ionization mass spectrometry, *Anal. Chem.* **48:**2098–2105.
19. Hunt, D. F., and Sethi, S.K., 1978, Analytical applications of positive and negative ion chemical ionization mass spectrometry, in *High Performance Mass Spectrometry: Chemical Applications* (M. L. Gross, ed.), ACS Symposium Series 70, Washington, D. C., pp. 150–178.
20. Beckey, H. D., 1977, *Principles of Field Ionization and Field Desorption Mass Spectrometry,* Pergamon, Oxford.
21. Baldwin, M. A., and McLafferty, F. W., 1973, Direct chemical ionization of relatively involatile samples. Application to underivatized oligopeptides, *Org. Mass Spectrom.* **7:**1353–1356.
22. Sundqvist, B., and Macfarlane, R. D., 1985, $^{252}$Cf-Plasma desorption mass spectrometry, *Mass Spectrom. Rev.* **4:**421–460.
23. Cotter, R. J., Chen, L., and Wang, R., 1991, Plasma desorption mass spectrometry of peptides and peptide conjugates, in *Mass Spectrometry of Peptides* (D. M. Desiderio, ed.), CRC Press, Boca Raton, FL, pp. 17–40.
24. Jonsson, G., Hedin, A., Håkansson, P., Sundqvist, S. U. R., Bennich, H., and Roepstorff, P., 1989, Compensation for non-normal ejection of large molecular ions in plasma-desorption mass spectrometry, *Rapid Commun. Mass Spectrom.* **3:**190–191.
25. Roepstorff, P., 1991, Analysis of peptides and proteins by plasma desorption mass spectrometry, in *Mass Spectrometry of Peptides* (D. M. Desiderio, ed.), CRC Press, Boca Raton, FL, pp. 65–86.
26. Benninghoven, A., Rudenauer, F. G., and Werner, H. W., 1987, *Secondary Ion Mass Spectrometry–Basic Concepts, Instrumental Aspects, Applications and Trends,* Wiley, New York.
27. Barber, M., Bordoli, R.S., Sedgwick, R. D., and Tyler, R. N., 1981, Fast atom bombardment of solids (FAB): a new ion source for mass spectrometry, *J. Chem. Soc. Chem. Commun.* 325–327.
28. Aberth, W., Straub, K., and Burlingame, A. L., 1982, Secondary ion mass spectrometry with cesium ion primary ion beam and liquid target matrix for analysis of bio-organic compounds. *Anal. Chem.* **54:**2029–2034.
29. Barofsky, D., 1985, Liquid metal ion sources, in *Desorption Mass Spectrometry—Are SIMS and FAB the Same?* (P. A. Lyon, ed.), ACS Symposium Series 291, Washington, D. C., pp. 113–124.
30. Pachuta, S., and Cooks, R. G., 1985, Molecular secondary ion mass spectrometry, in *Desorption Mass Spectrometry—Are SIMS and FAB the Same?* (P. A. Lyon, ed.), ACS Symposium Series 291, Washington, D. C., pp. 1–42.
31. Dass, C., 1990, Fast atom bombardment combined with mass spectrometry for characterization of polycyclic aromatic hydrocarbons, *J. Am. Soc. Mass Spectrom.* **1:**405–412.
32. Gower, J. L., 1985, Matrix compounds for fast atom bombardment mass spectrometry, *Biomed. Mass Spectrom.* **12:**191–196.
33. DePauw, E., 1990, Matrix selection for LSIMS and FAB-MS, in *Methods in Enzymology* (J. A. McCloskey, ed.), Vol. 193, Academic Press, New York, pp. 201–204.
34. DePauw, E., Liquid matrices for secondary ion mass spectrometry, *Mass Spectrom. Rev.* **5:**191–196.
35. Cook, K. D., Todd, P. J., and Frier, D. H., Physical properties of matrices used for fast atom bombardment, *Biomed. Environ. Mass Spectrom.* **18:**492–497.
36. Dass, C., and Desiderio, D. M., 1988, Particle beam–induced reactions between peptides and liquid matrices, *Anal. Chem.* **60:**2723–2729.
37. Busch, K. L., 1991, Sample preparation and matrix selection for analysis of peptides by FAB

and liquid SIMS, in *Mass Spectrometry of Peptides* (D. M. Desiderio, ed.), CRC Press, Boca Raton, FL, pp. 173–200.

38. Naylor, S., Findeis, A. F., Gibson, B. W., and Williams, D. H., 1986, An approach toward the complete FAB analysis of enzymic digests of peptides and proteins, *J. Am. Chem. Soc.* **108**:6359–6363.
39. Ligon, W. V., and Dorn, S. B., 1986, Mass spectrometry determination of amines after formation of a charged surface-active derivative, *Anal. Chem.* **58**:1889–1892.
40. Caprioli, R. M., Fan, T., and Cottrell, J.S., 1986, Continuous-flow sample probe for fast atom bombardment mass spectrometry, *Anal. Chem.* **58**:2949–2954.
41. Caprioli, R. M., 1988, Analysis of biochemical reactions with molecular specificity using fast atom bombardment mass spectrometry, *Biochemistry* **27**:513–520.
42. Caprioli, R. M., Moore, W. T., DaGue, B., and Martin, M., 1988, Microbore high-performance liquid chromatography–mass spectrometry for the analysis of proteolytic digests by continuous-flow fast-atom bombardment mass spectrometry, *J. Chromatogr.* **443**:355–362.
43. Mosely, M. A., Deterding, L. J., Tomer, K. B., and Jorgenson, J. W., 1989, Coupling of capillary zone electrophoresis and capillary liquid chromatography with coaxial continuous-flow fast atom bombardment tandem sector mass spectrometry, *J. Chromatogr.* **480**:197–209.
44. Caprioli, R. M., Moore, W. T., Martin, M., DaGue, B., Wilson, K., and Moring, S., 1989, Coupling of capillary zone electrophoresis and continuous-flow fast-atom bombardment mass spectrometry for the analysis of peptide mixtures, *J. Chromatogr.* **480**:247–257.
45. Mosely, M. A., Deterding, L. J., Tomer, K. B., and Jorgenson, J. W., 1989, Capillary-zone electrophoresis/fast-atom bombardment mass spectrometry: Design of an on-line coaxial continuous-flow interface, *Rapid Commun. Mass Spectrom.* **3**:87–93.
46. Caprioli, R. M., and Lin, S. N., 1990, On-line analysis of penicillin blood levels in the live rat by combined microdialysis/fast-atom bombardment mass spectrometry, *Proc. Natl. Acad. Sci. U.S.A.* **87**:240–243.
47. Caprioli, R. M., 1990, *Continuous-Flow Fast Atom Bombardment Mass Spectrometry*, Wiley, New York.
48. Hillenkamp, F., and Karas, M., 1990, Mass spectrometry of peptides and proteins by matrix-assisted ultraviolet laser desorption/ionization, in *Methods in Enzymology* (J. A. McCloskey, ed.), Vol. 193, Academic Press, New York, pp. 280–295.
49. Beavis, R. C., and Chait, B. T., 1989, Factors affecting the ultraviolet laser desorption of proteins, *Rapid Commun. Mass Spectrom.* **3**:233–237.
50. Tanaka, K., Waki, H., Ido, H., Akita, S., and Yoshida, T., 1988, Protein and polymer analysis up to *m/z* 100,000 by laser ionization time-of-flight mass spectrometry, *Rapid Commun. Mass Spectrom.* **82**:151–153.
51. Blakley, C. R., and Vestal, M. L., 1983, Thermospray interface for liquid chromatography/mass spectrometry, *Anal. Chem.* **55**:750–754.
52. Willoughby, R. C., and Browner, R. F., 1984, Monodisperse aerosol generation interface for combining liquid chromatography with mass spectrometry, *Anal. Chem.* **56**:2626–2631.
53. Dole, M., Mack, L. L., and Hines, R. L., 1968, Molecular beams of microions, *J. Chem. Phys.* **49**:2240–2249.
54. Mack, L. L., Kralik, P., Rheude, A., and Dole, M., 1970, Molecular beams of microions. II, *J. Chem. Phys.* **52**:4977–4986.
55. Yamashita, M., and Fenn, J. B., 1984, Electrospray ion source. Another variation on the free-jet theme, *J. Phys. Chem.* **88**:4671–4675.
56. Mann, M., and Fenn, J. B., 1992, Electrospray mass spectrometry: principle and methods, in *Mass Spectrometry: Clinical and Biomedical Applications* (D. M. Desiderio, ed.), Plenum Press, New York, pp. 1–35.
57. Aleksandrov, M. L., Gall, L. N., Krasnov, V. N., Nikolaev, V. I., Pavlenko, V. A., and

Shkurov, V. A., 1984, Ion extraction from solutions at atmospheric pressure—a method for mass spectrometric analysis of bioorganic substances, *Dokl. Akad. Nauk. SSSR* **277**:379–383.

58. Covey, T. R., Bonner, R. F., Shushan, B. I., and Henion, J. D., 1988, The determination of protein, oligonucleotide, and peptide molecular weights by ion-spray mass spectrometry, *Rapid Commun. Mass Spectrom.* **2**:249–256.
59. Bruins, A. P., Covey, T. R., and Henion, J. D., 1987, Ion-spray interface for combined liquid chromatography/atmospheric pressure ionization mass spectrometry, *Anal. Chem.* **59**:2642–2646.
60. Green, B. N., Oliver, R. W. A., Falick, A. M., Shackleton, C. H. L., Rutman, E., and Witkowska, H. E., 1990, Electrospray MS, LSIMS and MS/MS for the rapid detection and characterization of variant hemoglobins, in *Biological Mass Spectrometry* (A. L. Burlingame and J. A. McCloskey, eds.), Elsevier, Amsterdam, pp. 129–146.
61. Dass, C., Kusmierz, J. J., Desiderio, D. M., Jarvis, S. A., and Green, B. N., 1991, Electrospray mass spectrometry for the analysis of opioid peptides and for the quantification of endogenous methionine enkephalin and β-endorphin, *J. Am. Soc. Mass Spectrom.* **2**:149–156.
62. Lee, E. D., Muck, W., Henion, J. D., and Covey, T.R., 1989, Capillary zone electrophoresis/tandem mass spectrometry for the determination of sulfonated azo dyes, *Biomed. Environ. Mass Spectrom.* **18**:253–257.
63. Olivares, J. A., Nguyen, N.T., Yonker, C.R., and Smith, R. D., 1987, On-line mass spectrometric detection for capillary zone electrophoresis, *Anal. Chem.* **59**:1230–1232.
64. Smith, R. D., Olivares, J. A., Nguyen, N. T., and Udseth, H. R., 1988, Capillary zone electrophoresis-mass spectrometry using an electrophoresis ionization interface, *Anal. Chem.* **60**:436–441.
65. Loo, J. A., Edmonds, C. G., and Smith, R. D., 1991, Tandem mass spectrometry of very large molecules: serum albumin sequence information from multiply charged ions formed by electrospray ionization, *Anal. Chem.* **63**:2488–2499.
66. Roboz, J., 1968, *Introduction to Mass Spectrometry,* Interscience, New York.
67. McDowell, C. A., 1963, *Mass Spectrometry,* McGraw–Hill, New York.
68. Cottrell, J. S., and Evans, S., 1987, Characteristics of multichannel electro-optical detection system and its applications to the analysis of large molecules by fast atom bombardment mass spectrometry, *Anal. Chem.* **59**:1990–1995.
69. Gross, M. L., 1990, Tandem mass spectrometry: multisector magnetic sector instruments, in *Methods in Enzymology* (J. A. McCloskey, ed.), Vol. 193, Academic Press, New York, pp. 131–153.
70. Cotter, R. J., 1992, Time-of-flight mass spectrometry for the analysis of biological molecules, *Anal. Chem.* **64**:1027A–1039A.
71. Mamyrin, B. A., Karataev, V. I., Schmikk, D. V., and Zagulin, V. A., 1973, The mass reflection, a new magnetic time-of-flight mass spectrometer with high resolution, *Sov. Phys. JETP* **37**:45–48.
72. Gross, M. L., and Rempel, D. L., 1984, Fourier transform mass spectrometry, *Science* **226**:261–268.
73. March, R. E., and Hughes, R. J., 1989, *Quadrupole Storage Mass Spectrometry,* Wiley–Interscience, New York.
74. McLafferty, F. W., 1983, *Tandem Mass Spectrometry,* Wiley–Interscience, New York.
75. Busch, K. L., Glish, G. L., and McLuckey, S. A., 1988, *Mass Spectrometry/Mass Spectrometry: Techniques and Applications of Tandem Mass Spectrometry,* VCH Publishers, New York.
76. Dass, C., and Desiderio, D. M., 1987, Fast atom bombardment mass spectrometry analysis of opioid peptides, *Anal. Biochem.* **163**:52–66.

77. Beynon, J. H., Caprioli, R. M., and Ast, T., 1971, Effect of deuterium labeling on the width of a 'metastable peak,' *Org. Mass Spectrom.* **5**:229–234.
78. Cooks, R. G., Beynon, J. H., Caprioli, R. M., and Lester, G. R., 1973, *Metastable Ions*, Elsevier, Amsterdam.
79. Gross, M. L., 1989, Mass spectrometry: an interplay between ion chemistry, instrumental development, and applications, *Mass Spectrom. Rev.* **8**:165–197.
80. Jennings, K. R., and Mason, R. S., 1983, Tandem mass spectrometry utilizing linked scanning of double focusing instruments, in *Tandem Mass Spectrometry* (F. W. McLafferty, ed.), Wiley–Interscience, New York, pp. 197–222.
81. Dass, C., and Desiderio, D. M., 1989, Characterization of neuropeptides by fast atom bombardment and **B/E** linked-field scan techniques, *Int. J. Mass Spectrom. Ion Proc.* **92**:267–287.
82. Dass, C., and Gross, M. L., 1990, Structures of decomposing and nondecomposing gas-phase $[C_4H_6O]^{+\cdot}$ radical cations, and their $[C_2H_2O]^{+\cdot}$ and $[C_3H_6]^{+\cdot}$ product ions, *Org. Mass Spectrom.* **25**:24–32.
83. Biemann, K., 1990, Sequencing of peptides by tandem mass spectrometry and high-energy collision-induced dissociation, in *Methods in Enzymology* (J. A. McCloskey, ed.), Vol. 193, Academic Press, New York, pp. 455–479.
84. Yost, R. A., and Enke, C. G., 1979, Triple quadrupole mass spectrometry for direct mixture analysis and structure elucidation, *Anal. Chem.* **51**:1251A–1264A.
85. Cody, R. B., and Freiser, B. S., 1982, High-resolution detection of collision-induced fragments by Fourier transform mass spectrometry, *Anal. Chem.* **54**:1431–1433.
86. Louris, J. N., Cooks, R. G., Syka, J. E. P., Kelley, P. E., Stafford, Jr., G. C., and Todd, J. F. J., 1987, Instrumentation, applications, and energy deposition in quadrupole ion trap tandem mass spectrometry, *Anal. Chem.* **59**:1677–1685.
87. Denis, P. O., Wesdemiotis, C., and McLafferty, F. W., 1983, Neutralization–reionization mass spectrometry (NRMS), *J. Am. Chem. Soc.* **105**:7454–7456.
88. Dekrey, M. J., Mabud, Md. A., Cooks, R. G., and Syka, J. E. P., 1985, Applications of linked scan procedures in investigating polyatomic ions/surface interactions. *Int. J. Mass Spectrom. Ion Proc.* **67**:295–303.
89. Cody, R. B., and Freiser, B. S., 1987, Electron impact excitation of ions in Fourier transform mass spectrometry, *Anal. Chem.* **59**:1054–1056.
90. Tecklenburg, Jr., R. E., and Russell, D. H., 1990, An evalution of the analytical utility of the photodissociation of fast ion beams, *Mass Spectrom. Rev.* **9**:405–451.
91. Bateman, R. H., and Bott, P. A., 1987, A new high efficiency high mass ion detector, *Proc. 35th ASMS Conf. Mass Spectrom. Allied Topics*, pp. 253–254.
92. Biemann, K., 1990, Applications of tandem mass spectrometry to peptides and protein structures, in *Biological Mass Spectrometry* (A. L. Burlingame and J. A. McCloskey, eds.), Elsevier, Amsterdam, pp. 179–196.
93. Pesch, R., Jung, G., Rost, K., and Tietje, K.-H., 1989, A versatile array detection system, *Proc. 37th ASMS Conf. Mass Spectrom. Allied Topics*, pp. 1079–1080.
94. Biemann, K., Bommer, P., and Desiderio, D. M., 1964, Element-mapping, a new approach to the interpretation of high resolution mass spectra, *Tetrahedron Lett.* 1725–1731.

2

# Characterization of Diacylglycerylphospholipids by Fast Atom Bombardment Tandem Mass Spectrometry

*Douglas A. Gage, Zhi-Heng Huang, and Charles C. Sweeley*

## 1. INTRODUCTION

Fast atom bombardment desorption/ionization combined with collisionally activated dissociation tandem mass spectrometry (FAB-CAD-MS/MS) is a powerful method for directly characterizing underivatized glycerophospholipids, even in mixtures. The FAB mass spectrum in either the positive or negative ion mode contains high-mass ions indicative of the molecular weight of each of the molecular species present in the sample, as well as fragments derived from elimination of the head group and acyl groups. However, it is difficult to directly characterize the structure of any particular molecular species in a mixture from the FAB mass spectrum alone because the molecular ions ($[MH]^+$, $[M\text{-}H]^-$, or $[M\text{-}CH_3]^-$ for phosphatidylcholine) of isomeric molecular species

*Douglas A. Gage, Zhi-Heng Huang, and Charles C. Sweeley* • MSU–NIH Mass Spectrometry Facility, Department of Biochemistry, Michigan State University, East Lansing, Michigan 48824.

*Mass Spectrometry: Clinical and Biomedical Applications, Volume 2*, edited by Dominic M. Desiderio. Plenum Press, New York, 1994.

overlap and distinguishing fragments may be attributable to more than one molecular species. Only tandem mass spectrometric analysis of specific fragment or molecular ions permits the head group and acyl groups of the individual molecular species to be unambiguously determined. In addition, the relative abundance of specific fragment ions in the product ion spectrum can, in most cases, be used to assign the position of the acyl groups on the glycerol backbone. FAB-CAD-MS/MS of the carboxylate anions found in the low-mass region of the negative FAB spectrum has also been successfully used to localize the positions of double bonds and branch points in the acyl groups by analysis of charge-remote fragmentation patterns. Besides product ion scans, other tandem MS scan modes (namely, constant neutral loss scans and precursor ion scans) are useful in identifying, in mixtures, the molecular species of a particular lipid class or those containing a certain acyl group.

## 2. *STRUCTURAL ANALYSIS OF DIACYLGLYCERYLPHOSPHOLIPIDS*

In order to fully characterize a diacylglycerylphospholipid (PL), it is first necessary to identify the nature of the head group attached to the glycerol backbone. These groups provide the basis for the classification of phospholipids and include the phosphoesters of diacylglycerylphosphoric acid with choline (phosphatidylcholine, PC), glycerol (phosphatidylglycerol, PG), inositol (phosphatidylinositol, PI), ethanolamine (phosphatidylethanolamine, PE), and serine (phosphatidylserine, PS), as well as the parent compound, diacylglycerylphosphoric acid (phosphatidic acid, PA) (see Table 2.1). In addition to the head group, the two esterified fatty acyl groups must be characterized. The structural characterization should include not only the carbon chain length, but also the degree and position of unsaturation or branching in the chain. Finally, the relative positions of the two fatty acyl groups on the glycerol backbone must be distinguished. In naturally occurring compounds, the head group is always located on the terminal carbon of the glycerol moiety, whereas the acyl groups are found at the remaining two positions, designed *sn*2 (β to the esterified head group) and *sn*1 (γ to the head group). In this chapter, diagnostic features of the fast atom bombardment (FAB) mass spectra of various PL classes are discussed, and methods are described employing fast atom bombardment collisionally activated dissociation tandem mass spectrometry (FAB-CAD-MS/MS) for the direct structural characterization of underivatized phospholipids. This approach is particularly useful in mixture analysis because it does not require the prior separation of lipid classes or individual molecular species.

### *2.1. Traditional Methods and Alternative Mass Spectral Techniques for Phospholipid Analysis*

Diacylglycerylphospholipids have traditionally been characterized by enzymatic degradation with specific phospholipases and subsequent chromatographic and mass spectral analyses of the reaction products after chemical derivatization (1). Gas chromatography-mass spectrometry (GC-MS) is often used to identify the released fatty acids as their methyl esters. Direct analysis of intact phospholipids (even as their trimethylsilyl derivatives, TMS) by electron ionization mass spectrometry (EI-MS) is not generally successful because of their thermal lability. Pyrolysis products of the degradation of lysophospholipids have been characterized by EI-MS and chemical ionization mass spectrometry (CI-MS) (2); however, this approach does not seem to be generally applicable to intact diacylated compounds. The combination of enzymatic degradation, and gas-phase mass spectral analysis has proven to be more successful. For example, treatment of phosphatidylcholine or phosphatidylethanolamine with phospholipase C, which removes the polar head group, followed by derivatization as the dinitrobenzoate, yields a product (diacylnitrobenzoylglycerol) amenable to analysis by electron capture negative ion desorption chemical ionization mass spectrometry (3).

Other related derivatization and mass spectral analysis procedures for these and other subclasses of phospholipids have been reported (4–7). The enzymatic degradation and combined enzymatic degradation/derivatization approaches both involve multiple steps and may require chromatographic separation of individual lipid classes and molecular species prior to analysis. Other "soft ionization" techniques such as field desorption (FD), plasma desorption (PD), and laser desorption (LD) for direct PL mass spectral analysis are not addressed in this chapter. The interested reader should consult other reviews (8–10) for those leading references.

### *2.2. Direct Analysis of Phospholipids by FAB Mass Spectrometry*

Fast atom bombardment mass spectrometry (FAB-MS) provides an improved alternative for the direct analysis of phospholipids without requiring previous sample modification. In the FAB-MS experiment, 1–2 μl of the diacylglycerylphospholipid solution (0.1–1μg/μl in the appropriate solvent, e.g., $CHCl_3$:MeOH, 1:1 or $CHCl_3$:MeOH:$H_2O$, 86:14:1) is mixed with 1–2 μl of viscous liquid matrix on the probe tip. The common matrices employed include *m*-nitrobenzyl alcohol (NBA), thioglycerol, dithiothreitol:dithioerythritol (5:1), glycerol, diethanolamine, or triethanolamine. Because of solubility considerations, special precautions must be observed when glycerol is used as the matrix (11,12). For most applications in our laboratory, NBA is used for positive FAB-MS and triethanolamine or NBA for negative FAB-MS. Bombardment of the sample/matrix solution with 6–8 keV Xe atoms (or a high-energy $Cs^+$ beam) yields ions related to the intact molecules, as well as fragments indicative of the head group and esterified fatty acids.

Positive and negative ion mode analysis can both give complementary structural information and, because the sensitivity for different lipid classes varies between modes, it is often worthwhile to obtain spectra of the sample in each mode, particularly when different PL classes are present in a mixture. For example, it has been reported that in crude PL mixtures from rabbit and dog heart, PC, PE, and PI can be detected by positive FAB MS, whereas PS and sphingomyelin in the samples were not observed (10). Even PL classes that yield a good response in the positive mode may exhibit differential desorption in mixtures; the responses of PE (8) and acidic PL classes, such as PI, are often suppressed in the presence of PC [Huang and Gage, unpublished manuscript]. The converse is usually true in negative FAB MS. That is, the detectability of basic PL classes (including PC, with its fixed positive charge) is suppressed in the presence of acidic PLs. Nevertheless, mixtures of PL classes can be successfully analyzed by tandem MS procedures (Section 3.3), but analysis in the opposite mode is recommended for confirmation of the classes present in the sample.

The same structurally informative fragment ions in the FAB mass spectrum are also observed in collisionally activated dissociation (CAD) mass spectra, although they may appear in greater abundance. Because interpretation of these fragments is the key to PL characterization of head groups and acyl group composition in tandem mass spectrometric analysis, we will begin with a discussion of the structures of the ions formed from each of the PL classes.

The FAB mass spectra of diacylglycerylphospholipids can conveniently be divided into three regions (11,13): the high-mass region, containing ions related to the intact molecule, as well as others formed by cleavages within the head group; the middle-mass region, containing fragments from loss of one or the other acyl group; and the low-mass region, where head group–specific ions are found. In the negative FAB spectrum, the carboxylate anions of the fatty acyl groups are prominent in the low-mass region. Interpretation of these spectral regions will be discussed in turn.

### *2.2.1. The High-Mass Region in the FAB Spectra of Diacylglycerylphospholipids*

The positive mode FAB-MS spectra of some diacylglycerylphospholipid classes display $[MH]^+$ ions for each of the molecular species present. Phospholipids containing a fixed positive charge (PC) or a basic site (PE) give the strongest response in this mode (14,15). PS spectra also display $[MH]^+$ ions (16), although these ions may be weak (17). In contrast, the positive spectra of PI and PG often contain only weak ions in this region. Sodium or potassium adducts ($[MNa]^+$ or $[MK]^+$) are often observed in the positive mode; if the salt content of the sample is high, then the adduct ions may dominate the high-mass region (130). At least one report states that sodium adducts of PA and PI can be detected in the positive FAB mode (15), but high-mass peaks of these lipids are usually not observed. Although the relative abundances of the $[MH]^+$ are generally indicative of the composition of the molecular species present in the sample (18), matrix background peaks may obscure minor constituents. This effect can be minimized by using NBA as the matrix. Another concern is the misinterpretation of ions appearing at $[MH-2H]^+$ (particularly in the spectra of PCs) as a separate molecular species containing an acyl group with an additional degree of unsaturation (12,18,19). In glycerol matrix, this peak can be 10–20% of the corresponding $[MH]^+$ ion. The use of NBA as the matrix can reduce this artifact to 3–5% of the $[MH]^+$ ion (18,19). Positive mode FAB-MS analysis is most sensitive for phospholipid classes that are basic (PE, PS) or contain a preformed cationic center (PC). It is least sensitive for acidic lipids, such as PA.

Fragments present in the high-mass region of the positive FAB mass spectrum provide information about the nature of the head group. These fragments result from the cleavage of the phosphoester bond between the diacylglycerol moiety and the phosphate ester and consequent loss of neutral phosphoric acid with the attached head group. Charge is retained on the diglyceride fragment. For PC, this process involves loss of the neutral zwitterion, phosphocholine, so that the fragment is found at the *m/z* value of $[MH-183]^+$ (14). Analogous fragments for PS, PE, PG, and PI are found at $[MH]^+$ minus 185, 141, 172, and 260, respectively (Table 2.1). Note that the fragment ion formed by loss of the head group in the positive FAB spectrum PI overlaps what is described here as the middle-mass region. Careful inspection of this spectral region is required to identify this fragment correctly.

Because all phospholipids contain a phosphoric acid moiety, they can readily form stable anions in negative ion mode FAB analysis. All classes except PCs form prominent $[M-H]^-$ ions. The largest fragment in the negative spectra of PCs instead represents $[M-CH_3]^-$ (17,20). However, because N,N-dimethylphosphophatidylethanolamine yields high-mass ions identical to those produced by an analogous PC, confirmatory analysis by positive mode FAB-MS is advisable. Sodium adduct ions are much less prominent in the negative FAB spectra of diacylglycerylphospholipids, making the high-mass region often simpler and easier to interpret than in the positive FAB spectrum (but see Ref. 21). Using triethanolamine as the matrix, the background is often low enough in this region that even minor molecular species can be detected.

*Table 2.1*
*Characteristic Neutral Losses in the Positive Ion Mode*

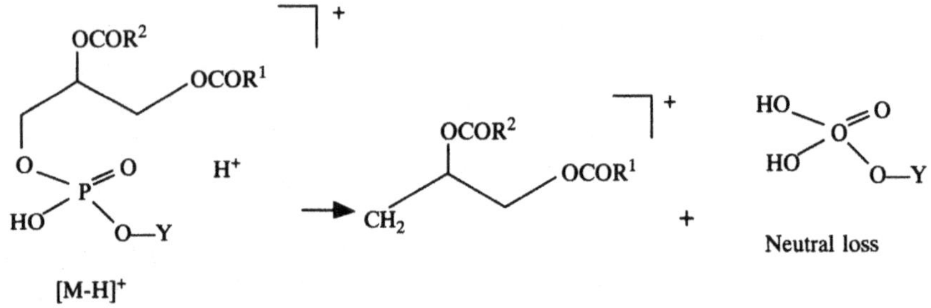


| Phospholipid | Y | Mass lost from [M-H]$^+$ |
|---|---|---|
| PC | $CH_2CH_2\overset{+}{N}(CH_3)_3$ | (183) |
| PE | $CH_2CH_2NH_2$ | 141 |
| PS | $CH_2CH(NH_2)CO_2H$ | 185 |
| PI | inositol | 260 |
| PG | $CH_2CH(OH)CH_2OH$ | 172 |
| (PA) | H | 98 |

In addition to the [M-H]$^-$ ions, the high-mass region in the negative mode spectrum contains a number of fragment ions diagnostic for individual phospholipid classes present in the sample (Table 2.2, Fig. 2.1). For example, the spectra of PCs display a characteristic triplet consisting of [M-15]$^-$, [M-60]$^-$ (loss of a methyl and dimethylamine), and [M-86]$^-$ (loss of dehydrated serine) (17). PE spectra contain fragments from the loss of vinylamine [M-H-43]$^-$. PE negative FAB spectra also sometimes display higher mass peaks due to adduct formation with triethanolamine matrix (22). The high-mass region of the negative spectrum of PG contains a diagnostic fragment at [M-H-74]$^-$ from loss of dehydrated glycerol with hydrogen transfer from the glycerol moiety (22). Another phospholipid class, PI, can be recognized by the prominent fragment appearing 162 mass units below the [M-H]$^-$, formed by elimination of neutral dehydroinositol. The mass of each of these fragments is equivalent to that of the [M-H]$^-$ ion of the analogous PA molecular species.

The identification of the lipid class from the high-mass fragments described above, together with the intact molecular weight derived from the [MH]$^+$ or [M-H]$^-$ (or [M-$CH_3$]$^-$ in the case of PC) ions, permits the total composition of the combined fatty acyl groups (number of carbon atoms and

*Table 2.2*
*Characteristic Fragments in the Negative FAB Spectrum*

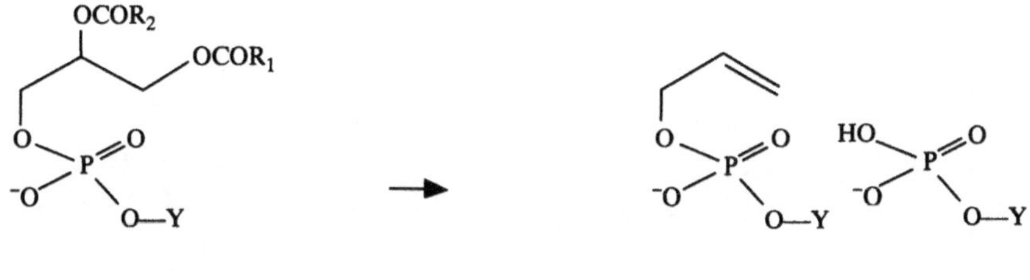


$[M\text{-}H]^-$

| Phospholipid | *Y* | Mass lost from $[M\text{-}H]^-$, (Y-H) | *a* | *b* |
|---|---|---|---|---|
| PC (DMPE)[a] | $CH_2CH_2\ N(CH_3)_2$ | 71 | 208 | 168 |
| PE | $CH_2CH_2\ NH_2$ | 43 | 180 | 140 |
| PS | $CH_2CHCO_2H$<br>\|<br>$NH_2$ | 87 | (224) | (184) |
| PI | inositol | 162 | 299 | 259 |
| PG | $CH_2CHCH_2OH$<br>\|<br>OH | 74 | 211 | 171 |

[a]PC undergoes N-demethylation to give dimethyl-PE (DMPE) in the FAB ionization process; thus, PC and naturally occuring DMPE cannot be distinguished in the negative ion mode.

double bonds or substituents) to be determined. However, because several combinations of different fatty acyl groups can yield molecular species of the same mass, it is necessary to characterize the individual acyl groups present. For purified single molecular species, this information can be obtained from fragments in the middle- and low-mass regions of the positive and negative FAB spectra.

### 2.2.2. *The Middle-Mass Region in the FAB Spectra of Diacylglycerylphospholipids*

The middle-mass region, as defined in this chapter, is composed of ions resulting from the loss of one or the other of the two fatty acyl groups with charge retention on the lysoglycerylphospholipid moiety. These fragments can be formed in two ways: either by loss of the equivalent of a free fatty acid, $RCH_2COOH$, or by loss of the ketene analog of one of the fatty acyl groups, R—CH═C═O. If the acyl groups are different, then this process can result in the appearance of four significant peaks in this region of the spectrum involving two pairs of fragments, each separated by 18 mass units. Depending upon the

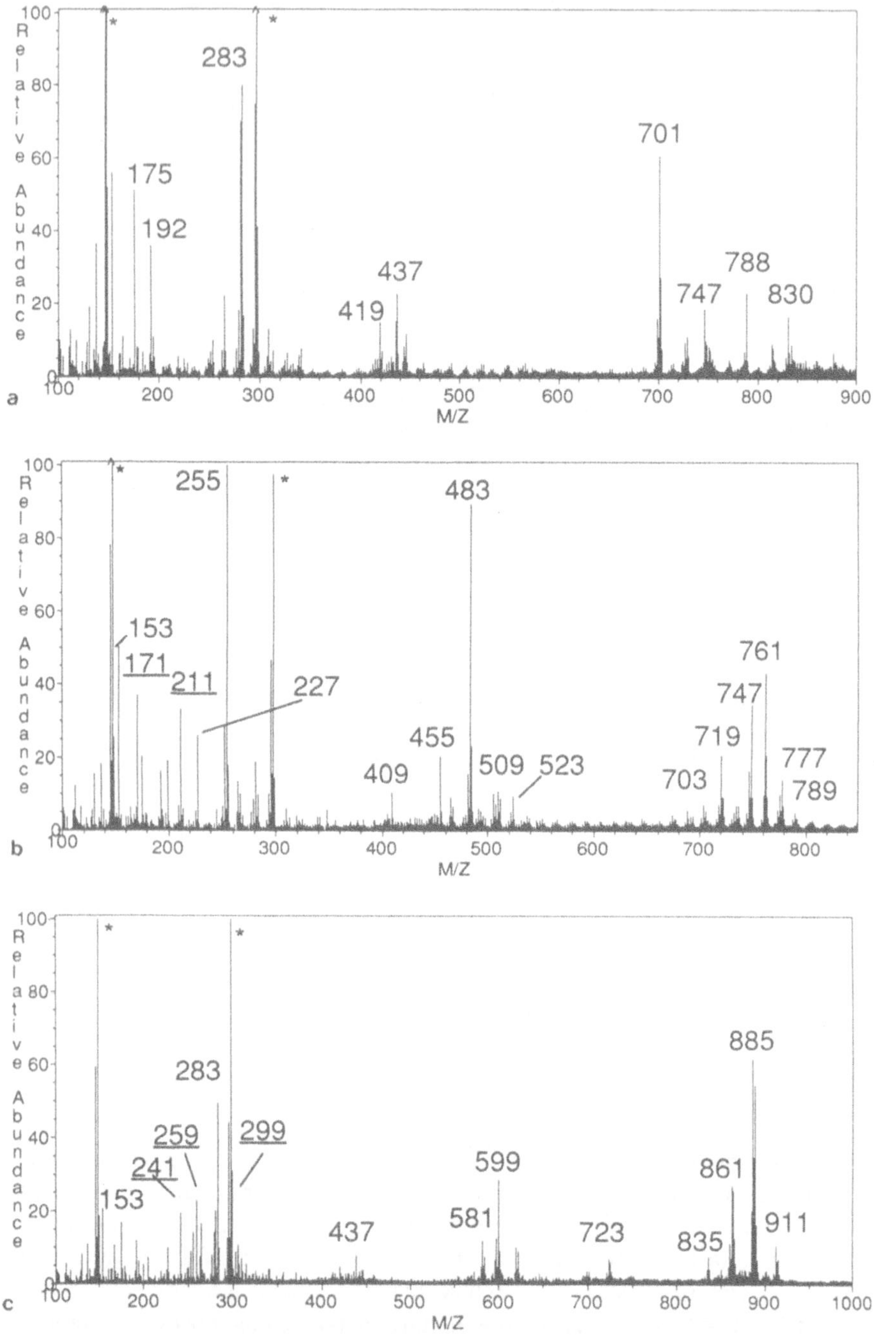


*Figure 2.1.* **Negative FAB mass spectra of (a) PS, (b) PG, and (c) PI. Low-mass fragments indicative of the headgroup are underlined. Note that in the spectra of PS, the low mass ions can be weak. Triethanolamine matrix peaks are designated by an asterisk.**

head group and the mode of analysis, only some of these fragments may be present in the spectrum. The sensitivity of positive or negative mode analysis for a particular lipid class influences how prominent these fragments are. Thus, all four fragments are seen in the positive FAB spectra of unsymmetrical PC and PE compounds (14), but they are weak in the spectrum of PS and PG (Ref. 22, but see Ref. 23 also). In the negative FAB spectra of all classes, losses of the ketene analogs or the free fatty acids from the $[M-H]^-$ ions (or $[M-15]^-$, $[M-60]^-$, and $[M-86]^-$ of PC) can be observed. Again, the abundance of these ions depends upon the concentration of the analyte in the matrix or the presence of other phospholipid classes in the sample (see Section 2.3). However, the nature of the acyl group can also influence the relative abundances of the lysoglycerylphospholipid ions, so that in some cases only the fragment(s) formed by loss of one of the acyl groups may be detected. In either the positive or negative mode, when the molecular weight and head group are known, the appearance of these fragments can allow the molecular formulas of the individual acyl groups to be determined. Nevertheless, the analysis can be complicated when multiple molecular species are found together in a sample. Unambiguous linkage of a particular acyl group to a particular molecular species requires that tandem mass spectrometry be performed (Section 2.3).

### *2.2.3. The Low-Mass Region in the FAB Spectra of Diacylglycerylphospholipids*

Fragments in this region of this spectrum provide confirmatory information about the nature of the head group, as well as the acyl groups. The low-mass region can appear rather complex, because ions derived from a variety of fragmentation processes have masses in this region. These ions include head group–specific fragments in both the positive and negative mode, some of which contain the dideacylated glyceryl moiety (24). In the low-mass region of the negative FAB spectrum, the most prominent fragments observed are usually the carboxylate anions of the two acyl groups. In principle, the absence of certain carboxylate anions would allow isomeric molecular species to be distinguished, (e.g., 18:0/20:0-PL from 16:0/22:0-PL or 1-0-alkenyl plasmalogens from their isomeric diacyl PLs). The carboxylate anions could also be used to determine whether an $[MH]^+$ or $[M-H]^-$ ion observed two mass units below the protonated or deprotonated molecule of another molecular species represents an oxidation artifact or a molecular species containing an acyl group with an additional degree of unsaturation. However, in mixtures, each molecular species contributes carboxylate anions, so that linking particular acyl groups to a specific molecular species is not usually possible. Another complication is that the relative abundance of these fragments can vary significantly, depending on their chain length and degree of unsaturation, as well as their relative position on the glycerol backbone (11,15,25). Some long-chain polyunsaturated acyl groups (e.g., 22:6) yield weak carboxylate anions that may be lost in the matrix background. The presence of contaminating free fatty acids in the sample can also lead to incorrect assignments. Thus, direct interpretation of acyl group composition from the negative FAB spectrum must be viewed with caution.

Specific fragments for each PL class in the low-mass region of the negative ion mode FAB spectrum are found in pairs that differ by 40 mass units (see Ref. 24). The lower-mass ion results from cleavage of the phosphoester bond between the glycerol moiety and the phosphate group containing the esterified head group (or H in the case of PA). The higher-mass ion contains, in addition to the phosphoester, the dideoxyglycerol group (Table 2.2). For N-modified PEs, the fragment pair shifts in mass accordingly. The FAB spectrum of N-palmitoyl-PE is shown in Fig. 2.2. In the spectrum of PI, the two head group fragments are accompanied by a prominent ion at *m/z* 241, presumably representing a dehydrated phosphoinositol product. In the spectra of PS samples, the low-mass fragments can be weak (see Fig. 2.1a). The low-mass fragments of PC differ from other classes because the head group is N-demethylated. Thus, the fragments at *m/z* 208 and 168 do not permit PC to be distinguished from N,N-dimethylphosphatidylethanolamine.

The low-mass region of the positive FAB spectrum is less useful for characterizing the head groups of different PL classes. Only PC and PE (and its related N-methyl and N,N-dimethyl derivatives) usually yield diagnostic fragments in this part of the spectrum (but see Ref. 19). A series of prominent

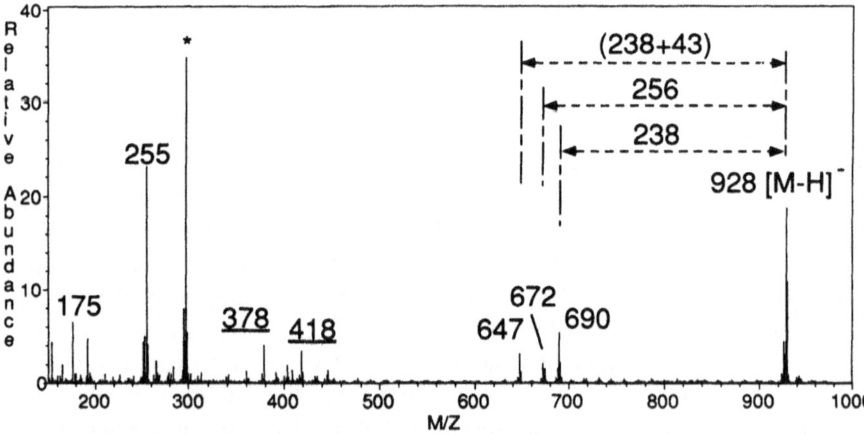


*Figure 2.2.* Negative FAB mass spectrum of N-palmitoyl PE. Structurally diagnostic ions are shifted by the mass of the N-acyl group relative to their *m/z* value in an analogous PE sample.

ions is found in the spectrum of PC at *m/z* 258 (glycerophosphocholine), 224 (dideoxyglycerophophocholine), and 184 (phosphocholine). Analogous fragments in the spectrum of PE appear at *m/z* 182 (dideoxyglycerophosphocholine) and 142 (phosphoethanolamine). The latter fragments shift by 14 and 28 mass units, respectively, in the N-methyl and N,N-dimethyl PE analogs (19). Information about the acyl group composition of the total sample can be found from the monoacyl glyceride fragments in the low-mass region of the positive FAB spectrum, but these ions are often found in low abundance.

## *2.3. Analysis of Diacylglycerylphospholipids by FAB-CAD-MS/MS*

As described above, fragments characteristic of the head group and fatty acyl groups of different PL classes are present in the normal (i.e., *B* scan) FAB spectrum. For the analysis of highly purified individual molecular species, interpretation of this spectrum might be sufficient for the determination of these structural features, although the relative positions and structures of the fatty acyl groups can not be confirmed. However, PLs isolated from biological sources are rarely composed of a single molecular species or lipid class. The multiplicity of molecular ions and fragments in the spectra of such samples usually precludes the unambiguous connection of a particular molecular species to particular fragment ions. Rigorous purification to obtain single molecular species for each of the PL classes present would be extremely tedious and time consuming. Thus, tandem mass spectrometry (MS/MS) is an obvious alternative to obtain structural information for individual molecular species occurring in a mixture. Another advantage of this approach is that additional structural details unavailable from interpretation of the FAB spectrum can also be obtained. For example, the relative positions of the acyl groups on the glycerol backbone and the location of double bonds or branch points in the acyl chains can be directly determined only by FAB-CAD-MS/MS analysis.

Tandem mass spectrometry can be defined as the detection of ions that, after their initial formation from a precursor in the source, have undergone a change in mass during the course of their analysis in a mass spectrometer (26). Implied in this definition is the ability to select and separate ions of a particular mass-to-charge ratio from all others produced during the ionization process. This process allows one particular fragment (or protonated or deprotonated molecule) to be analyzed whether it is found in the spectrum of a purified molecular species or that of an impure mixture containing many molecular species of several different lipid classes. Ideally, the ions produced are directly related to only the selected precursor ion. The fragments from the precursor can form from metastable decomposition or be promoted by energy deposition from collision with a neutral gas (collisionally activated dissociation, CAD). In the MS/MS analysis of PLs, head group and acyl group fragments can be observed with or without CAD, but their abundance and relative intensity will differ between the two processes. As will be discussed below, the relative intensities of certain fragments can be critical in assigning the positions of the two acyl groups on the glycerol backbone. In our laboratory, we have found that CAD-MS/MS is preferable over metastable analysis for PLs because the diagnostic fragment ions are usually more abundant and the relative intensities of these fragments are more reproducible and predictive of the structure. However, CAD is required for the acyl group structure to be characterized by charge remote or other fragmentation processes (11,17,27–29) (Section 3.1).

Several different FAB-CAD-MS/MS experiments have been profitably employed for the characterization of PLs. The most widely utilized technique is the product ion scan. Here, the precursor ion is selected and induced to fragment by CAD. The resulting product ions are mass-analyzed (Fig. 2.3). The precursor ion selected can be located in any region of the PL spectrum, including the molecular ion, a high-mass fragment, or a low-mass fragment such as the carboxylate anion. Different instrumental configurations have been used for the acquisition of product ion spectra of phospholipids. Much of the initial work was performed on reverse geometry (*B*/*E*/*B*) (15) and forward geometry (*E*/*B*/*E*) triple sector instruments (17,22), as well as hybrid (*B*/*E*/*qQ*) mass spectrometers (11) ($B$ = magnetic sector, $E$ = electric sector, $q$ = rf-only quadrupole, $Q$ = quadrupole mass analyzer). With each of these instruments, the first two sectors can be used to select the precursor ion with unit resolution; CAD occurs in the third field-free region (or in an rf-only quadrupole in the hybrid after deceleration of the precursor ion); and the product ions are mass-analyzed by the third sector or quadrupole. One disadvantage of the *E*/*B*/*E* instrument (or in any case where $E$ is used to mass-analyze collision products, e.g., MIKES analysis, see Sherman *et al.*, Ref. 15) is the poor product ion resolution obtained (see Ref. 26). For example, in the published reports (11,15,17) it was not possible to distinguish unambiguously the carboxylate fragments of two fatty acyl groups differing only by one double bond.

Other MS/MS studies of PLs have been conducted with forward geometry *E*/*B* double focusing mass spectrometers (13,25,30). With these instruments, collisional activation occurs in the first field-free region. Product ion spectra are acquired by scanning the electric sector and the magnetic sector in a fixed ratio (*B*/*E* linked scans) (31). This instrumental approach suffers from poor precursor ion resolution, but product ions are measured at unit resolution. Again, this limited resolution can lead to problems in spectral interpretation when other molecular species are present in the sample that differ by only two mass units from the molecular species being analyzed. The product ion spectrum of the selected molecular ion or fragment is likely to contain product ions from the adjacent ions of the other molecular species.

Two alternative instruments have been used to overcome these limitations.

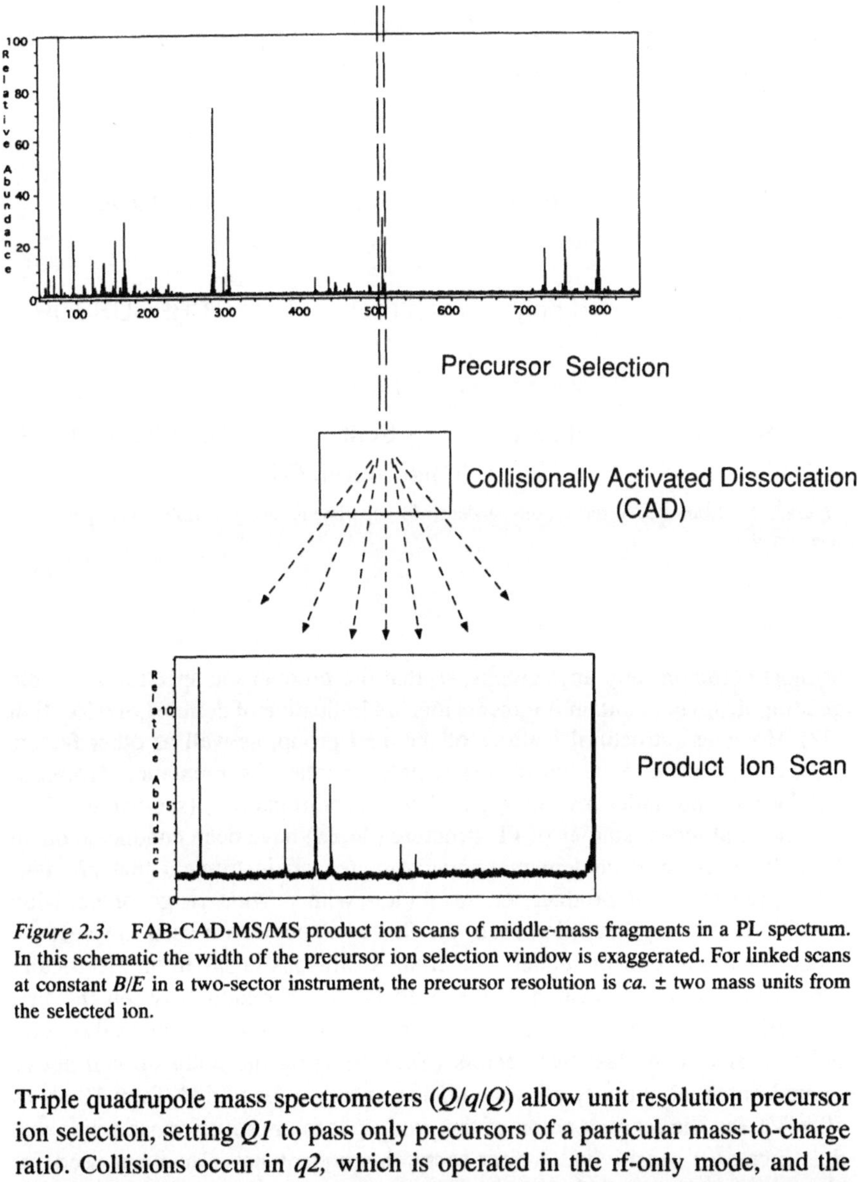


*Figure 2.3.* FAB-CAD-MS/MS product ion scans of middle-mass fragments in a PL spectrum. In this schematic the width of the precursor ion selection window is exaggerated. For linked scans at constant *B/E* in a two-sector instrument, the precursor resolution is *ca.* ± two mass units from the selected ion.

Triple quadrupole mass spectrometers (*Q/q/Q*) allow unit resolution precursor ion selection, setting *Q1* to pass only precursors of a particular mass-to-charge ratio. Collisions occur in *q2,* which is operated in the rf-only mode, and the resulting product ions are mass-analyzed with unit resolution by scanning *Q*3 (Fig. 2.4). One significant difference between the triple quadrupole instrument and the instruments described above (except for the hybrid) is that collisional activation occurs at much lower energies (10V–100eV) compared to the keV collisions of the sector mass spectrometers. The lower energies are generally not sufficient to induce charge-remote fragmentation of the carboxylate anions

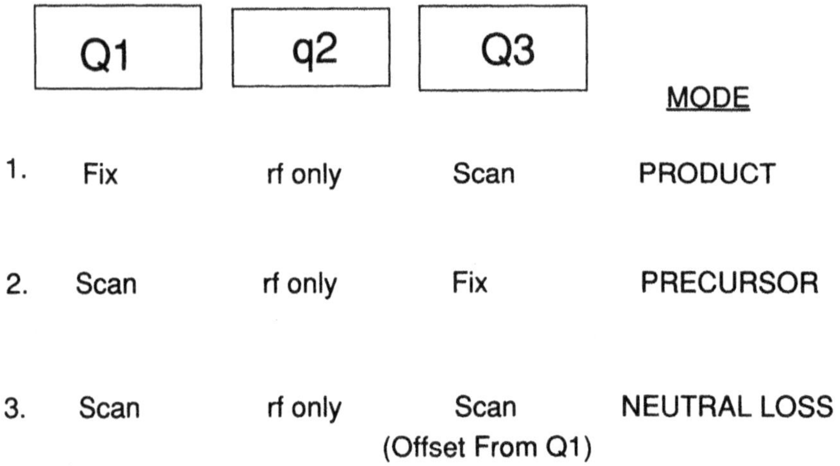


*Figure 2.4.* Scan modes for various tandem MS experiments using a triple quadrupole mass spectrometer.

of most common fatty acyl groups, so that the product ion spectra on a triple quadrupole do not contain fragments that are indicative of double bond location (32). However, structural features of the acyl group, as well as other factors (33), can influence whether charge-remote or other fragmentation processes will be induced under low-energy collision conditions (29) (see Section 3.1).

Several recent studies of PL structure (34,35) have been conducted on an *E/B/E/B* four-sector tandem mass spectrometer, an instrument that provides good precursor and product ion resolution with a wide range of collision conditions. For a product ion scan, the first two sectors are used to select the precursor ion at unit or higher resolution. Collisions occur in an electrically floated collision cell located in the third field-free region between the first magnetic sector and the second electric sector. The products are analyzed by linked scans of the last two sectors (*E/B*). Floating the collision cell above ground potential not only allows collision energy to be varied (from 10 eV to 10 keV), but the reacceleration of the product ions improves transmission and thus increases sensitivity. A four-sector instrument has also been used for higher-order tandem MS (MS/MS/MS) analyses of PLs (34,35). If the precursor ion (e.g., the molecular ion, $[M-H]^-$) undergoes CAD in the first field-free region, then a specific product (e.g., a carboxylate anion) can be selected by setting the first magnetic sector and electric sector at the appropriate *B/E* ratio. CAD of this product ion in the collision cell located in the third field-free region produces a second generation of products that are mass-analyzed by scanning the second electric and magnetic sectors at a constant *B/E* ratio.

Another useful MS/MS technique for PL characterization is the precursor ion scan. In this experiment, all the precursors of a specific CAD (or metastable) product ion are analyzed. These data can be used in the analysis of a mixture where one wishes to identify all the components that share a specific structural feature. Precursor ion scans can be conducted on all the instruments described above. For the sector instruments, this experiment involves linked scanning of the electric and magnetic sectors in a constant ratio of $B^2/E$, with CAD occurring in the first field-free region (or in the third field-free region of a four-sector instrument) (see Ref. 26 and 31 for a more detailed discussion). The precursor ion scan in a triple quadrupole instrument is performed by fixing *Q*3 to pass only the product ion of interest (e.g., a specific head group fragment). The first quadrupole mass analyzer (*Q*1) is scanned across the desired mass range, and CAD takes place in *q*2, as before (Fig. 2.4). For hybrid instruments, the quadrupole mass analyzer is similarly set to pass only a selected product ion, and the sectors are scanned normally. The spectrum obtained will display peaks only for the precursors of the product ion selected to pass through *Q*3 (or the quadrupole mass analyzer of hybrid instruments). The selectivity of this experiment allows minor components to be detected in complex mixtures that might otherwise be missed.

Two groups (32,36) have demonstrated the value for PL mixture analysis of another tandem MS experiment, the constant neutral loss scan. In this case, all ions are detected that can lose a specific neutral fragment. Sector instruments can perform this analysis by a complicated linked scan algorithm, $B^2(1 - E)/E^2$ (26,31). Conceptually, this experiment is much easier to understand when it is conducted on a triple quadrupole or hybrid instrument. Here, both *Q*1 and *Q*3 (or the sector and quadrupole mass analyzers in a hybrid) are simultaneously scanned in the normal mode, but the mass axes of the two quadrupoles are offset by the mass of the selected neutral fragment that is lost. Thus, to identify all ions that lose a neutral mass *X*, during scanning *Q*1 will pass an ion at $m/z_n$ while *Q*3 simultaneously passes all ions at $m/z_n - X$ (Fig. 2.4). CAD can occur in *q*2, or metastable decompositions can be monitored without CAD. As will be discussed below, this technique has been used to identify specific PL classes in mixtures by constant neutral loss scanning for characteristic head group losses (32,36). One caveat for this analysis is that other components will be incorrectly identified if they can lose an unrelated fragment of the same mass. It should be noted that constant neutral gain experiments can also be performed easily in the same manner by changing the offset between *Q*1 and *Q*3 so that only ions that acquire a specific increase in mass in *q2* will be detected. This technique has been used by one group to study selective ion–molecule reactions of different PL classes (37).

## 3. APPLICATIONS OF TANDEM MASS SPECTROMETRY FOR DIACYLGLYCERYLPHOSPHOLIPID CHARACTERIZATION

After the initial investigations of PLs by both positive and negative FAB-MS (10,12,14,21), several groups applied this ionization method with collisionally activated dissociation tandem mass spectrometry (15,17, 22). Product ion scans of high-mass ions in both the positive and negative ion mode were used in these studies to confirm the PL class-specific head group fragments and to identify the two acyl groups of particular molecular species (Fig. 2.5). In addition, product ion scans in the negative ion mode of carboxylate anions were employed to characterize the structures of the individual acyl groups by their charge-remote fragmentation patterns (Section 3.1). In a study by Sherman (15) of PI, negative ion mode analysis was found to be preferable for MS/MS analysis of the high-mass ions, because $[M\text{-}H]^-$ ions were prominent, and only cationized species were present in the positive ion spectrum. The comprehensive investigation of other PL classes by Jensen *et al.* (17,22) suggested that negative ion FAB-CAD-MS/MS would be most generally useful for the analysis of unknown PL samples. Even classes that reliably provide good-quality positive ion spectra such as PC and PE, give readily interpretable product ion spectra in the negative ion mode. However, other authors (11) found positive mode FAB-MS useful, at least for an initial survey of the molecular species present in PC samples, because only $[MH]^+$ ions were significant (although this pattern may not hold when salt contamination is high), whereas the negative mode contained three high-mass fragments ($[M-15]^-$, $[M-60]^-$, and $[M-86]^-$) for each molecular species. It should be noted, though, that only $[M-15]^-$ is at an even mass. Precursor ion scans were used in several studies of PC to investigate the formation of the three high-mass fragments observed in the negative FAB spectra of this PL class. It was determined that these fragments are formed from the dissociation of a matrix adduct to the intact PC (13,20).

Although negative ion mode analysis for PLs is today the most common procedure (25,30,34; but see Ref. 16 and 35), it has been found that, for characterization of mixtures of different lipid classes by constant neutral loss scanning of head group specific fragments, the positive mode may often be preferable (32,36). For other applications, particularly acyl group characterization, however, negative mode analysis is more informative. For example, during the course of the MS/MS studies of the $[M-15]^-$ ions (and other high-mass ions) of PC molecular species, patterns in carboxylate product ion relative abundances emerged, which suggested to several groups (11,13,17) that these patterns were indicative of the relative positions of the two acyl groups. Later studies indicated that, although this pattern was not universal, other FAB-CAD-MS/MS procedures could be used for reliable acyl group position analysis (Section 4). Another powerful experiment, available in the

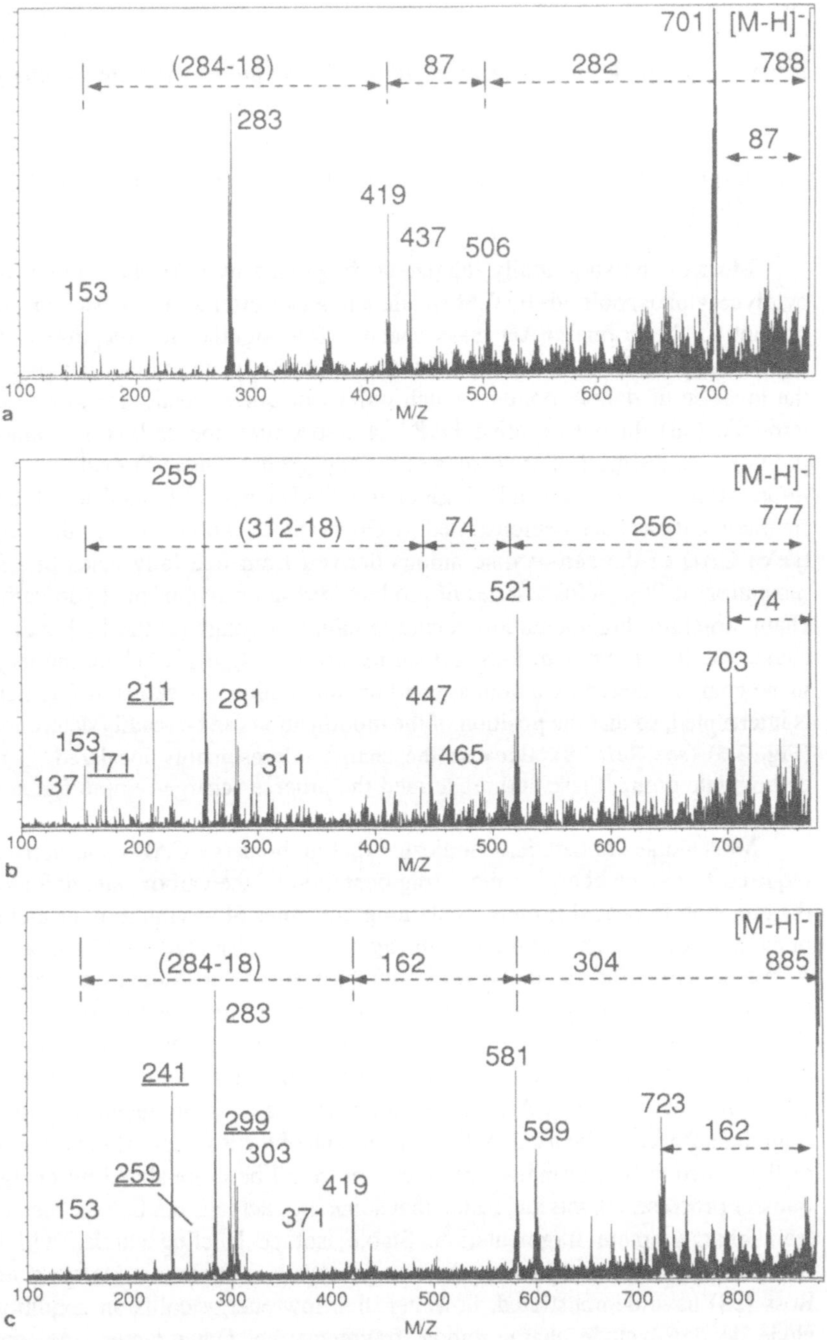


*Figure 2.5.* **Negative FAB-CAD-MS/MS of (a) PS, (b) PG, and (c) PI showing diagnostic fragmentation patterns indicative of the head group (Table 2.2) as well as low-mass and middle-mass fragments representative of the acyl groups. See Figure 2.1 for full (b) scans of these samples.**

negative ion mode, is the determination of acyl group structure by charge-remote fragmentation of the carboxylate anion fragments.

### 3.2. Characterization of Acyl Group Structure by Charge-Remote Fragmentation

Most of the structurally diagnostic fragments used to characterize diacylglycerylphospholipids by CAD tandem mass spectrometric analyses are also present in the normal FAB mass spectra. One significant exception is the high-energy CAD-induced fragmentation of carboxylate anions that permits the location of double bonds, branch points, or other substitutions to be determined (38). In the negative FAB mass spectrum, the carboxylate anions provide only the molecular weights of the acyl groups, but additional structural information can be obtained by high-energy CAD tandem MS analysis of these fragments. It has been demonstrated by Gross and co-workers that high-energy (keV) CAD of the carboxylate anions derived from free fatty acids or fragmentation of PLs yields a series of product ions indicative of the hydrocarbon chain structure. Fragmentation occurs randomly throughout the hydrocarbon chain, yielding prominent ions formed by losses of $C_nH_{2n+2}$. Where the methylene chain is altered by a double bond or substitution, the series of fragments is interrupted, so that the position of the modification can be readily determined (Fig. 2.6) (see Ref. 38). Because the charge is presumably localized on the carboxylate group, these authors termed the process "charge-remote fragmentation."

Most studies to date have indicated that high-energy CAD conditions are required to induce charge-remote fragmentation of the carboxylate anions of the common fatty acyl groups, containing only sites of unsaturation or methyl or cyclopropyl branch points along the hydrocarbon chain (13,17,27,28,32,38). A recent report indicated, however, that for some substituted acyl groups low-energy CAD of their carboxylate anions can provide structurally information fragment ions. Bernstrom *et al.* (29) used a triple quadrupole mass spectrometer to characterize PC and PE species containing isomeric epoxyeicosatrienoyl acyl groups. CAD of the carboxylate anions produced diagnostic fragments that could be used to distinguish each of the isomers. Although many of the structurally informative product ions could be rationalized by charge-transfer processes, it was suggested that some product ions could have resulted from charge-remote fragmentation. Stable isotope labeling studies will be required to determine which of the two processes is operating. Wysocki and Ross (33) have demonstrated, however, that low-energy collision conditions alone do not exclude charge-remote fragmentation. Other factors including structure of the precursor ion, collision gas, and collision-gas pressure may influence the type of fragmentation observed. It is likely that the epoxy-

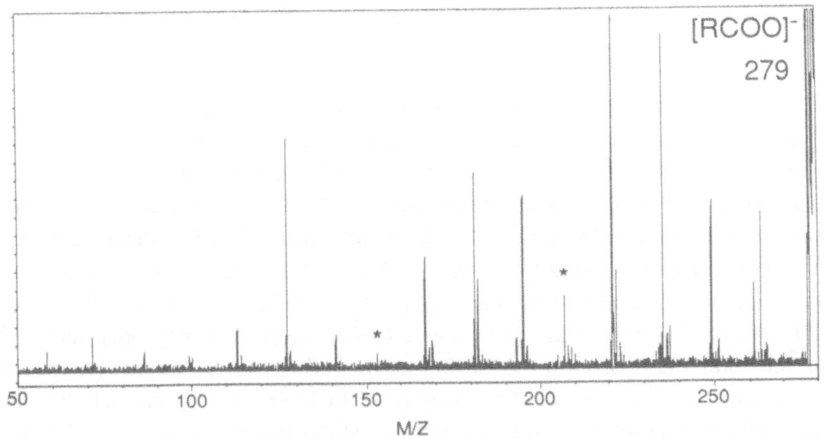


*Figure 2.6.* Negative FAB-CAD-MS/MS of the carboxylate anion of an oleoyl-containing PL showing charge-remote fragmentation pattern. These data indicate that the double bonds are at C-9 and C-12.

eicosatrienoyl acyl groups have a significant influence in promoting formation of the observed fragments from the carboxylate anions because low-energy CAD of other acyl group carboxylates has not generally been successful (32).

Although for most cases FAB-CAD-MS/MS analysis of the carboxylate anions can provide definitive structural information about the two acyl groups in an individual PL species, in mixtures it is possible that two different molecular species could contain isomeric acyl groups that yield carboxylate anions at the same *m/z* value. Tandem MS analysis would yield a mixed spectrum representing the two acyl groups, which even if it could be deconvoluted so that the two acyl groups could be identified, would not allow the acyl groups to be assigned to a particular molecular species. Recently, a group has reported the use of higher-order (MS/MS/MS, Section 2.3) experiments on a four-sector mass spectrometer, which overcomes this potential limitation (34,35). In these experiments, a high-mass ion of an individual molecular species is induced to fragment by CAD; one particular product ion, a fatty acyl carboxylate, is selected and undergoes CAD to produce charge-remote product ions indicative of its structure. Although this experiment appears to be similar to a simple product ion scan of a carboxylate fragment produced in the source during the ionization process, it has the advantage of directly relating the second-generation product ions to a specific initial precursor ion.

### 3.2. *Identification of Molecular Species Containing Specific Acyl Groups*

In a study of the phospholipid precursors involved in the liberation of arachidonic acid for leukotriene and prostaglandin biosynthesis, as well as the formation of lyso-platelet activating factor (1-O-alk-1′-enyl-*sn*-glycerophosphocholine, LPAF), Kayganich and Murphy (39) investigated PC molecular species that contained arachidonoyl (20:4) acyl groups in a PC preparation from human neutrophils. Using precursor ion scans of the arachidonate carboxylate anion on a triple quadrupole mass spectrometer (Fig. 2.7), these authors were able to identify more than eight molecular species (comprising only about 5% of the total PC species present) that contained an arachidonoyl group. The composition of the second acyl group (or 1-O-alk-1′-enyl group) can be calculated by difference from the inferred molecular weight $[M-15]^-$ of the arachidonoyl PC species. The precursor ion spectrum appears complicated, but only the even-mass peaks represent the $[M-15]^-$ ions of the PC molecular species; other, odd-mass fragments correspond to the $[M-60]^-$ and $[M-86]^-$ ions of the molecular species. Subsequent product ion scans of the identified $[M-15]^-$ anions were used to confirm the acyl group and head group composition, and in some cases indicated that several isobaric molecular species overlapped the arachidonoyl species (e.g., 18:0/20:4-PC, 18:2/20:2-PC, and 16:0/22:4-PC all have an $[M-15]^-$ ion at *m/z* 794). Carboxylate anion relative abundances in the product ion spectra were also used in this study to assign relative positions of the two acyl groups. These results will be discussed below (Section 4).

The same group has recently extended the use of precursor ion scans to distinguish arachidonoyl-containing PE plasmalogens (1-O-alk-1′-enyl-*sn*-glycerophosphoethanolamine) from isomeric 1-O-alkenyl species containing a double bond elsewhere in the hydrocarbon chain (40). Comparison of precursor ion scans of *m/z* 303 (arachidonate anion) before and after treatment of the sample with dilute mineral acid allowed the molecular species of the acid-labile plasmalogens to be distinguished from those of the acid-stable 1-O-aklenyl PEs.

### 3.3. *Identification of PL Classes in Crude Lipid Extracts*

Fenselau and co-workers (36) developed the use of constant neutral loss scans for the different PL classes and their molecular species in crude lipid extracts of lysed bacterial cells. These PL profiles can be used to identify rapidly the bacterial species in a small sample without culturing (32). Because this method takes advantage of the preferential desorption/ionization of PLs in these extracts by FAB, the samples do not require purification prior to analysis. The head group–specific neutral fragment losses observed in both the positive

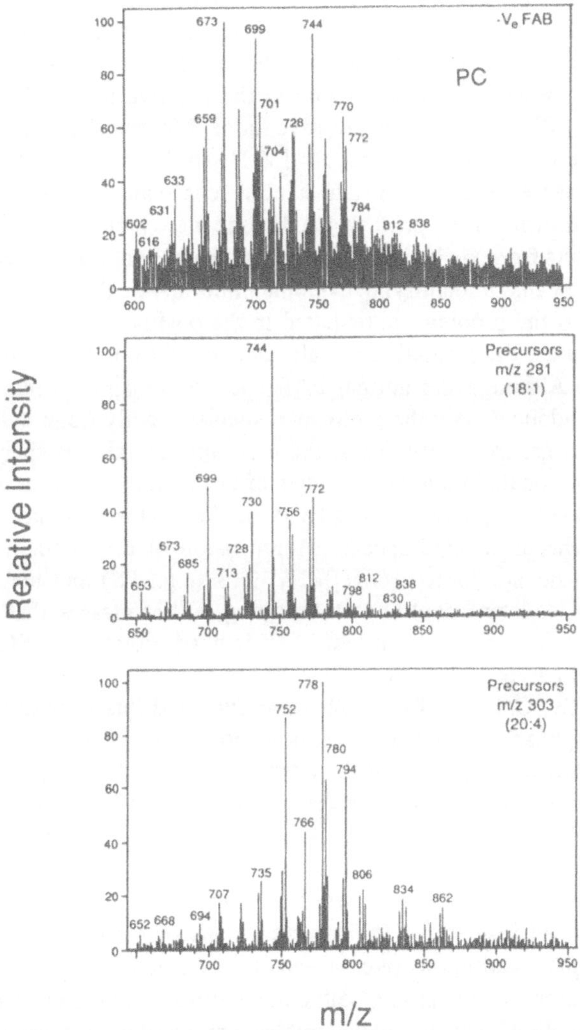


*Figure 2.7.* Negative FAB-CAD-MS/MS analysis of PC molecular species from human neutrophils. Precursor ion scans of the carboxylate anions permit the molecular species containing specific acyl groups to be identified in a mixture. [Reprinted from Kayganich and Murphy, 1991, with permission from *J. Am. Soc. Mass Spectrom.*]

and negative FAB mass spectra (see Section 2.2.1) were considered for the constant neutral loss experiments, because in principle both could be used to selectively identify the molecular species of a particular class. Preliminary product ion scans (linked scanning at constant *B/E* ratio) of the $[MH]^+$ and $[M\text{-}H]^-$ ions of the major molecular species of each of the main PL classes present (PE, PG, and PI) indicated that the ions formed by the expected neutral losses were in every case detectable in the positive ion mode, but were not observed for PE and PG in the negative mode. Moreover, in the negative mode the characteristic neutral loss for PI, $[M\text{-}H\text{-}162]^-$, is shared by other glycolipids. For these reasons, constant neutral loss scans [$B^2(1-E)/E^2$ on an *E/B* sector instrument] in the positive FAB mode were employed. An examination of the extracts from six bacterial species indicated that this procedure provides a sensitive means of identifying even the minor molecular species of each lipid class. One of the problems anticipated in the positive FAB analysis of PLs in crude mixtures, the formation of salt adducts, did not cause any appreciable difficulties. Although the normal FAB spectra contained a number of peaks from these adducts, only the protonated species readily fragmented to lose the neutral head group moiety. Thus, the constant neutral loss spectra contained primarily protonated molecular species of each class. However, salt-free samples would still improve the sensitivity of the analysis by increasing the ion current of the protonated species. A limitation of this procedure is that PL classes that do not form strong $[MH]^+$ ions (e.g., PS) and are not as easily detected by constant neutral loss scanning. Other classes that yield a low-abundance fragment from the loss of their head group (e.g., PC) can also be difficult to analyze by this technique.

Recently, Cole and Enke (32) have extended this approach for bacterial and fungal phospholipid analysis with a triple quadrupole mass spectrometer. These investigators again used the positive FAB mode and analyzed the head group–specific neutral losses. For the constant neutral loss scan of PC mixtures, it was reported that the molecular species detected were all sodium adducts. Because of the potential problems associated with interpretation, precursor ion scans of the characteristic low-mass fragment for PC at *m/z* 184 (phosphocholine) were used instead for PC detection. One of the primary advantages of using a triple quadrupole over a sector instrument in these studies is the computer control of its tuning, collision conditions, and scan mode switching. This control allowed an automated analysis protocol to be developed, whereby all the constant neutral loss scans for each PL class were sequentially performed. The instrument was programmed to switch to the negative ion mode after these scans were completed, and conduct product ion scans on each of the ten major molecular species detected for each PL class, in order to determine their acyl group compositions (Fig. 2.8). A complete PL profile of the bacterial sample could be obtained in 10 min. However, because high-energy collision

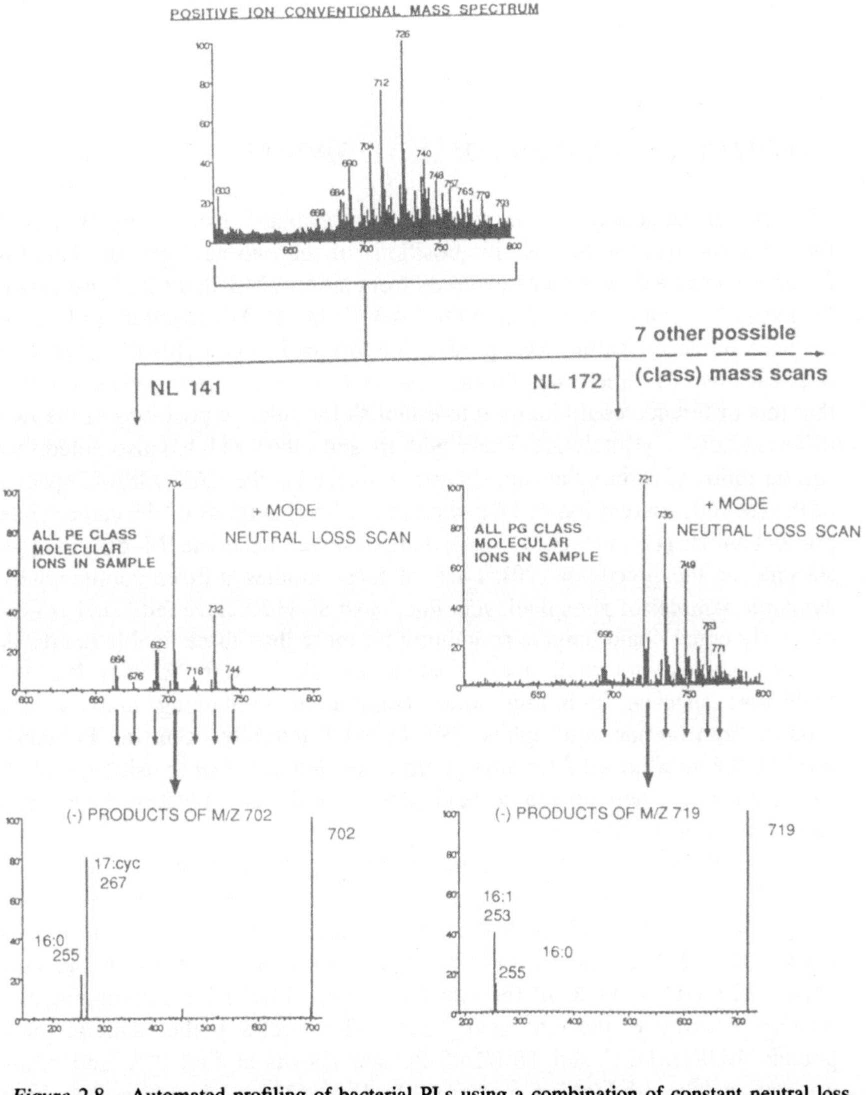


*Figure 2.8.* **Automated profiling of bacterial PLs using a combination of constant neutral loss scans and product ion scans with a triple quadrupole mass spectrometer. [Reprinted from Cole and Enke, 1991 with permission from *J. Anal. Chem.*]**

conditions are not available on this instrument, these researchers did not pursue characterization of the acyl group structure (double bond location, etc.) by additional product ion scans of the carboxylate anion fragments (see Section 3.2).

## *4. POSITIONAL ANALYSIS OF ACYL GROUPS*

One of the structural features of PLs not available from interpretation of the FAB spectrum is the relative positions of the two acyl groups. This information can be directly obtained only from tandem MS analysis. Jensen *et al.* (17) were the first to report that in the FAB-CAD-MS/MS spectrum of PC, the intensity of the *sn*2 carboxylate product ion formed from the $[M-15]^-$ precursor is always two to three times greater than that of the *sn*1 carboxylate ion, and that this difference could be used to establish the relative positions of the two different fatty acyl moieties. These authors and others (11,13) also noted that similar ratios of carboxylate anions were observed in the CAD-MS/MS spectra of the $[M-60]^-$ parent ion of PC, whereas the inverse ratios of the carboxylate product anions (i.e., *sn*1/*sn*2 = 2 to 3) were observed when the $[M-86]^-$ ion was selected as the precursor (20). Each of these studies utilizes commercially available samples of phosphatidycholine, most of which have fatty acyl groups of nearly equal chain lengths containing no more than three double bonds. A recent study also indicated that this pattern was observed in the analysis of PC molecular species, including some compounds containing arachidonoyl groups, from human neutrophils (39). In most naturally occurring PC compounds, the unsaturated fatty acyl group is located at the *sn*2 position, though PLs with long-chain polyenoic acyl groups at the *sn*1 position have been identified in rat brain (41).

The influence of fatty acyl group structures on the fragmentation pattern under FAB-CAD-MS/MS was systematically investigated in our laboratory using a series of 24 synthetic and naturally occurring PCs (25). The results of FAB-CAD-MS/MS analysis of the $[M-15]^-$ ions of a number of the PCs examined were found to fit the empirical rules (11,17) for determining the relative positions of the fatty acyl groups. The spectra of the isomeric compounds 16:0/18:1-PC and 18:1/16:0-PC are shown in Fig. 2.9a and b for illustration. The product ion spectra of the $[M-15]^-$ ion of each isomer show that the *sn*2-derived carboxylate ions (*m/z* 281 and 255 for 18:1 and 16:0, respectively) are more abundant (Fig. 2.9c and d). Although the ratio of intensities of the *sn*2 to *sn*1 carboxylate ions approached 3:1 in the 16:0/18:1-PC isomer (Fig. 2.9c), this ratio was less than 2:1 in the 18:1/16:0-PC isomer (Fig. 2.9d).

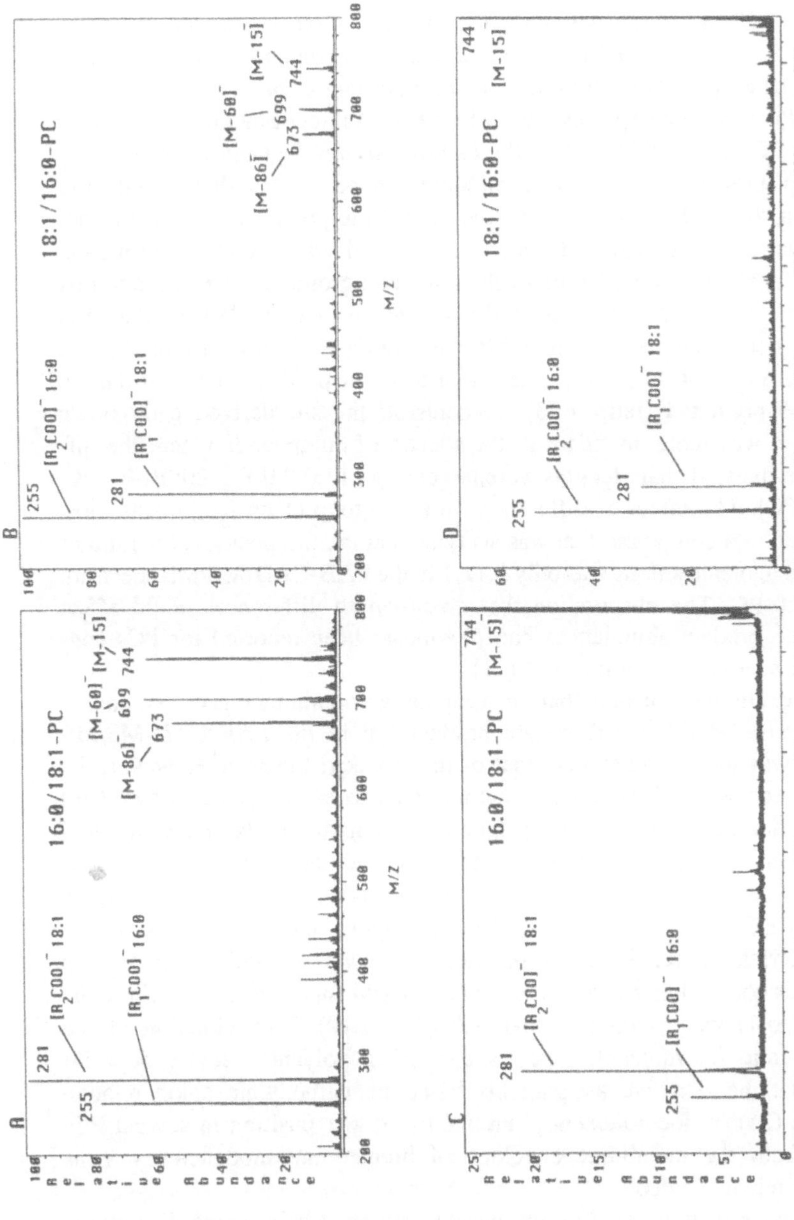


***Figure 2.9.*** Negative FAB mass spectra of two isomeric PC molecular species, (a) 16:0/18:1-PC and (b) 18:1/16:0-PC, show the three characteristic high-mass fragments in the spectra of PCs and prominent fragments representing the carboxylate anions of the two acyl groups. Panels (c) and (d) are the FAB-CAD-MS/MS product ion spectra of the $[M-15]^-$ ion of the two isomers. [Reprinted from Huang *et al.*, 1992, with permission from *J. Am. Soc. Mass Spectrom.*]

However, several examples were encountered where the empirical rules for positional analysis were directly violated. Errors in positional assignments based on carboxylate product ion intensities occurred for compounds in two categories: PC compounds with polyunsaturated (> three or four double bonds) fatty acyl groups at the *sn*2 position and those that contain a much shorter (> six methylene unit) fatty acyl group at *sn*2 than that at *sn*1.

For PC molecular species with a longer-chain acyl function at *sn*2 (e.g., 14:0/16:0-PC and 14:0/20:0-PC), the carboxylate product ion from the fatty acyl group at this position was substantially more intense than that from the *sn*1 position in the FAB-CAD-MS/MS spectrum. This predicted pattern did not hold, however, in the analyzed samples that contained a shorter acyl function at the *sn*2 position. As the length of the fatty acyl group at *sn*1 relative to that at *sn*2 was increased, the ratio of the *sn*2 to *sn*1 carboxylate product ion intensities decreased. With a small difference in chain length between the two acyl groups (e.g., 14:0/12:0-PC), the *sn*2 carboxylate product ion was still more intense, as predicted (ratio 1.73). In contrast, the *sn*2-derived carboxylate product ion was more abundant in the spectra of compounds where the differences in the acyl chain lengths were larger (e.g., 14:0/8:0-PC, 20:0/14:0-PC, 16:0/5:0-PC). The effect of differences in acyl group chain length was most pronounced in a compound that was acetylated at the *sn*2 position; the ratio of the acetate to oleate anions was only 0.12:1 in the FAB-CAD-MS/MS spectrum of 18:1/2:0-PC. The observation that chain-length differences could affect carboxylate product abundances had previously been reported for PCs containing an acetoxy function at *sn*2 (11).

Earlier studies indicated that for compounds containing one or two double bonds, the *sn*2-derived carboxylate product ion in the FAB-CAD-MS/MS spectrum was more intense than that of the *sn*1 acyl function. However, the ratio of intensities of these two product ions (*sn*2 carboxylate/*sn*1 carboxylate) tended to decrease with increasing number of double bonds in the *sn*2 acyl group. When the number of double bonds in the *sn*2 fatty acyl group was four or greater, the *sn*1 carboxylate ion was more abundant in all but one example (18:0/20:4-PC, ratio 1.88) (see Ref. 39). The significantly larger peak for the *sn*1 carboxylate ion relative to the *sn*2 carboxylate was particularly apparent in compounds containing docosahexenoic (22:6) acyl functions at *sn*2 where the ratio (*sn*2/*sn*1) was much less than 1.0 (Fig. 2.10). This violation of the predicted ratio for molecular species containing polyenoic acyl groups indicates that the tentative assignments based on carboxylate anion relative intensities (35) of docosahexenoyl groups to the *sn*1 position in several PCs isolated from the membrane envelope of human immunodeficiency virus should be reinvestigated.

It appears that *sn*2 acyl group unsaturation and chain-length differences

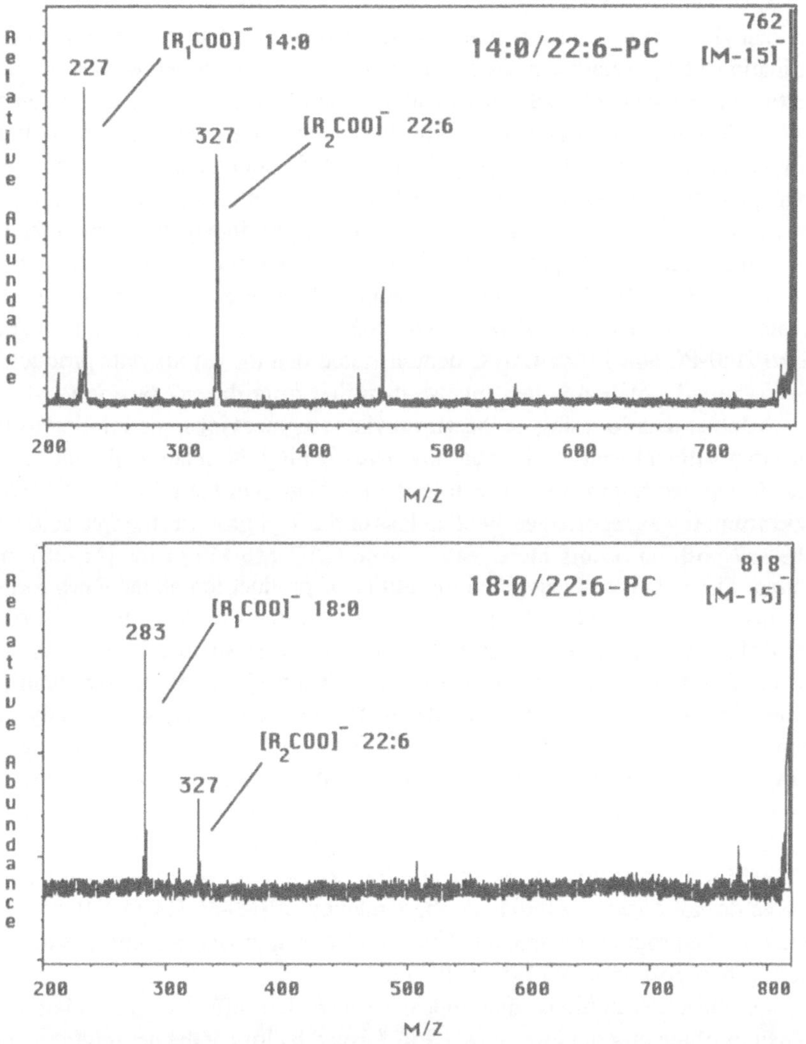


*Figure 2.10.* Negative FAB-CAD-MS/MS spectra of the $[M-15]^-$ ion of two PC molecular species containing docosahexenoyl groups at the *sn*2 position. [Reprinted from Huang *et al.*, 1992, with permission from *J. Am. Soc. Mass Spectrom.*]

between the two acyl groups can act together to influence carboxylate-ion abundances. The MS/MS spectrum of the $[M-15]^-$ ion of 20:0/18:3-PC displays a more intense *sn*1-derived carboxylate fragment.

The dependence on acyl group structural features demonstrates that the relative intensity of the carboxylate product ions from the $[M-15]^-$ precursor of diacyl PC derivatives cannot always be relied on for unambiguous determination of the relative position of the two acyl groups. However, an alternative FAB-CAD-MS/MS experiment does permit acyl position assignments to be made with confidence, even for the exceptional cases discussed. If the high-mass ion $[M-86]^-$ (PC – (choline–water), equivalent to the analogous phosphatidic acid) is used as the precursor ion, then a significantly different product ion spectrum is obtained. As mentioned previously, the ratio of the two prominent carboxylate product ions' intensities is usually reversed (i.e., *sn*1 > *sn*2) (11,13,17,20). Yet, there are compounds where this prediction is also violated. For example, FAB-CAD-MS/MS of the $[M-86]^-$ ion of both 14:0/20:0-PC and 14:0/24:1-PC demonstrated that the carboxylate product ion from the *sn*1 position is less intense than that from the *sn*2 position.

Rather than focusing on the carboxylate fragment ions, the solution to acyl group positional analysis is found by investigating the relative abundances of the deacylated fragments in the middle-mass region of the FAB-CAD-MS/MS spectrum. It was reported earlier that loss of the acyl group as the free acid from the *sn*2 position occurs more readily from CAD-MS/MS of the $[M-86]^-$ precursor (13,17). We investigated the utility of product ion abundances formed by loss of the *sn*1 and *sn*2 free fatty acids from the $[M-86]^-$ precursor ion of 24 different PC molecular species (25). In every case so far examined, the loss of the free fatty acid from the *sn*2 position is always more frequent than that from the *sn*1 position. For example, in the PC containing a 14:0 fatty acyl function at *sn*1 and a 22:6 fatty acyl function at *sn*2, the loss of 22:6 as the free fatty acid from the $[M-86]^-$ parent is significantly more intense than the loss of myristic acid (Fig. 2.11). This relationship is in marked contrast to the violation of the expectation of carboxylate product ion abundances from the $[M-15]^-$ parent of this compound (cf. Fig. 2.10). Further, chain-length differences do not appear to influence the tendency of $[M-86-R^2COOH]^-$ to dominate in the product ion spectra. The use of this approach reliably predicts the acyl group positions in 14:0/8:0-PC (25).

Collision conditions also appear to have less effect on the relative intensities of the product ions of $[M-86]^-$ formed by loss of the neutral fatty acids than the relative abundances of the carboxylate products of this precursor. The metastable decomposition spectrum of the $[M-86]^-$ ion shows preferential neutral loss of the fatty acid from the *sn*2 position (e.g., 14:0/18:3-PC) (25), whereas the *sn*2 carboxylate product is more abundant in this particular case

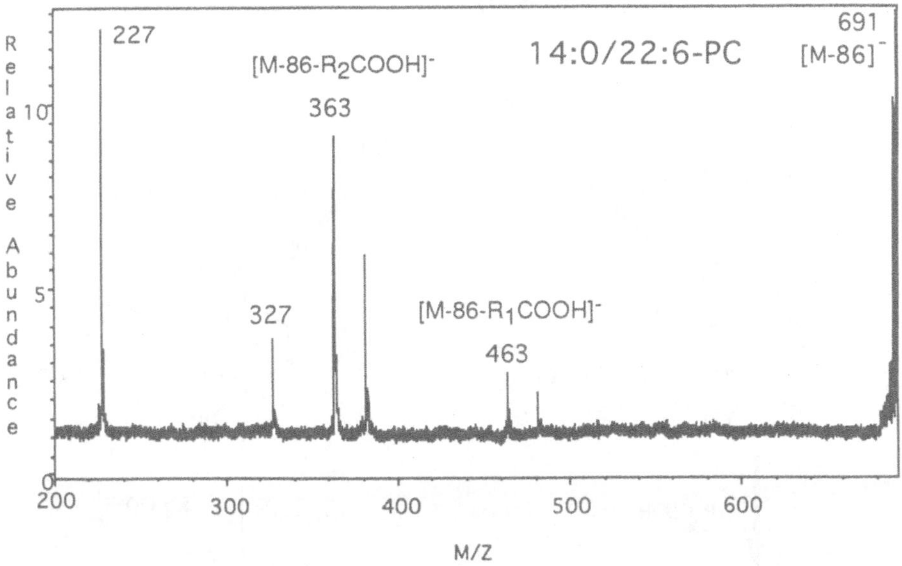


*Figure 2.11.* Negative FAB-CAD-MS/MS spectrum of the [M−86]⁻ ion of 14:0/22:6-PC. [Reprinted from Huang *et al.*, 1992, with permission from *J. Am. Soc. Mass Spectrom.*]

(even though the *sn*1 carboxylate product is predicted to be more abundant than the *sn*2 ion from [M−86]⁻ (13,17). Note, however, that in the high-energy CAD-MS/MS of the [M−86]⁻ ion of this compound, the *sn*1 carboxylate is relatively more intense than the *sn*2 carboxylate, as expected from earlier reports (13,17).

The observed relative abundances of the product ions formed by loss of the free fatty acids may involve the stability of the product ion formed. A mechanism for the formation of these fragments is shown in Fig. 2.12. The loss of the secondary ester function is favored over the elimination of the primary ester because the latter results in the formation of an isolated terminal double bond (Fig. 2.12, structure B). In comparison, conjugation with the phosphate moiety of the internal double bond formed by the loss of the *sn*2 acyl group (Fig. 2.12, structure A) stabilizes this product. Because the fatty acyl group at *sn*2 is lost as a neutral fragment, subsequent fragmentation or reactions of the fatty acyl group do not influence the abundances of the lysophosphatidic acid product ions. These subsequent reactions may account for differences in carboxylate product ion abundances in PCs containing polyunsaturated fatty acyl functions. As shown in Fig. 2.13, equimolar quantities of saturated and unsat-

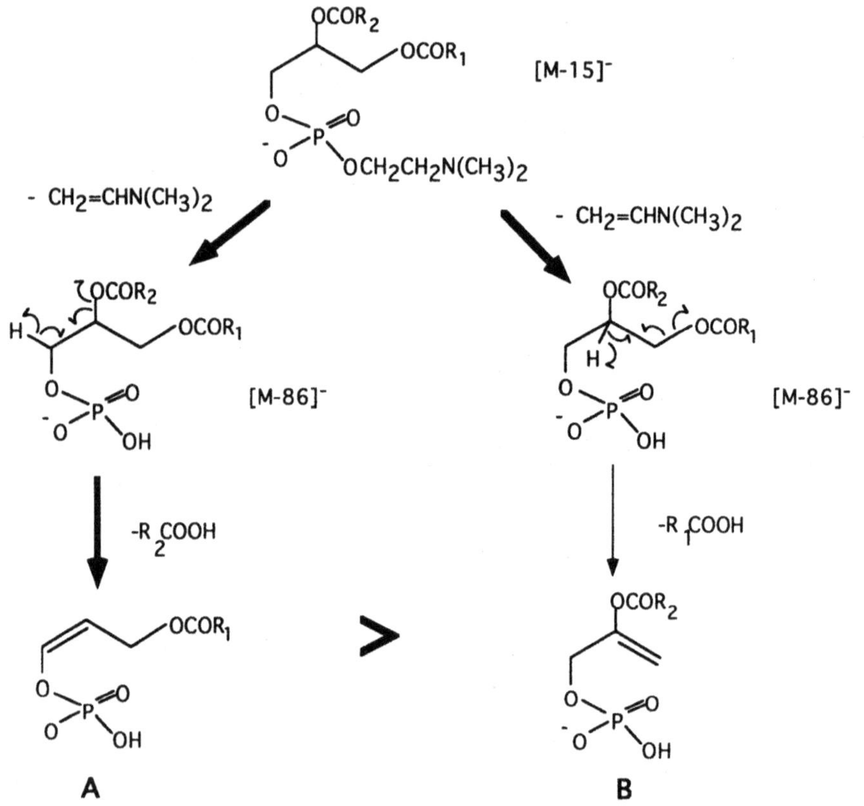


*Figure 2.12.* Proposed mechanism for the formation of the middle-mass fragments of PC molecular species: structures A and B (see text). > Represents "more abundant than." [Reprinted from Huang *et al.*, 1992, with permission from *J. Am. Soc. Mass Spectrom.*]

urated fatty acids may not display $[M-H]^-$ ions in equal abundance. The enhanced response of saturated and long-chain carboxylate anions may provide an additional explanation as to why the carboxylate products cannot be reliably used to determine PC fatty acyl group positions.

The selection of the $[M-86]^-$ precursor ion of diacyl PC compounds for positional analysis indicates that this CAD-MS/MS approach may have general applicability to other classes of PL. Because this ion is equivalent to phosphatidic acid, the same ion found in the spectra of other PL classes can be subjected to CAD-MS/MS analysis. Preliminary studies with diacylglycerylphosphoserine (PS) and diacylglycerylphosphoinositol (PI) compounds have demonstrated that the free fatty acid from the *sn*2 position is also preferentially

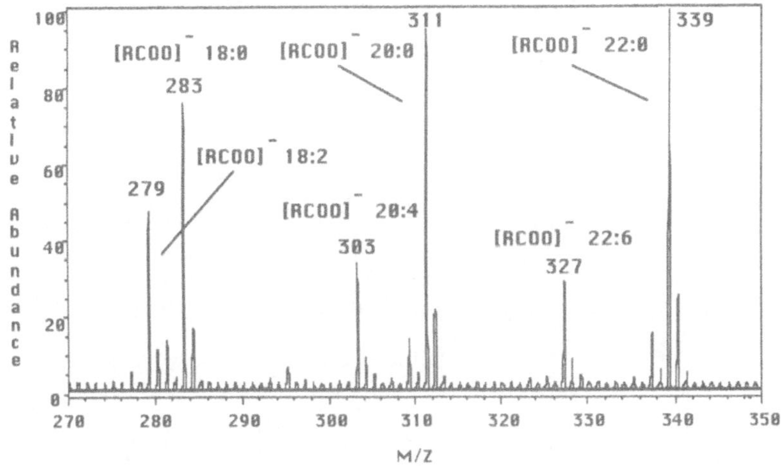


*Figure 2.13.* Negative FAB-MS of an equimolar mixture of different fatty acids. [Reprinted from Huang *et al.*, 1992, with permission from *J. Am. Soc. Mass Spectrom.*]

lost from the $[M-88]^-$ and $[M-162]^-$ ions, respectively (equivalent to the phosphatidic acid fragment), found in the spectra of these compounds (Fig. 2.14). However, the proposed mechanism for the observed differences in the monoacylated fragments' relative intensities suggests that the selection of the phosphatidic acid analog of each PL class as the precursor ion is not essential. We have found that FAB-CAD-MS/MS analysis of any of the high-mass fragments in the negative FAB spectra of different PLs can successfully predict acyl group positions from the abundance of the lyso PL fragment ions, as long as these deacylated product ions are detectable.

## 5. CONCLUSIONS

Fast atom bombardment combined with tandem mass spectrometry has now become a standard method for diacylglycerylphospholipid characterization. It is possible to identify the head group, acyl composition, and relative positions, as well as their individual structures. No other competing methods offer the sensitivity, speed, and structural detail obtained by using these techniques. Moreover, these analyses can be conducted on relatively crude sample preparations. These powerful techniques should allow new structural insight into these molecules and the roles they play in biological systems.

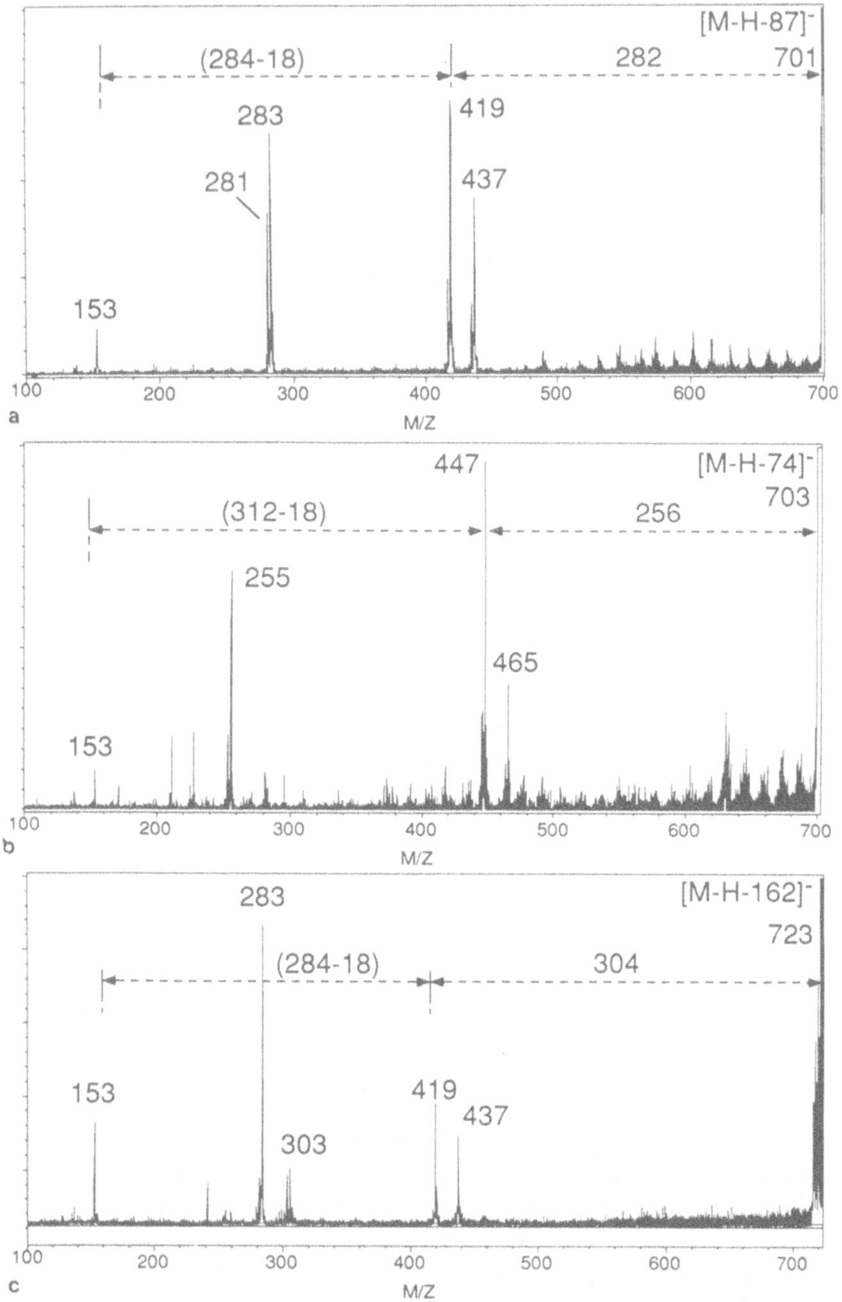


*Figure 2.14.* Negative FAB-CAD-MS/MS spectra of the phosphatidic acid fragment in the spectra of (a) 18:0/18:1—PS, (b) 20:0/16:0—PG, and (c) 18:0/20:4—PI. The relative abundances of the middle-mass fragments formed by the loss of the two fatty acyl groups can be used to characterize their relative positions (Section 4). The PG sample, isolated from *C. butyricum,* was kindly provided by Professor Howard Goldfine, University of Pennsylvania.

## *REFERENCES*


1. Christie, W. W., 1982, *Lipid Analysis*, Pergamon Press, New York.
2. Tokumura, A., Handa, Y., Yoshioka, Y., and Tsukatani, H., 1983, Mass spectrometric analyses of biologically active choline phospholipids and their lyso derivatives, *Chem. Pharm. Bull.* **31**:4425–4435.
3. Haroldsen, P. E., and Murphy, R. C., 1987, Analysis of phospholipid molecular species in rat lung as dinitrobenzoate diglycerides by electron capture negative chemical ionization mass spectrometry, *Biomed. Environ. Mass Spectrom.* **14**:573–578.
4. Blank, M. L., Robinson, M., Fitzgerald, V., and Snyder, F., 1984, Novel quantitative method for determination of molecular species of phospholipids and diglycerides, *J. Chromatog.* **298**:473–482.
5. Weintraub, S. T., Lear, C. S., and Pinkard, R. N., 1990, Analysis of platelet-activating factor by GC-MS after direct derivatization with pentafluorobenzoyl chloride and heptafluorobutyric anhydride, *J. Lipid Res.***31**:719–725.
6. Satsangi, R. K., Ludwig, J. C., Weintraub, S. T., and Pinkard, R. N., 1989, A novel method for the analysis of platelet-activating factor: direct derivatization of glycerophospholipids, *J. Lipid Res.* **30**:929–937.
7. Knörr, W., and Spiteller, G., 1990, Simple method for the analysis of glycerol enol ethers from plasmalogens in complex lipid mixtures and subsequent determination of the aldehydic components by gas chromatography-mass spectrometry, *J. Chromatog.* **526**:303–318.
8. Fenselau, C., Heller, D. N., Olthoff, J. K., Cotter, R. J., Kishimoto, Y., and Uy, O. M., 1989, Desorption of ions from rat membranes: selectivity of different ionization techniques, *Biomed. Environ. Mass Spectrom.* **18**:1037–1045.
9. Jensen, N. J., and Gross, M. L., 1988, A comparison of mass spectrometry methods for structural determination and analysis of phospholipids, *Mass Spectrom. Rev.* **7**:41–70.
10. Benfenati, E., and Reginato, R., 1985, A comparison of three methods of soft ionization mass spectrometry of crude phospholipid, extracts, *Biomed. Mass Spectrom.* **12**:643–651.
11. Münster, H., and Budzikiewicz, H., 1988, Structural and mixture analysis of glycerophosphoric acid derivatives by fast atom bombardment tandem mass spectrometry, *Biol. Chem. Hoppe–Seyler* **369**:303–308.
12. Fenwick, G. R., Eagles, J., and Self, R., 1983, Fast atom bombardment mass spectrometry of intact phospholipids and related compounds, *Biomed. Mass Spectrom.* **10**:382–386.
13. Hayashi, A., Matsubura, T., Morita, M., Kinoshita, T., and Nakamura, T., 1989, Structural analysis of choline phospholipids by fast atom bombardment mass spectrometry and tandem mass spectrometry, *J. Biochem.* **106**:264–269.
14. Ohashi, Y., 1984, Structure determination of phospholipids by secondary ion mass spectrometric techniques: differentiation of isomeric esters, *Biomed. Mass Spectrom.* **11**:383–385.
15. Sherman, W. R., Ackermann, K. E., Bateman, R. H., Green, B. N., and Lewis, I., 1985, Mass analyzed ion kinetic energy spectra and $B_1$–E–$B_2$ triple sector mass spectrometric analysis of phosphoinositides by fast atom bombardment, *Biomed. Mass Spectrom.* **12**:409–413.
16. Chen, S., Mariot, R., Kirschner, G., Farretto, D., and Traldi, P., 1990, Analysis of arachidonic-acid-containing molecular species in glycerophospholipid classes from rat kidney by fast atom bombardment mass spectrometry, *Rapid Commun. Mass Spectrom.* **4**:495–497.
17. Jensen, N. J., Tomer, K. B., and Gross, M. L., 1986, Fast atom bombardment and tandem mass spectrometry of phosphatidylserine and phosphatidylcholine, *Lipids* **21**:580–588.
18. Gasser, H., Strohmaier, W., Schlag, G., Schmid, E. R., and Allmaier, G., 1991, Characterizationof phosphotidylcholines in rabbit lung lavage fluid by positive and negative ion fast atom bombardment mass spectrometry, *J. Chromatog. Biomed. Appl.* **562**:257–266.
19. Allmaier, G., Schmid, E. R., Gasser, H., Strohmaier, W., and Schlag, G., 1990, Methodological

approach to the characterization of diacylphosphatidylcholines in rabbit lung lavage fluid by fast atom bombardment mass spectrometry, *Rapid Commun. Mass Spectrom.* **4**:19–23.
20. Zirrolli, J. A., Clay, K. C., and Murphy, R. C., 1991, Tandem mass spectrometry of negative ions from choline phospholipid molecular species related to platelet activating factor, *Lipids* 2bf26:112–116.
21. Münster, H., Stein, J., and Budzikiewicz, H., 1986, Structure analysis of underivatized phospholipids by negative ion fast atom bombardment mass spectrometry, *Biomed. Environ. Mass Spectrom.* **13**:423–427.
22. Jensen, N. J., Tomer, K. B., and Gross, M. L., 1986, FAB-MS/MS for phosphatidylinositol, -glycerol, -ethanolamine and other complex phospholipids, *Lipids* **22**:480–489.
23. Chen, S., Kirschner, G., and Traldi, P., 1990, Positive ion fast atom bombardment mass spectrometric analysis of the molecular species of glycerophosphotidylserine, *Anal. Biochem.* **191**:100–105.
24. Chen, S., Benfenati, E., Fanelli, R., Kirschner, G., and Pregnolato, F., 1989, Molecular species analysis of phospholipids by negative ion fast atom bombardment mass spectrometry: application of surface precipitation technique, *Biomed. Environ. Mass Spectrom.* **18**:1051–1056.
25. Huang, Z-H, Gage, D. A., and Sweeley, C. C., 1992, Characterization of diacylglycerylphosphocholine molecular species by FAB-CAD-MS/MS: a general method not sensitive to the nature of the fatty acyl groups, *J. Am. Soc. Mass Spectrom.* **3**:71–78.
26. Busch, K. L., Glish, G. L., and McLuckey, S. A., 1988, *Mass Spectrometry/Mass Spectrometry Techniques and Applications of Tandem Mass Spectrometry*, VCH Publishers, Inc., New York.
27. Dasgupta, A., Ayanoglu, E., Wegmann-Szente, A., Tomer, K. B., and Djerassi, C., 1986, Mass spectral behavior and HPLC of some unusual molecular phospholipid species, *Chem. Phys. Lipids* **41**:335–347.
28. Dasgupta, A., Ayanoglu, E., Tomer, K. B., and Djerassi, C., 1987, High performance liquid chromatography and fast atom bombardment mass spectrometry of unusual branched and unsaturated phospholipid molecular species, *Chem. Phys. Lipids* **43**:101–111.
29. Bernstrom, K., Kayganich, K., and Murphy, R. C., 1991, Collisionally induced dissociation of epoxyeicosatrienoic acids and epoxyeicosatrienoic acid–phospholipid molecular species, *Anal. Biochem.* **198**:203–211.
30. Van Breeman, R. B., Wheeler, J. J., and Boss, W. F., 1990, Identification of carrot inositol phospholipids by fast atom bombardment mass spectrometry, *Lipids* **25**:328–334.
31. Jennings, K. R., and Mason, R. S., 1983, Tandem mass spectrometry utilizing linked scanning of double focusing instruments, in *Tandem Mass Spectrometry* (F. W. McLafferty, ed.), Wiley, New York.
32. Cole, M. J., and Enke, C. G., 1991, Direct determination of phospholipid structures in microorganisms by fast atom bombardment triple quadrupole mass spectrometry, *Anal. Chem.* **63**:1032–1038.
33. Wysocki, V. H., and Ross, M. M., 1991, Charge-remote fragmentation of gas-phase ions: mechanistic and energetic considertions in the dissociation of long-chain functionalized alkanes and alkenes. *Int. J. Mass Spectrom. Ion Proc.* **104**:179–211.
34. Bryant, D. K., and Orlando, R., 1991, Location of unsaturated positions in phosphatidylcholines by consecutive-reaction monitoring, *Rapid Commun. Mass Spectrom.* **5**:124–127.
35. Bryant, D. K., Orlando, R. C., Fenselau, C., Sowder, R. C., and Henderson, L. E., 1991, Four-sector tandem mass spectrometric analysis of complex mixtures of phosphatidylcholines present in a human immunodeficiency virus preparation, *Anal. Chem.* **63**:1110–1114.
36. Heller, D. N., Murphy, C. M., Cotter, R. J., Fenselau, C., and Uy, O. M., 1988, Constant neutral loss scanning for the characterization of bacterial phospholipids desorbed by fast atom bombardment, *Anal. Chem.* **60**:2787–2791.
37. Cole, M. J., and Enke, C. G., 1991, Fast atom bombardment/ion molecule reactions for the differentiation of phospholipid classes, *J. Am. Soc. Mass Spectrom.* **2**:470–475.

38. Jensen, N. J., and Gross, M. L., 1987, Mass spectrometry methods for structural determination and analysis of fatty acids, *Mass Spectrom. Rev.* **6**:497–536.
39. Kayganich, K., and Murphy, R. C., 1991, Molecular species analysis of arachidonate containing glycerophosphocholines by tandem mass spectrometry, *J. Am. Soc. Mass Spectrom.* **2**:45–54.
40. Kayganich, K., and Murphy, R. C., 1992, Fast atom bombardment and tandem mass spectrometric analysis of diacyl, alkylacyl and alk-1-enyl-acyl glycerophosphoethanolamine in human polymorphonuclear leucocytes. *Anal. Chem.,* **64**:2965–2971.
41. Robinson, B. S., Johnson, D. W., and Poulos, A., 1990, Unique molecular species of phosphatidylcholine containing very-long-chain ($C_{24}$–$C_{38}$) polyenoic fatty acids in rat brain, *Biochem. J.* **265**:763–767.

# 3

# DNA Modifications

## Investigations by Mass Spectrometry

*Curt B. Norwood and Paul Vouros*

### 1. INTRODUCTION

DNA, deoxyribonucleic acid, is crucial to life. With its triplet coding, DNA serves as the template for messenger RNA, and is therefore responsible for the myriad proteins that ensure the ongoing health and life of the current cell or organism. Future generations are similarly dependent on the DNA code, which cellular mechanisms help propagate. DNA strands must remain intact and undamaged to guarantee that the code functions with high fidelity. If DNA becomes covalently damaged, repair mechanisms can excise the affected region and replace it with the correct sequence. Indications of this biochemical transformation can be found among the excretion products of the cell or organism. However, the repair processes do not always work perfectly, and certain modifications escape the repair system. These covalent modifications may be of critical importance to the causes of disease states, including carcinogenesis. In addition, certain medications, known to generate covalent modifications of DNA, are used therapeutically in an effort to combat an observed disease state. These deliberately produced DNA modifications also need to be studied. In this review, the term *DNA modification* includes not only DNA adducts (xenobiotic compounds covalently bound to DNA) but also other aberrant forms of DNA, such as ring-opened products and dimers.

*Curt B. Norwood* • U.S. Environmental Protection Agency—Environmental Research Laboratory, Narragansett, Rhode Island 02882. *Paul Vouros* • Department of Chemistry and Barnett Institute, Northeastern University, Boston, Massachusetts 02115.

*Mass Spectrometry: Clinical and Biomedical Applications, Volume 2,* edited by Dominic M. Desiderio. Plenum Press, New York, 1994.

The great attraction of mass spectrometry (MS) lies in its ability to offer both specificity (i.e., identification) and sensitivity—i.e., improved limits of detection (LOD). Nuclear magnetic resonance (NMR) techniques may offer a higher degree of structural elucidation but require considerable amounts of material. Other currently used techniques, e.g., $^{32}P$ postlabeling–autoradiography and high-performance liquid chromatography (HPLC)–fluorescence detection (1), offer superior sensitivity, but do not have the identification power that can be obtained with MS. In many cases, MS has provided confirmation or identification of a specific DNA modification, albeit at sometimes fairly high levels. On the other hand, the recently introduced technique of accelerator MS has reached levels of detection sensitivity that rival or surpass those attained with non-MS techniques, but does not possess the identification ability inherent to most types of MS. MS is well positioned to occupy an important niche between the high information–low sensitivity technique of NMR and the high sensitivity–low information techniques of, e.g., fluorescence detection or $^{32}P$ postlabeling. It is in this context that the future role of MS methods in DNA adduct research needs to be assessed.

MS has already played an important role in both the detection and the identification of DNA modifications, as detailed in two recent reviews. In 1992, McCloskey and Crain (2) discussed much of the recent research (through mid-1991) of the analysis of xenobiotic modifications of DNA and some emerging techniques for investigating larger fragments of DNA. Lay's 1992 article (3) included a historical perspective and a section devoted to the recent use of tandem MS techniques for the analysis of carcinogen–DNA adducts. Also of importance is Crain's review (4), which detailed MS techniques utilized in nucleic acid research (not specific to modified DNA), and a similar review by Schram (5), which delved into many sample preparation and chemical derivatization techniques. The present chapter is organized according to the mode of MS analysis; for each method, we will discuss its advantages, its difficulties, and its applicability to the form of modified DNA to be investigated.

DNA that is subjected to analysis may be in any one of a number of forms. For example, the DNA may exist as a double-stranded segment. However, cleavages of the interstrand hydrogen bonds yield an oligomeric sequence of single-stranded DNA. Hydrolysis of the phosphodiester backbone of the strand produces individual nucleotides, each containing the purine or pyrimidine base, the five-carbon 2′-deoxyribose sugar, and the 3′ or 5′ phosphate. Removal of the phosphate produces the nucleoside (base plus sugar), and removal of the sugar produces the base. In many instances, the adduct can be excised (hydrolyzed) from the base (or other part of the DNA) and investigated as a separate compound. Or the excretion process(es) of the cell or organism can be investigated to provide some indication of the nature of the DNA modification(s) and of the repair processes. Under certain conditions, the base itself can undergo ring opening (although in some cases, at least, this modification seems to result from the analytic method). Furthermore, DNA can form covalent bonds between adjacent bases (dimers) or crosslinks with residues on proteins, and these modifications also can be investigated. Different sample-preparation techniques must be used in order to analyze for different forms of DNA. Similarly, different sample-preparation methods must be employed depending on the chosen MS mode of analysis. This chapter will detail the amount and nature of the sample processing required prior to actual MS analysis.

DNA modification research often considers the number of modifications present for every $10^N$ unmodified nucleotides. The exponent $N$ may reach values close to 1 for certain *in vitro* reactions, or can be 12 or greater for environmental samples and low-level-dosimetry experiments. When the number of modifications is large (when $N$ is small), the MS analysis is accomplished more easily. However, when $N$ is large, high sensitivity (low LOD) analyses are required, as is shown by the following example. DNA usually comprises approximately 0.1% of the mass of a given tissue. A calculation of the mass of tissue needed for an MS analysis is based on a number of factors. Let the assumptions be that a small adduct nucleotide is present at a level of 1 in $10^6$, that the analysis will consume 100 pg of the modified nucleotide, that the determination will be near the LOD, and that ~20% of the sample is consumed during the analysis. (This 20% represents a reasonable compromise between repeatability and LOD). Under these conditions, 500 mg of tissue would be required. The amount of tissue that can be dedicated to analysis is often very limited and may be much less than 500 mg (particularly with human tissue). Our review details current LODs so that the reader can assess the applicability of a given mode of MS analysis to the problem at hand.

Throughout this chapter, emphasis will be placed not only on the detection of the molecular ion, but also on its observed fragmentation. In many instances, especially when the identification of unknown modifications is desired, the presence of a molecular ion may be essential. In other cases, the fragmentation pattern itself may provide additional information about the type of modification, the position of modification, or the nature of the base that had been modified. The generally accepted nomenclature for these spectra uses the terms M, B, S, and A to indicate, respectively, the molecular, base, sugar, and aglycone (M–S) species. However, when the actual instrumental hardware is discussed, B, E, and Q denote, respectively, magnetic, electrostatic, and quadrupole analyzers.

Chemical derivatization techniques are often used to enhance the detection of a significant ion or to increase the molecular weight so that the ions of interest are no longer interfered with by the chemical background signal at a lower mass-to-charge ratio (*m*/*z*). Derivatization is also used to provide increased fragmentation, or to augment chromatographic separations of various constituents prior to MS analysis. Similarly, derivatization can provide the various components with improved peak shapes or increased volatility to aid in the chromatography. In many cases, a variable amount of functional group incorporation occurs, so that the ionization is spread over a number of molecular ions. This extra range of ions can be detrimental because the LOD deteriorates correspondingly; however, the extra ions can help determine the molecular species. Derivatization may also be a difficult process when dealing with a small amount of sample.

Certain sections of this chapter will discuss the presence of various alkali metal ions, specifically $Na^+$ and $K^+$, in the sample(s) under consideration. Often, these ions substitute for $H^+$ in the molecular ion, so that the ion current for the molecular species is distributed over a number of different *m/z*. This spread decreases the analyte signal strength at a particular *m/z* and thus the LOD deteriorates. As Russell and co-workers have observed (6), $H^+$, $Na^+$, and $K^+$ can have different binding sites on the molecule and can aid in the identification of the molecular species. However, most researchers today try to eliminate these two ions during sample preparation. If a cation is required, then the ammonium ion may often be successfully utilized for either of two reasons: ammonia can easily be volatilized prior to sample analysis, or if left in the sample, it doesn't tend to replace $H^+$ in the MS.

In many cases, mass spectrometry/mass spectrometry (MS/MS) techniques (also known as tandem MS) have been used to increase significantly the degree of specificity and/or detectability. MS/MS uses are discussed in this chapter when appropriate to the MS mode under consideration. Many types of MS/MS experiments exist, and the exact analytical details depend on the presence or absence of a collision gas and the type of instrumentation used. But all MS/MS experiments rely on the fragmentation of the selected precursor ion and result in spectra, of which there are three common types: the product ion spectrum, which contains all ions that result from a selected precursor ion; the precursor ion spectrum, which contains all precursor ions that produce a specific product ion; and the constant neutral loss (CNL) spectrum, which contains product or precursor ions formed when a neutral fragment of the selected mass was lost from a precursor ion. Ions used in MS/MS determinations may result from metastable transitions in ion traps or in the field-free regions of sector instruments, or may be generated by collisions with an inert gas, in a process known as collision-induced dissociation (CID). MS/MS analyses, which can be performed on a wide variety of MS systems, afford a higher degree of selectivity that can permit significant reductions in background noise or chemical interference. Thus, even if less total signal is being collected, the signal-to-noise ratio may be increased and the LOD improved. Alternatively, this specificity can sometimes reduce the amount of sample cleanup required.

## 2. *FIELD DESORPTION (FD)*

The technique of FD involves coating a sample onto an FD emitter wire and placing the wire in a strong electric field. Subsequent desorption of the ionized species may be enhanced by passing a current through the wire. FD/MS, although used more extensively in the past, continues to be used today. FD was employed by Shaulsky *et al.* (7) to investigate the binding of aflatoxins to DNA. Aflatoxins are structurally related to furocoumarins (psoralens), which are used therapeutically for skin cancer, because photoactivation of the psoralens leads to DNA binding. Aflatoxin–base adducts were obtained from *in vitro* reactions and the following briefly described methods. The DNA was acid-hydrolysed to the base level; the bases were separated by HPLC and collected into fractions; the volume was reduced; the residue was dissolved in methanol and coated onto the FD emitter wire. One of the resulting spectra (Fig. 3.1) clearly displays the molecular ion as well as several fragments, including those ions corresponding to the guanine base and the adduct, which had been cleaved from it. No details are given with respect to the amount of material

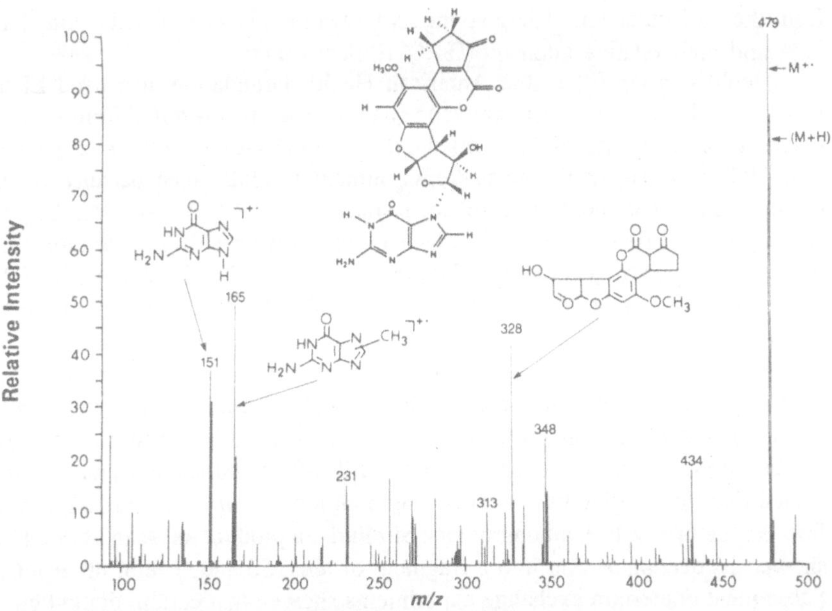


*Figure 3.1.* Field-desorption positive-ion full-scan mass spectrum of the photoadduct of DNA and aflatoxin $B_1$. Structures of the guanine base adduct and selected fragment ions are shown. Modified from Ref. 7; used with permission.

subjected to MS analysis. These researchers were unsuccessful in obtaining spectra when they utilized fast atom bombardment (FAB) with a variety of matrices.

## *3. DIRECT INSERTION PROBE (THERMAL DESORPTION) TECHNIQUES*

### *3.1. Electron Ionization*

With direct insertion probe MS, the sample is first placed near, or just within, the source region of the MS and thermally vaporized. Because of this heating process, compounds that are thermally labile may not be analyzed by this approach. In combination with this desorption mode, electron ionization (EI) and chemical ionization (CI) have both been used extensively in the past 5 yr. With EI, base and nucleoside adducts have both been investigated. For example, positive-ion EI spectra aided in the investigation of the 5-(hydroxymethyl) chrysene guanine and adenine base adducts formed *in vitro* in calf thymus and collected in HPLC fractions (8). The most intense ion observed was the *m/z* 239 fragment, corresponding to the methylchrysene moiety cleaved from the molecular ion. The guanine and adenine adduct molecular ions had 20% and 60% relative intensity (%RI), respectively.

Hecht's group (9) at the American Health Foundation also used EI to investigate the much smaller exocyclic guanine bases that resulted from *in vitro* reactions of nitrosopyrrolidine with single-strand DNA and with deoxyguanosine (dG). Nitrosopyrrolidine has environmental significance because of its presence in cured meats and tobacco smoke. UV, NMR, and MS helped identify three different isomeric adducts collected from HPLC separations. EI positive-ion spectra generated in the presence of $NH_4Cl$ exhibited a 100 %RI molecular ion peak at *m/z* 221. The spectrum also showed considerable fragmentation associated with ions with charge retention on the guanine base.

Positive-ion EI was used as part of a project to investigate similar small ethyl and dimethyl guanine adducts, all of which have a molecular ion of *m/z* 179 (10). A sample isolated from human urine was subjected to high-energy source acceleration and subsequent CID. MS/MS precursor ion analysis revealed that the guanine base–related fragment ions of *m/z* 150 and 151 derive from an ion of *m/z* 179, among others. Similarly, a product ion scan of *m/z* 179 showed its decomposition into a fragment of *m/z* 150 and several other ions. Subsequent deuterium exchange experiments showed that certain dialkyl guanines could be readily distinguished. That research demonstrated some of the potential of MS/MS techniques to determine more specifically the presence of a suspected analyte.

### 3.2. Chemical Ionization

EI and CI have both been extensively utilized by Solomon *et al.* at the New York University Medical Center, usually for relatively small base or nucleoside adducts. A recent article (11) described isopropyl modifications. Both ionization techniques yielded spectra with extensive fragmentation, indicative of modification at the base rather than at the sugar. CI was used to determine the structure of both nucleoside and base modifications. The intensity of the $[M+H]^+$ ion ranged from 20 %RI for a nucleoside adduct to 70 %RI for a base adduct. All nucleoside spectra showed a prominent $[AH_2]$ ion, corresponding to the loss of the deoxyribose moiety, and an intense ion corresponding to the base. Their more recent reports (12,13) use similar EI and CI MS techniques to investigate *in vitro*–formed ethyl and hydroxyethyl adducts of various bases and nucleosides, sometimes with dimethoxytrityl derivatization to protect the 5′ end of the nucleoside monomer intended for incorporation into an oligomeric sequence. Their current research involves the rapid hydrolytic deamination of alkylated deoxycytidine and the possible influence of uracil glycosylase repair.

CI of nucleosides and bases has been employed extensively by various groups at the American Health Foundation (14–16). Spectra were not presented, but isobutane CI tabular data for a pyridyloxobutyl nucleoside adduct (15) gave a $[M+H]^+$ ion of <2 %RI.

A phenyl–glycidyl ether nucleoside adduct was investigated by van den Eckhout *et al.* (17). The ammonia CI tabular data indicated an $[M+H]^+$ ion of 100 %RI. Ions corresponding to the loss of sugar and to the loss of sugar plus the adduct's phenyl group were strong, as were the ions corresponding to the sugar itself. This group conducted considerably more FAB research that will be mentioned below, in Section 8.3.

Styreneoxide (alkyl) adducts were analyzed by triple quadrupole (TQ)/MS using CI with either isobutane or ammonia as the reacting gas (18). Tabular MS data showed that the site of adduction, as well as the identity of a specific nucleoside base, caused a dramatic shift in the intensity of the $[M+H]^+$ ion, which ranged between 10 and 100 %RI. The abundances of other fragment ions indicative of specific cleavage pathways also varied widely among the various adducts.

Base adducts of benzoquinone were similarly investigated with CI by Pongracz and Bodell (19). With high-resolution MS, they established the nature of the chemical modifications of the bases. They postulated that some of the adducts were changed by ring-opening during the mild acid hydrolysis used to generate the bases from the nucleotides. FAB analysis of the nucleotides was also included in their research.

Excellent LODs for small alkylated nucleosides were recently attained by

Cooks's group (20). Their *in vitro* reactions of methyl methanesulfonate and 1-methyl-1-nitrosourea with calf thymus DNA resulted in quantifiable levels of adduction (3-methylthymidine and $O^4$-methylthymidine) as low as 1 in $10^5$ unmodified nucleotides. Enzymatic hydrolysis of the DNA to nucleosides was followed by microfiltration to remove the proteinaceous material. Subsequent HPLC fraction collections (performed twice to eliminate unwanted interferences that resulted from the first separation) were analyzed by CI MS/MS with CID with argon collision gas in the central quadrupole of their TQ/MS. The actual MS analysis was completed in a few seconds, and the MS/MS product ion scans added required specificity to permit improved LODs. The first quadrupole was centered on the *m/z* 257 $[M+H]^+$ of the nucleoside adduct and detuned sufficiently to allow quantification with a deuterated analog. Figure 3.2 shows the desorption profile for $O^4$-methylthymidine and ~1 ng of its deuterated analog, the structure of the nondeuterated nucleoside adduct, and the MS/MS CID product ion spectrum. This CI probe methodology avoids the matrix-associated interferences present with some other MS methodologies. Their methods should permit investigations of mutagenesis and cytotoxicity relative to adduct formation and repair as a result of cell culture exposures of various alkylating agents.

## *3.3. Electron Ionization with Chemical Derivatization*

The two well-established chemical derivatization techniques of silylation and acetylation have been used recently with EI for the determination of base and nucleoside modifications (21–25). Similarly, CI of derivatized modified DNA has helped in certain research efforts. Chadha *et al.* (26) examined the acetylated and the underivatized dibenzanthracene nucleoside. Derivatization has played a more prominent role in the MS research of Solomon's group. EI with trimethylsilyl (TMS) derivatization and CI with acetylation of modified nucleosides (as well as high-resolution MS) were both investigated (27). That work established a different approach to investigating various modified nucleosides that result from oxidative damage to DNA. Their methods relied on gentle enzymatic hydrolysis of the DNA, acetylation with tritium labeling, and HPLC separation of the various oxidative products. As a result, oxidative damage caused by acid hydrolysis of the DNA was avoided, and the analysis was applicable to a wider suite of components than those amenable to HPLC–electrochemical detection. Their *in vivo* research involved exposure of mice to the tumor promoter 12-O-tetradecanoylphorbol 13-acetate. However, their MS research involved *in vitro*–produced standards that were used for comparison with the products that had been formed *in vivo.* Figure 3.3 depicts three spectra and their corresponding structures. Note that in Fig. 3.3a and b, the same compound was subjected to both TMS and acetyl derivatization and analyzed

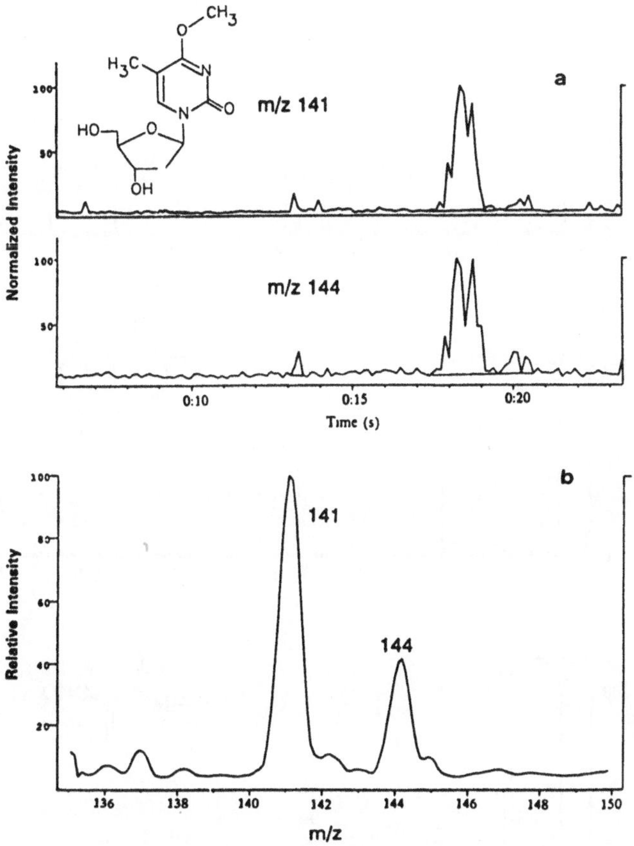


*Figure 3.2.* Thermal-desorption positive-ion CI MS/MS analysis of $O^4$-methylthymidine and 1 ng of the deuterated internal standard analog. CID TQ/MS analysis used a detuned first quadrupole centered at the nucleoside $[M+H]^+$ ion (*m/z* 257), collisions in the central Q, and product ion scans to analyze the *m/z* region of the modified base. (a) Desorption profiles for the protonated, modified bases. (b) Product ion spectrum. Modified from Ref. 20; used with permission.

under EI and CI conditions, respectively. The molecular ion does not occur at greater than 10 %RI. Variable amounts of incorporation of the functional groups were often observed. The spectra detail fragment ions corresponding to the sugar and the base moieties.

Schram's group at the University of Arizona reported (28) the use of TMS derivatization techniques. Two nucleotides important in anticancer research were investigated, and LODs of 1 µg were obtained. The primary focus of their report was the use of FAB to achieve more sensitive LODs, as will be discussed in Section 8.2.

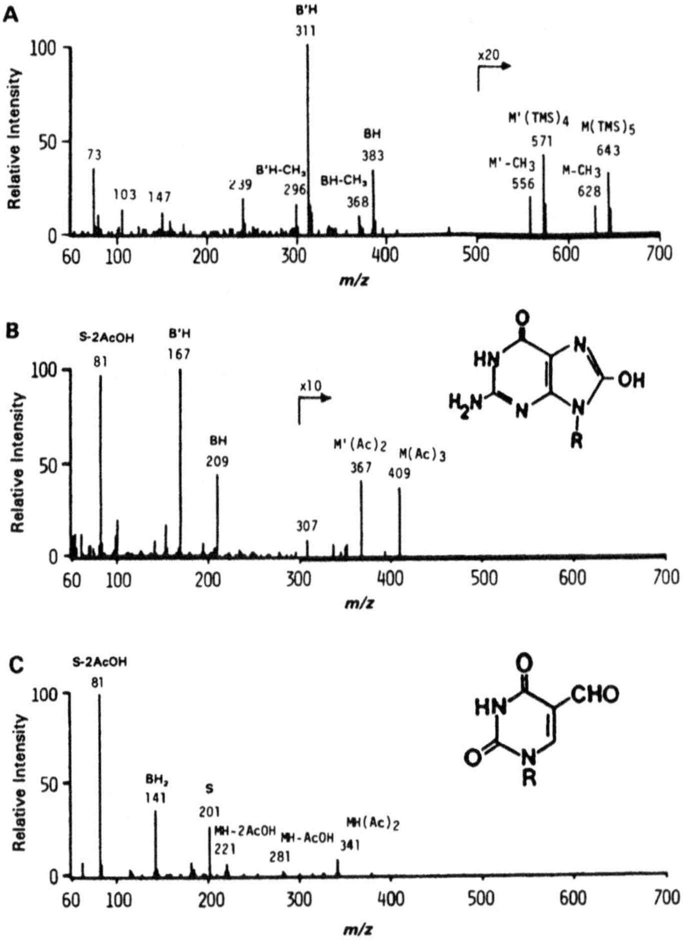


*Figure 3.3.* Structures and thermal-desorption positive-ion mass spectra. (A) TMS derivative of 8-OHdG under EI conditions. (B) Acetyl derivative of 8-OHdG under EI conditions. (C) Acetyl derivative of 5-formyl-2′-deoxyuridine under CI conditions. Modified from Ref. 27; used with permission.

## 3.4. Chemical Ionization with Chemical Derivatization

More extensive derivatization investigations with CI and EI were performed by researchers at Northeastern University (29). Research groups directed by Giese and Vouros sought to improve the LOD for nucleosides through the use of pentafluorobenzyl (PFB), cinnamyl, or mixed derivatives. EI, CI, and electron-capture negative ionization (ECNI) spectra were compared. The positive-ion EI and CI spectra gave significant fragmentation for all the derivatives. The PFB derivative is noted for its chemical stability and its ability to yield structurally relevant fragmentation with EI and positive ion CI modes. In a later paper (30), these researchers investigated a wider variety of modified bases that were derivatized with PFB and analyzed by EI, CI, and gas chromatography–ECNI (GC–ECNI). This last technique will be discussed more fully with GC-MS in Section 4.3.

## 4. *COMBINED GAS CHROMATOGRAPHY-MASS SPECTROMETRY*

### 4.1. *Electron Ionization*

This chapter considers a number of on-line couplings of separation schemes with MS. Presented in this section is the first of these couplings, GC-MS. Parts of this presentation will be dedicated to EI and CI modes, usually with chemical derivatization of the sample in order to impart increased volatility and sensitivity, to avoid most of the problems due to thermal lability, and to allow improved GC resolution and peak shape. However, derivatization was not used when ion–molecule reactions were investigated in the ion source and the collision cell of a TQ/MS (31). Some of that research involved CI of a model DNA compound, pyridine, which was reacted with GC effluent components inside the MS so that formation of adducts could be monitored. The first quadrupole selected the pyridine molecular species. Subsequent low-energy collisions occurred in the second quadrupole between these ions and the GC effluent's neutral molecules. The third quadrupole determined those GC components that had reacted with the model DNA ion. The method is potentially of use in assessing the risk of adduct formation by environmental contaminants. LODs in the high-pg range are suggested. Only those electrophilic compounds that are direct-acting and unaffected by the sample preparation and analysis (i.e., GC injector and column) conditions are amenable to that method.

### 4.2. *Electron Ionization with Chemical Derivatization*

The vast majority of researchers who employ on-line GC techniques derivatize their samples for the abovementioned reasons. A number of groups have utilized this approach using EI (32–41). In many instances, selected ion monitoring (SIM) techniques were employed to achieve improved LODs for known compounds. In most cases, the base form of the DNA moiety was utilized. However, in a review article dealing with free radical–induced damage to DNA, Dizdaroglu (36) notes that sugar modifications and DNA–protein crosslinks can also be monitored. Although many research groups are now utilizing alternative methods of analysis that employ FAB or other desorption techniques, Schram's group continues to develop the usefulness of GC-MS with derivatization. As already noted, their review article (5) clearly portrays many of the advances that have been achieved with MS in DNA and RNA research and details many derivatization and sample preparation procedures. Their more recent article (42) describes the use of HPLC to separate human urine into fractions that were derivatized (TMS) and subjected to GC-MS analysis. The urine of a lung cancer patient was found to contain a 5′-deoxy-5′-methylthioguanosine. This compound is probably not derived from DNA, but may be important as an indicator of disease states because it is also found at elevated levels in the urine of patients with AIDS. Figure 3.4 shows the structure and spectrum of this compound, whose identification was facilitated by comparison with spectra of known TMS-derivatized methylthioadenosine and guanosine.

The TMS derivatization is a straightforward approach that results in a number of ion series that are useful in spectral interpretation. The molecular ion is usually obvious, in part because of the normally intense fragment ion at $[M-15]^+$ that results from the facile loss of a methyl group from $M^{+\cdot}$. Derivatization also helps in many of the ways previously mentioned. That research with biological samples has successfully combined two successive pre-MS separation schemes, HPLC and GC. LODs of 10 to 100 pg have been attained with full-scan spectra (i.e., spectra obtained when the analyzer is operated over the full pertinent range of *m/z* values) and with injection of only 2% of the sample. Variable amounts of functional group incorporation are sometimes encountered, but can readily be discerned by careful examination of the spectra.

### *4.3. Chemical Ionization with Chemical Derivatization*

Combined GC-MS has played a similar important role when CI, rather than EI, has been employed (43–45). Giese's group has developed ECNI techniques to attain improved adduct LODs. Their early work demonstrated that PFB derivatization was ideal for very sensitive measurements with ECNI. They have now focused on three broad classes of adducts. With small alkyl adducts, they can directly derivatize the base. With larger amino polynuclear aromatic hydrocarbon (PAH) adducts, hydrazinolysis (to excise the amino-PAH) is followed by derivatization (46). With similarly large diolepoxide PAH adducts, the adduct is cleaved from the base with mild acid hydrolysis, subjected to potassium superoxide treatment to form a diacid, and then derivatized (47). They showed impressive zeptomole ($10^{-21}$ mole) detection levels for representatives of these classes of adducts, with SIM focused on either the $M^{\cdot-}$

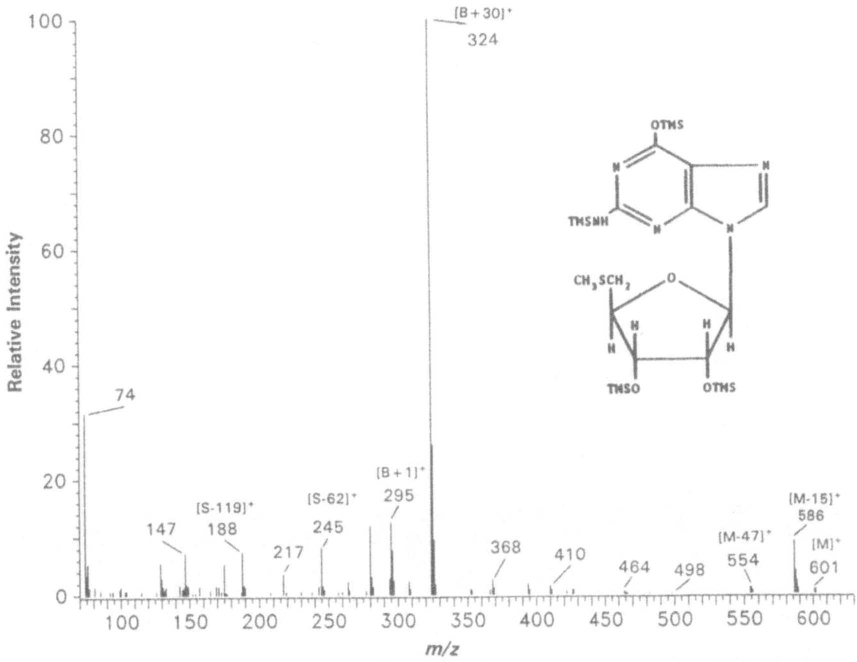


*Figure 3.4.* **Structure and EI mass spectrum of TMS-derivatized 5′-deoxy-5′-methylthioguanosine obtained by GC-MS analysis of urine from a lung cancer patient. Modified from Ref. 42; used with permission.**

or the $[M-181]^-$ ion (48). An example of their SIM results for a PAH diol-epoxide (47) is given in Fig. 3.5. In this case, ~100 pg of a BaP-tetrahydrotetrol was reacted to yield the final PFB derivative. After addition of the deuterated internal standard, 1% of the final volume was subjected to GC-ECNI-MS analysis. SIM traces for the blank and the analyte both reveal peak widths of ~15 sec. Most recently, Giese's group has investigated benzo(*a*)pyrene (B*a*P) adducts formed through the diolepoxide route in human lymphocyte cell cultures reacted with B*a*P. Although the SIM technique functions well for known analytes, wider *m*/*z*-range scans can detect unknown compounds, admittedly with some considerable loss in sensitivity.

## 5. PLASMA DESORPTION

With plasma desorption (PD), the first event in the ionization scheme is the nuclear fission of $^{252}$Cf, which generates two extremely high energy frag-

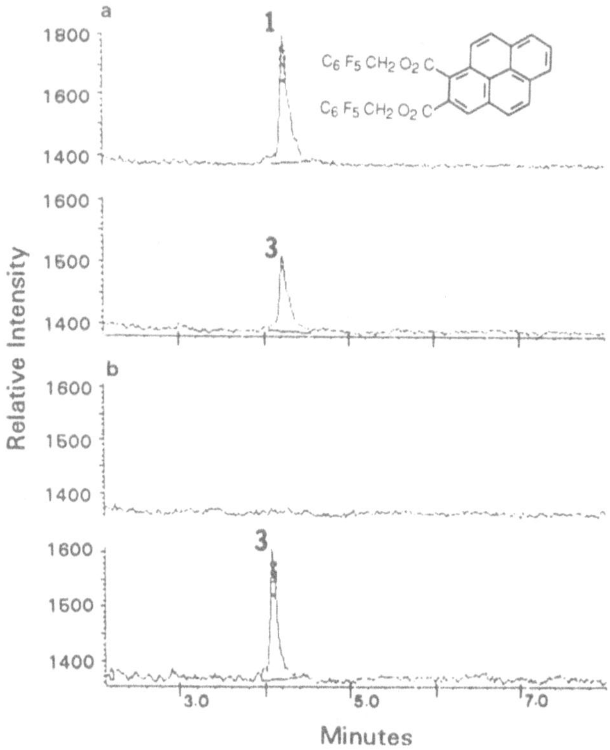


*Figure 3.5.* Structure and extracted ion-current profiles resulting from selected ion monitoring with GC-ECNI-MS. (a) Profile of the $[M-181]^-$ ion, *m/z* 469, for a benzo[*a*]pyrene tetrol that had been subjected to potassium superoxide treatment to form a diacid and further reacted to form the pentafluorobenzyl derivative. (b) Profile for the analysis of a blank. Both analyses included a deuteriated internal standard indicated as peak 3 in each of the lower traces. Modified from Ref. 47; used with permission.

ments. One of these fragments is directed to, and starts, the system clock. The second fragment is accelerated in the opposite direction to the back side of a Mylar® foil on which the sample has been deposited. Subsequent ionization and desorption generates the molecular species, which are subjected to MS analysis. This cycle is repeated approximately every 50 μsec. In contrast to the previously mentioned types of ionization, PD is of the pulsed rather than of the continuous type and is more compatible with either time-of-flight (TOF) or ion-cyclotron-resonance (ICR) mass analyzers.

Although New York University School of Medicine researchers investigated acrylamide- and acrylonitrile-formed nucleoside adducts using PD-MS in 1985 (49,50), in the last few years, the PD-MS analysis of modified DNA has been practiced primarily by a collaboration of French researchers (51–53). They used a TOF MS of their own manufacture to investigate methyl modifications and a number of photodimers formed *in vitro* in oligomeric DNA. Earlier work (54) by these researchers focused on benz(*a*)anthracene (B*a*A) adducts of poly(dG) and (B*a*P) nucleoside modifications. Positive and negative ion spectra were both recorded for a cyclobutane thymine nucleoside dimer (53). Their article included a structure and spectrum (Fig. 3.6), which is indicative of the results that can be obtained with this instrumentation and ionization. This spectrum shows the natriated molecular species, as well as the fragments, for a cyclobutane thymine nucleoside. No attempt was made to eliminate sodium, nor were the effects due to its presence discussed. However, it should be noted that the intensity of the molecular species is distributed over a number of different ions. With some other techniques, such as FAB, the presence of $Na^{+}$ or $K^{+}$ can either markedly increase or decrease the intensity of the observed ions. Note that this Fig. 3.6 spectrum also displays less than unit

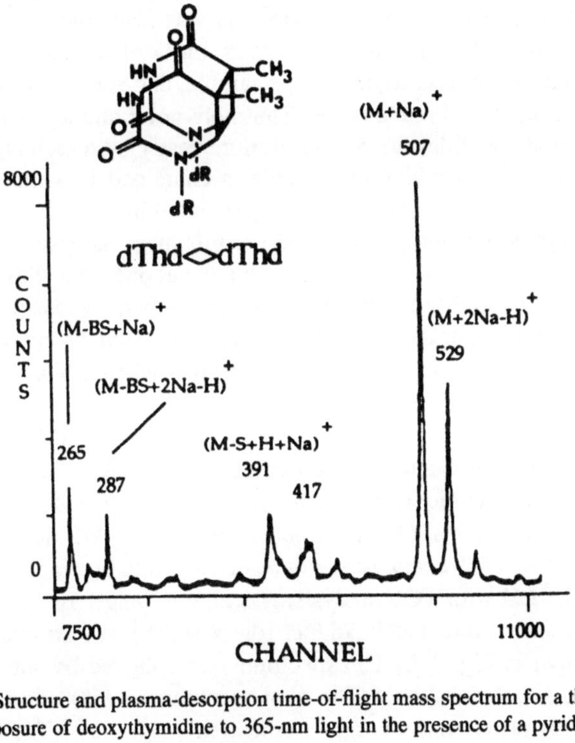


*Figure 3.6.* Structure and plasma-desorption time-of-flight mass spectrum for a thymidine dimer formed by exposure of deoxythymidine to 365-nm light in the presence of a pyridopsoralen drug. Modified from Ref. 53; used with permission.

resolution, so that identification of the molecular species would be further complicated if the sample contained unknown adducts, especially if a mixture of adducts were deposited on the foil. The report also mentions the use of FAB, but gives no details.

## 6. LASER DESORPTION

### 6.1. Time-of-Flight Spectra

A notable feature of both PD and laser desorption (LD) is their ability to bring into the vapor phase nonvolatile or thermally labile molecules. As a result, these techniques possess the ability to deal with large modified oligonucleotides in cases where the DNA hydrolysis has produced such species. Ionization by LD involves a laser focused on a small spot where the sample has been deposited. The laser's photon beam ionizes the sample through a number of different processes (55). Nanosecond-wide pulses help avoid formation of ions with a wider range of kinetic energies. Such a wide range would lead to poorer resolution. This pulsed operation, as was also true with PD, is more amenable to either TOF or ICR/MS analyses. Earlier LD analyses relied on the heat generated by the laser to desorb the sample, and such analyses functioned quite well for bases and nucleosides. The more polar and less volatile nucleotides and oligonucleotides are analyzed more easily by a technique known as matrix-assisted LD, in which the sample is contained in a matrix known to absorb radiation in the wavelength range generated by the laser. That absorbed energy aids the desorption process. LD permits many laser shots per loading, so that most or all of the sample could be loaded at one time while maintaining the experimenter's ability to perform repeated analyses, subsequent modifications on the probe tip, or recovery of the sample. The kinetic energy spread that results in loss of resolution seems to be lessened when UV as opposed to IR laser excitation is employed.

TOF and ICR have both been used recently with modified DNA. Lay's group at the National Center for Toxicological Research (56) contrasted two very different MS techniques for the analysis of the base and nucleoside forms of various DNA adducts. One MS method involved TMS derivatization followed by FAB-MS/MS analysis; it will be discussed in Section 8.4. The second MS technique, LD with TOF analysis, was able to detect 10 pg of a synthetic acetylaminofluorine nucleoside adduct [dg($N^2$)AAF]. (Its structure and spectrum are shown in Fig. 3.7.) The spectrum was obtained by summing 100 of 500 laser shots to the probe tip. The mass resolution is ~200, and the accuracy for compounds in this mass range is essentially unit mass. The most intense peak in the spectrum corresponds to the *m/z* 489 $[M+H]^+$ ion. The presence of sodium and potassium is indicated in the spectrum by the *m/z* 511 and 527 peaks, which correspond to $[M+Na]^+$ and $[M+K]^+$, respectively. The LOD for this nucleoside adduct is better than what can be achieved with derivatization and FAB/MS/MS, although the latter technique offers more information.

### *6.2. Ion Cyclotron Resonance*

ICR-MS studies of small alkyl and exocyclic modifications to DNA bases have recently been reported (57,58). That research is focused on the known chemical carcinogen diethylnitrosamine and its tumor-forming capabilities in fish. LD-ICR-MS, operating at very high resolution (~250,000), permits LODs in the low-ng range.

Hettich's group at Oak Ridge has employed the same LD-ICR-MS approach in studying a wide variety of forms of DNA, including methylated nucleosides (59), cyclobutane photodimers (60), various modified bases and nucleotides (61), and oligonucleotides with small alkyl groups on the phosphate linkage (62). The spectrum for a thymine dinucleotide with an ethyl

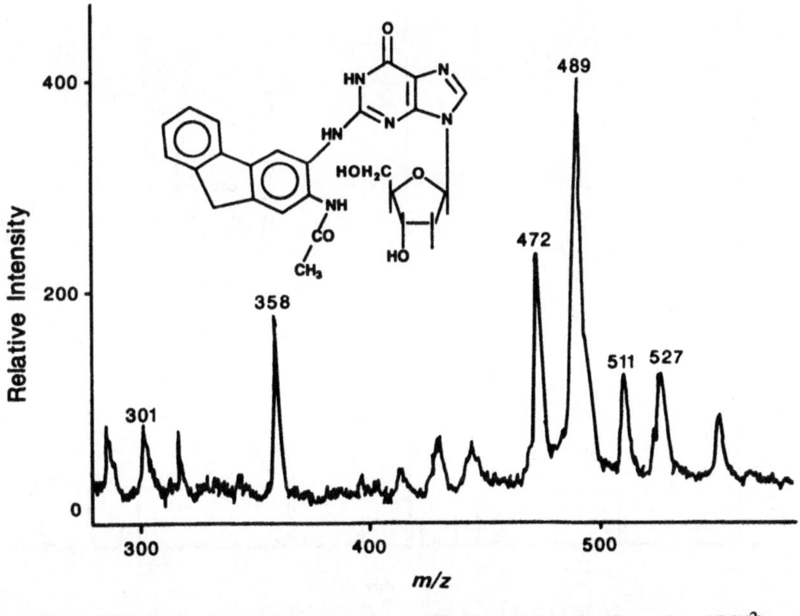


*Figure 3.7.* Structure and laser-desorption time-of-flight spectrum of ~10 pg of an dG($N^2$)acetylaminofluorene nucleoside adduct. Modified from Ref. 56; used with permission.

modification of the phosphate linking the two nucleotides is given in Fig. 3.8 (R.L. Hettich, personal communication). Nicotinic acid served as the matrix, and negative-ion detection clearly shows the molecular anion, loss of an ethyl group, and the *m/z* 349 fragment that results from the loss of one of the thymine nucleosides. For this analysis, approximately 10 ng of the analyte was loaded onto the probe tip, and resolution was approximately 1000, as determined by the full width of the peak at half its maximum height (FWHM). Subsequent CID of the *m/z* 349 ion showed further fragmentation that helped define the structure of the compound. Furthermore, higher resolution (5000 FWHM), combined with CID, revealed the presence of two separate fragments, both with nominal mass 125. Only a small fraction of the applied sample was consumed during the actual analysis, yet the recorded spectra were clearly well above the LOD. Unwanted ions can be ejected from the ICR, and CID MS/MS experiments can be performed. As a result of these instrumental factors, LD offers excellent sensitivity for low-level adduct research, coupled with identification ability. However, because of the high vacuum requirements of ICR, the coupling of LD-ICR to any on-line separation technique to obtain additional discriminating powers is problematic.

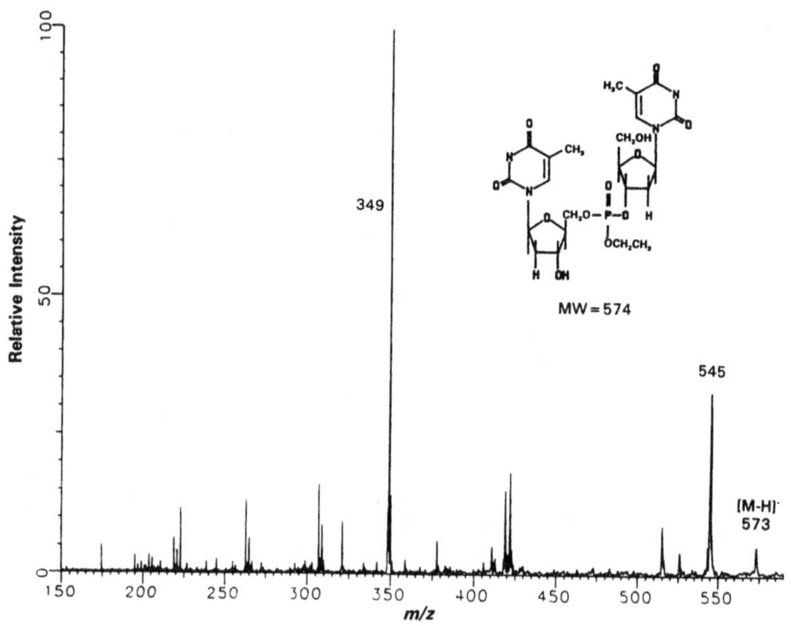


*Figure 3.8.* Structure and laser-desorption negative-ion cyclotron-resonance spectrum for a thymine dinucleotide with an ethyl modification of the phosphate linking the two nucleotides. Unpublished data from R. H. Hettich.

## 7. *ACCELERATOR MS (AMS)*

This recently adapted technique is based on determination of the $^{14}C/^{12}C$ ratios in organic compounds. AMS was previously utilized in carbon dating, but has more recently been employed by Turteltaub, Felton, and others to investigate, among other problems, the binding of heterocyclic amines to DNA (63,64). These types of adducts are thought to be primarily responsible for the mutagenic and carcinogenic activity of cooked meats. As currently practiced, AMS functions well only in dosimetry experiments because the heterocyclic amine must first be labeled with one or more $^{14}C$ atoms. After the dosing experiment, the intact DNA sample is pyrolyzed to form graphite prior to the actual AMS ratio measurement. If the starting sample consists of 1 mg of DNA, then levels of one adduct in $10^{12}$ unmodified bases can be determined (the observed ratio difference is ≈10%). Only ≈15% of the sample is used in the analysis. This sample size would correspond to an adduct–nucleotide with a mass of ~1 fg (~0.5 amol), which is approximately the same amount of adduct spotted on a thin layer chromatography (TLC) plate when $^{32}P$ postlabeling is employed near its LOD with a few μg of DNA at an adduct level of 1 in $10^9$. Conventional MS measurement of adducts at this low level can be approached only by the GC-MS derivatization techniques previously discussed. It should be noted that the sample preparation is relatively straightforward, because the DNA need not be hydrolyzed down to the monomer level or beyond. However, extreme care must be taken in the laboratory to assure the purity of the DNA and to avoid any $^{14}C$ contamination of the samples. The current limitation to dosimetry experiments can be lifted if researchers are successful in efforts to develop a postlabeling methodology. Indeed, they currently are improving their techniques for excising individual spots from $^{32}P$ TLC plates. One long-range research aim is to investigate adducts on specific genes.

## 8. *FAB AND SECONDARY ION MASS SPECTROMETRY*

### 8.1. *Standard FAB*

This section addresses ionization methods based on the impact of a beam of either ions or atoms generated by a gun and focused onto a sample target. More commonly, the techniques are known by the terms (liquid) secondary-ion mass spectrometry and fast atom bombardment (FAB), respectively. The "liquid" merely denotes that the sample has been dissolved in a liquid matrix. Utilization of the atom gun usually involves a similar matrix. These two particle-induced desorption techniques will be considered together because they result in ionization, i.e., spectra, that are nominally the same (65). Static FAB experimentation involves the use of a sample placed on the tip of a probe, which is inserted into the source area. A few articles that exemplify this basic ionization mode will be discussed first. Subsequent sections will deal with derivatization techniques, MS/MS methods, and finally a combination of these two aspects.

Recently, FAB has been used for modified DNA analyses much more frequently (19,21,24,66–76) than in the past. In part, this increased utilization is because the ionization is of high enough energy to induce some fragmentation yet still retain information about the molecular ion. Increased utilization is also due to the amenability of these techniques to more polar samples, such as nucleotides, which can be difficult to analyze by some of the direct insertion probe or GC-MS techniques. In fact, a wide range of forms of DNA can be analyzed by FAB. For example, ethyl glutathione base adducts were investigated (77). Small nucleoside modifications were similarly examined (78,79). Carbazole attachments to the phosphates of nucleotides were considered by McCloskey's group (80). Oligonucleotide–*cis*platin adducts were investigated (81), as were similar oligomers that had lost a central pyrimidine base (82). Spectra, which were given for most of the above publications, showed a profusion of matrix ions, which interfered with the peaks of interest and substantially degraded the LODs. When FAB is used, these matrix-ion contributions cannot be subtracted away unless the matrix (sometimes containing sample) is flowing continuously into the source, as will be discussed in Section 9. Without this feature, substantial amounts of sample must be used to overcome the background. Many of these research articles do not give actual spectra. For the most part, quantitative information is not detailed, although it generally appears that microgram quantities of material were often subjected to analysis. This technique is thus not among the most sensitive. Rather, the spectrum is designed to either establish or to confirm the molecular weight or to provide information about the site of attachment or nature of the modification.

### *8.2. FAB with Chemical Derivatization*

Derivatization of the sample has been recently utilized for the same reasons already presented in Section 3, i.e., to enhance pre-MS analysis separations or to improve detectability. For example Silk *et al.* (25) reported the detection of the silylated derivative of N-(deoxyguanosin-$O^6$-yl)-4-amino-3-chlorobenzyl alcohol. These researchers indicated that they had been unable to detect the underivatized nucleoside. Their report showed no spectral data and gave no quantitation details but did give some textual data that indicated the presence of $Na^+$ and $K^+$ ions, as well as the presence of molecular ions carrying different numbers of silyl groups. In the same year, Schram's group (28) investigated TMS derivatives of two clinically important drug–nucleotide adducts, tricyclic nucleoside-5′-monophosphate and 5-fluoro-2′-deoxyuridine-5′-monophosphate. Figure 3.9 contains the structure for one of these compounds, along with spectra obtained with two different amounts (500 and 2.5 ng). When the negative-ion spectrum was obtained in a 3-nitro-benzyl alcohol matrix, the LOD for these compounds was two orders of magnitude better than the LOD achieved without derivatization. The LODs for the derivatized compounds were, respectively, 20 and 100 times better than those obtained with EI and CI techniques.

### *8.3. FAB with MS/MS*

Considerable importance is rightfully placed on the use of MS/MS techniques with FAB ionization to investigate DNA modifications. Recent studies have employed a variety of instrument configurations, including four-sector (83–86), three-sector (85), two-sector (66,87,88), TQ (71), and hybrid EBQQ (10,89–92). Of these studies, several (66,83,84) deal with oligonucleotide modification, whereas others (10, 90–92) discuss bases. The remaining articles deal with nucleosides. Van Breemen's group at North Carolina State University details the binding of platinum, with two attached amino groups, to two adjacent bases in synthetic tetramers following *in vitro* exposures of the antitumor drug Cisplatin [*cis*-diamminedichloroplatinum(II)] (66). This type of platinum binding is responsible in part for the observed Cisplatin effects, which include inhibition of DNA repair, DNA replication, and cell division. The two-sector (EB) MS used for that research provided positive-ion FAB-MS/MS results somewhat better than those for the negative-ion case. The full-scan spectrum for the molecular-ion region of one of the adducted tetramers is shown in Fig. 3.10 (R. B. van Breeman, personal communication). (The structure of that platinated tetramer is given in Fig. 3.11.) Note that the ion clusters show a multitude of peaks due to the many isotopes of platinum. Such an isotope pattern spreads out the molecular-ion current over a range of *m*/*z* values and can lead to poor LOD and difficult spectral interpretation if multiple components were present. On the other hand, the existence of isotopic clusters in the spectrum can help establish the elemental composition of ions. MS/MS on a forward-geometry EB instrument is based on the preselection of a range of ions in the first field-free region. Because of the limited energy resolution of the electrostatic analyzer, the energy bandwidth may be as much as 1 or 2%, or more, relative to the selected precursor mass. Breeman's MS/MS approach

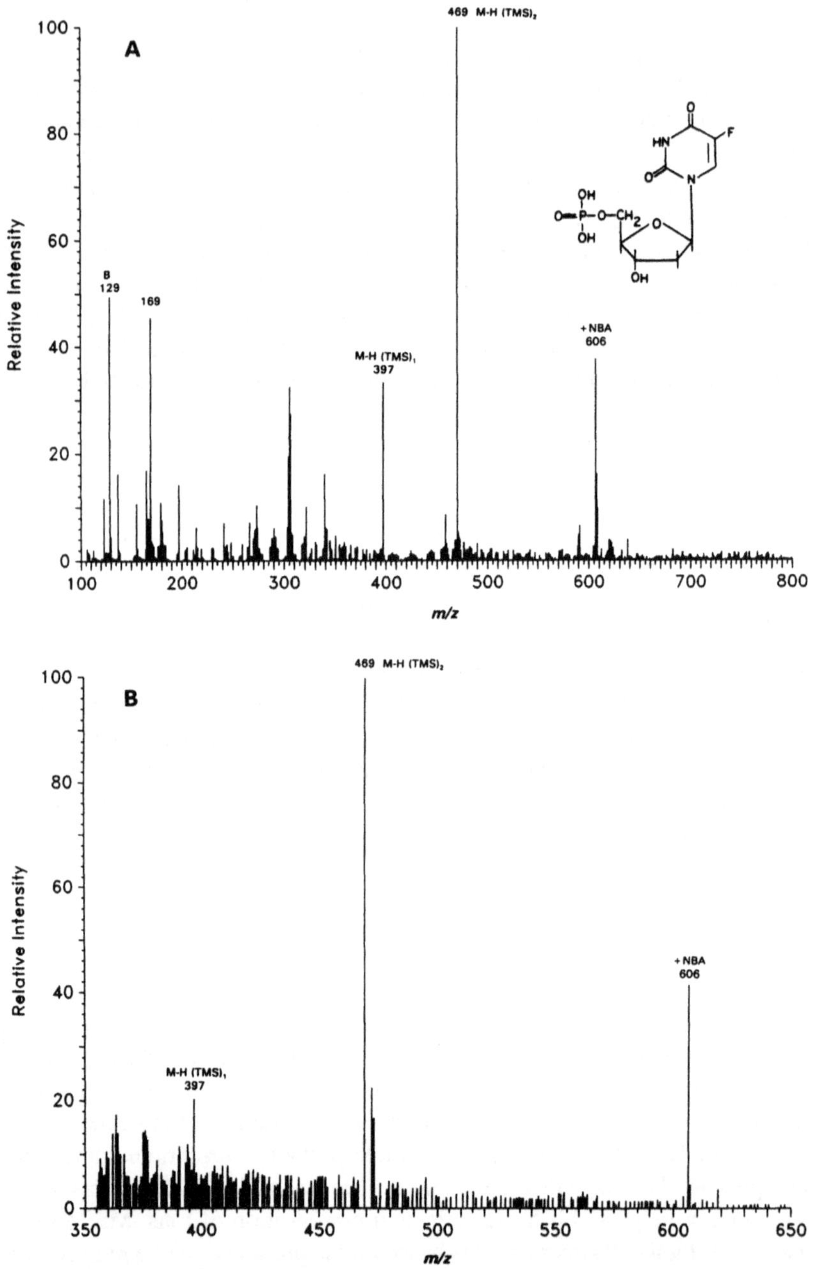


*Figure 3.9.* Structure and positive-ion fast atom bombardment spectra for (A) 500 ng and (B) 2.5 ng of the trimethylsilyl derivative of the drug fluoro-deoxyuridine-monophosphate. Nitrobenzyl alcohol (NBA) was utilized as the fast atom bombardment matrix. Modified from Ref. 28; used with permission.

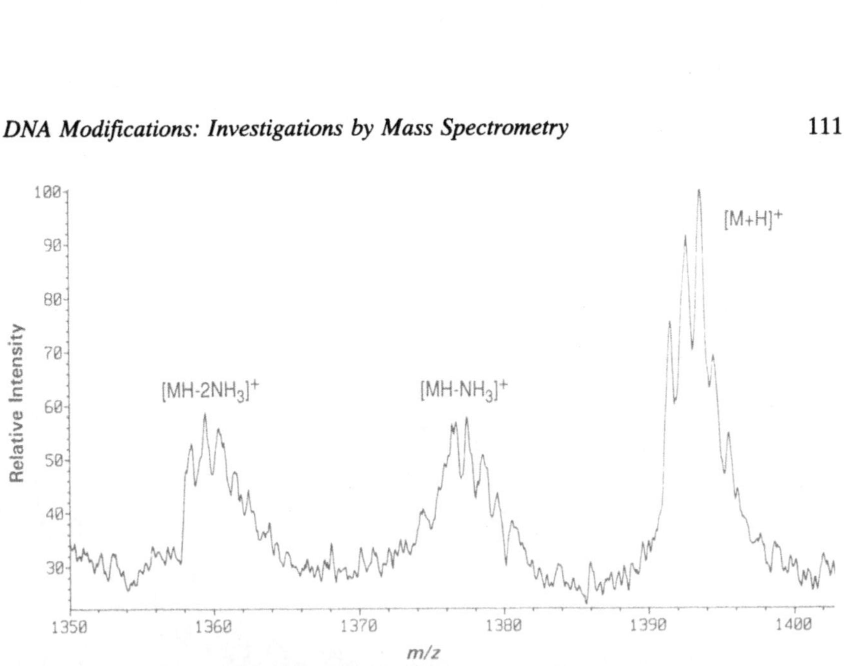


*Figure 3.10.* The molecular-ion region of the positive-ion fast atom bombardment full-scan spectrum of the synthetic tetramer TGCT containing platinum. The entire $[M+H]^+$ cluster with many platinum isotopes was selected in the MS/MS analysis shown in Fig. 3.11, which also displays the structure of the modified tetramer. Unpublished data from R. B. van Breeman.

preselects the entire molecular ion cluster. Utilization of this wide band is of great advantage, as long as the product ion spectrum of interest does not include any platinated species. Figure 3.11 shows such a MS/MS product ion spectrum, with its much simpler fragmentation pattern due to the loss of the two segments of the tetramer containing the bound platinum. These lower-*m/z*-fragment ions are not seen in the non-MS/MS spectrum. Thus, MS/MS more fully describes the nature of the binding of platinum to the tetramer. This same relatively wide energy band of preselection would be more problematic if contaminants existed in the chosen energy range.

MS/MS on a four-sector instrument was also recently employed by Gross's group at the University of Nebraska to investigate BaP adduct formation mechanisms (86). For many years, most biochemical researchers had believed that B*a*P mutagenic activity was mediated through a diolepoxide form of B*a*P. However, Cavalieri, also at the University of Nebraska, has long championed the importance of the more direct radical cation pathway, a pathway admittedly more difficult to evaluate because it relies on compounds that exist for only a short time. Indeed, the radical cations react only with double-stranded DNA, not with single-stranded DNA or RNA, and they are formed by cytochrome P450 enzyme activation, which for many years was also thought impossible. The diolepoxide pathway is important for some carcinogens, but

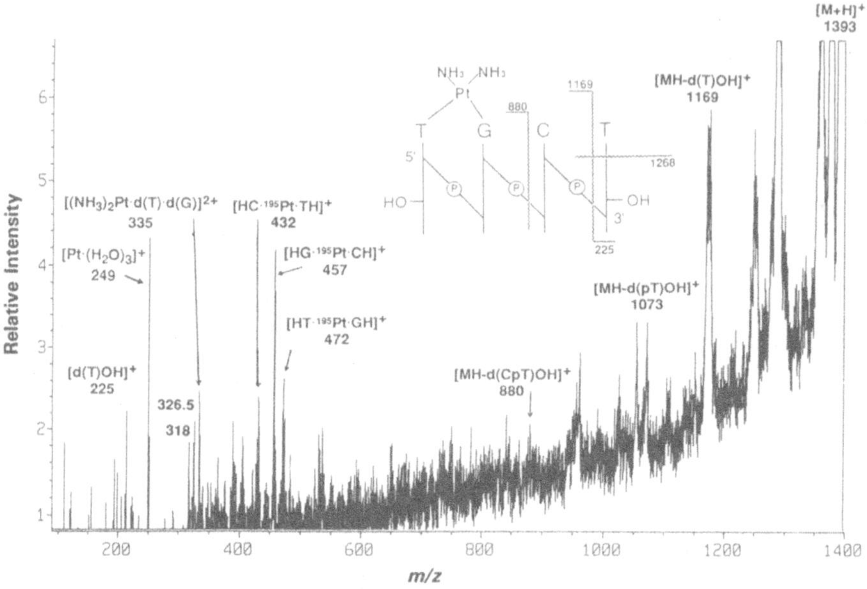


*Figure 3.11.* Structure and positive-ion fast atom bombardment MS/MS product ion spectrum obtained on a two-sector forward-geometry (EB) MS with collision-induced dissociation of the $[M+H]^+$ cluster shown in Fig. 3.10. Modified from Ref. 66; used with permission.

does not seem to be of major importance for, e.g., B*a*P and 7,12-dimethylbenzanthracene. Gross's four-sector MS/MS research was instrumental in helping to prove Cavalieri's hypothesis. Figure 3.12 shows the structure and the positive-ion FAB product ion MS/MS spectra for the N7 and C8 B*a*P–dG adducts formed by electrochemical oxidation. The *m/z* 402 source-produced fragment ion (corresponding to loss of the sugar moiety) was selected by the first MS and subjected to CID to generate the displayed product ion spectra. The CID product ion spectra are markedly dissimilar because of the different points of attachment of the adduct to the guanine base. Figure 3.13 (R. L. Cerny, personal communication) details the structure and the positive-ion–FAB CID-product ion spectra generated by three-sector and four-sector MS analysis of the dimethylbenzanthracene N7–guanine base adduct, which was again formed by the radical cation mechanism as opposed to the diolepoxide pathway. In both cases, the CID spectra resulted from collisions of the $[M+H]^+$ ions, but the energy analysis shown in Fig. 3.13a does not give the level of detail seen in the four-sector results in Fig. 3.13b. Indeed, only the extra resolving power provided by the four-sector analysis permitted determination of the 239/240 ratio and subsequent elucidation of the point of attachment.

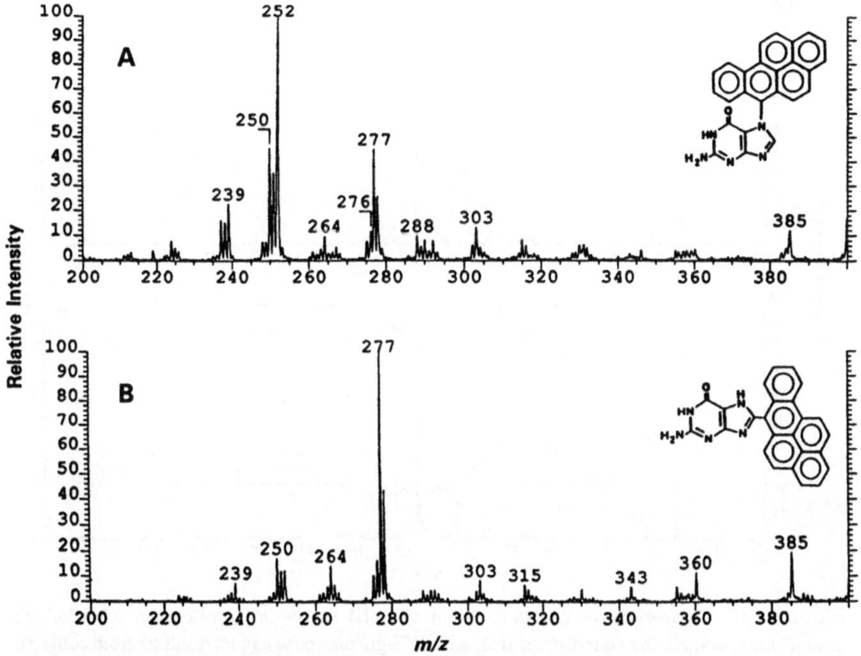


*Figure 3.12.* Two structures and positive-ion FAB product spectra (A,B) obtained on a four-sector MS with collision-induced dissociation of the aglycone ions, *m/z* 402. Benzo(*a*)pyrene nucleoside adducts were formed by the radical cation mechanism. Observed spectral variations are due to differences in the site of attachment to the guanine base: (A) N7, (B) C8. Modified from Ref. 86; used with permission.

## 8.4. FAB with Chemical Derivatization and MS/MS

Researchers at Northeastern University combined positive ion FAB-MS/MS techniques with TMS derivatization to investigate four amino-PAH-nucleoside adducts (93). In agreement with Schram's work (28), their data showed that TMS derivatization gave low-nanogram LODs, an order of magnitude better than LODs without derivatization. A predominant fragmentation pathway involved cleavage of the glycosidic bond to yield the $[AH_2]^+$ fragment. Metastable-ion MS/MS and CID-MS/MS results were comparable when the product ion spectrum resulted from the $[M+H]^+$. However, when the product ion spectrum resulted from the *m/z* 343 $[AH_2]^+$ ion, whose abundance had been increased by CID, striking differences could be seen between the spectra of two amino–phenanthrene nucleoside adducts, that differed only in their point of attachment to the guanine base, as seen in Fig. 3.14. The TMS derivatization

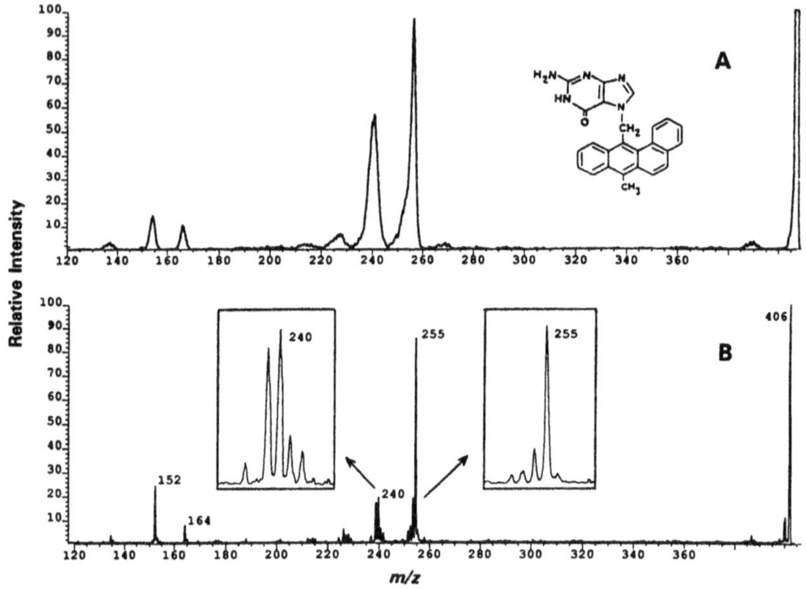


*Figure 3.13.* Comparison of (A) three-sector and (B) four-sector positive-ion FAB-MS/MS product ion spectra of the dimethybenzanthracene N7-guanine base adduct resulting from collision-induced dissociation of the $[M+H]^+$ ion, *m/z* 406. Unpublished data from R. L. Cerny.

for these two samples resulted in the addition of two TMS groups onto the sugar in a relatively clean manner; i.e., there was little addition of a different number of TMS groups. Furthermore, there was no confusion due to TMS addition onto other parts of the adducts. The obvious fragmentation differences seen in Fig. 3.14 facilitate spectral interpretation. For example, in the Fig. 3.14a spectrum, the *m/z* 218 and 203 ions support a C8 linkage, whereas in Fig. 3.14b, the $N^2$ linkage gives rise to an analogously formed ion at *m/z* 233, which represents the portion of the adduct that still contains a free amino residue.

TQ-MS/MS analyses of TMS derivatives of very similar amino–PAH compounds were reported more recently (56,94) by Lay's group. Improved LODs in the low-ng range were observed, perhaps in part due to the different type of MS used. Of note in achieving those improved LODs is the importance of sample derivatization—four different TMS schemes were compared in the latter article (94). Various isomeric forms of the derivatized components were observed and avoided by selecting for CID product ion analysis the $AH_2$ fragments, which did not contain any TMS modification. The resulting CID fragmentation was quite informative. However, the suspect stability of the TMS

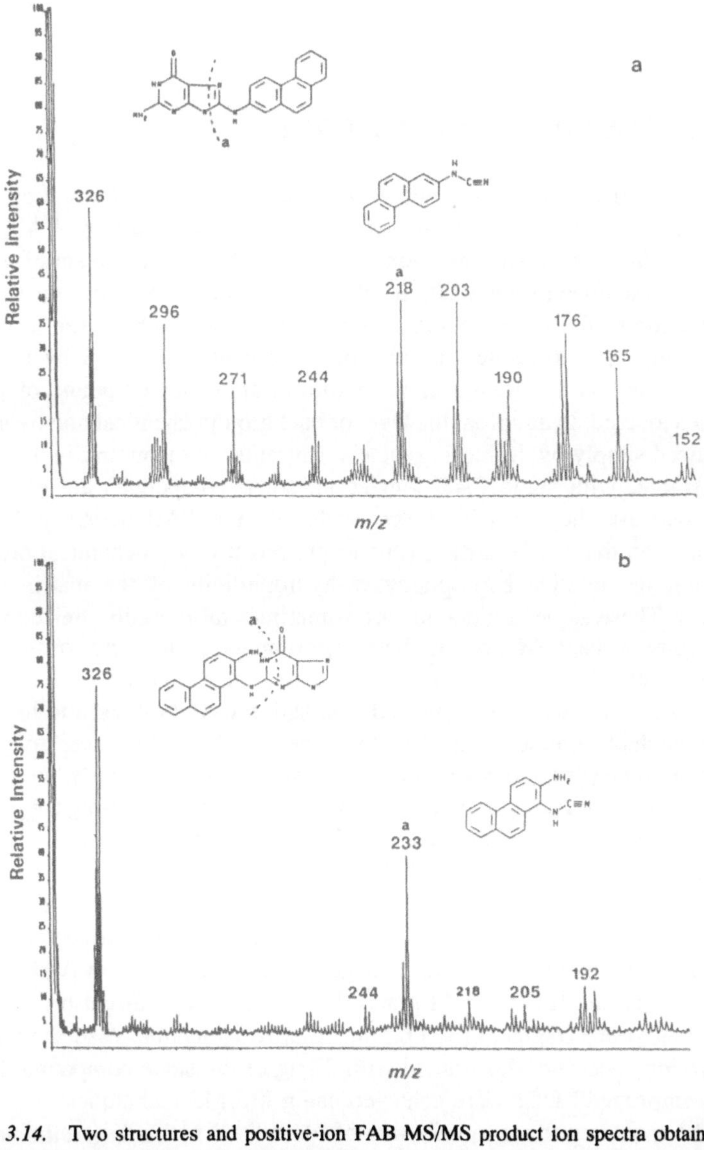


*Figure 3.14.* Two structures and positive-ion FAB MS/MS product ion spectra obtained on a two-sector MS with collision-induced dissociation of the aglycone fragment ions, *m/z* 343. Observed spectral variations are due to differences in the site of attachment to both the aminophenanthrene and the guanine base. Modified from Ref. 93; used with permission.

derivative is justly observed to be in part due to the derivatization process and to the FAB matrix itself. This propensity to hydrolysis of TMS derivatives makes their use with Flow-FAB very problematic.

## 9. FLOW-FAB MASS SPECTROMETRY

A modified FAB probe allows for the continuous introduction of the matrix and sample to the probe tip. This technique, known as Flow-FAB, is thus capable of handling direct injections of the sample into the stream of material headed for the tip, where ionization will occur. Several benefits accrue from this modification to the FAB technique. First, background subtraction techniques can be employed to reduce the chemical noise in a spectrum by removing contributions that are present either before or after the compound of interest becomes ionized. In addition, the level of background chemical noise can often be reduced simply by decreasing the concentration of the matrix in the stream. Furthermore, with static FAB ionization, some compounds are difficult to ionize because they dissolve more readily in the FAB matrix rather than remaining on the matrix surface (due to important physicochemical processes involving the relative hydrophobicity–hydrophilicity of the matrix–analyte mixture). These same compounds are sometimes more easily ionized when a flow regime is used. As a result, these experimental factors and considerations can substantially improve the LODs for particular analytes.

Positive and negative Flow-FAB ionization were both used to help identify the nucleotide adducts resulting from the use of the anticancer drug cyclophosphoramide (95). However, no spectra were shown. Similarly, Vouros and co-workers investigated a variety of nucleoside adducts (96) using the positive- and negative-ion modes, as well as MS/MS techniques. The full-scan mass spectra exhibited moderately detailed fragmentation patterns and low-nanogram LODs. Two different sample introduction schemes, "dipping" and "syringe injection," were contrasted, and the effects of various concentrations of glycerol were compared. Figure 3.15 gives the structure of an AAF guanine nucleoside adduct [dG(C8)AAF], and the extracted ion-current profile of the *m/z* 373 aglycone fragment for a 6.5-ng "dipped" injection. Also shown is the positive-ion spectrum, determined with 37 ng of the same compound. Significantly improved LODs were achieved using MS/MS techniques.

The most important contribution of Flow-FAB is that it permits the coupling of separation techniques with the FAB ionization mode. As a result, HPLC and capillary electrophoresis eluates can be directly interfaced to the stream of material flowing into the FAB source. These two techniques will be discussed in Sections 10.6 and 11.

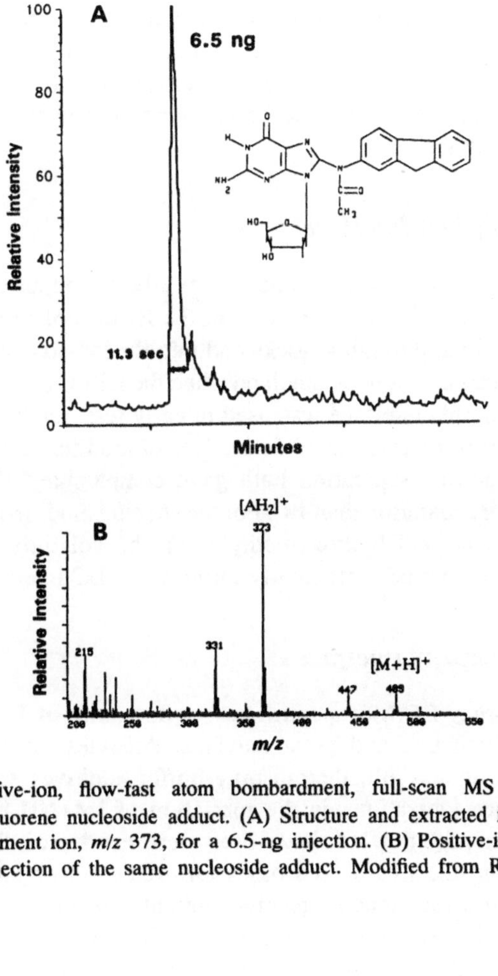


*Figure 3.15.* Positive-ion, flow-fast atom bombardment, full-scan MS analysis of the dG(C8)acetylaminofluorene nucleoside adduct. (A) Structure and extracted ion-current profile of the aglycone fragment ion, *m/z* 373, for a 6.5-ng injection. (B) Positive-ion full-scan spectrum for a 37-ng injection of the same nucleoside adduct. Modified from Ref. 96; used with permission.

## *10. LIQUID CHROMATOGRAPHY-MASS SPECTROMETRY*

A wide variety of interfaces has been introduced to handle eluates for a number of HPLC flow regimes, with or without splitting, into an array of MS systems. The primary importance of such schemes is to afford some degree of separation of the sample components prior to the actual MS analysis. Such separation should facilitate the identification of unknowns and remove possible interfering compounds (in part through the use of background subtraction). In contrast to GC, HPLC systems can be compatible with the nucleosides and nucleotides, which are more polar than the corresponding bases. Often, fewer cleanup steps are required during sample preparation, and derivatization steps may not be necessary. Of the various LC-MS modes of coupling, six types have been used for DNA adduct analysis. Those types include the moving belt, thermospray, particle beam, atmospheric pressure ionization, electrospray, and Flow-FAB methods.

### *10.1. Moving Belt Interface*

The moving belt interface, one of the earliest used, uses a narrow belt that is continuously moving from the exit of the HPLC column, where the HPLC eluate is deposited, through airlocks and into the ion-source region of the MS. EI and CI spectra can both be obtained when the HPLC eluate is interfaced with a moving belt. This interface was used in early research by Giese and Vouros for the analysis of perfluorinated derivatives of modified DNA bases (97). GC- and LC-MS on-line separation both gave comparable LODs, although GC provided better separation than LC. For the specific modified bases investigated (5-methylcytosine and hydroxymethylurea), the volatility of the derivatives provided optimal transfer efficiently through the LC interface.

### *10.2. Thermospray Interface*

Thermospray LC-MS operates best at flow rates of 1–2 ml/min and uses conventional columns and polar solvents. Analytes must be preionized in solution, and, as a result, thermospray buffer additives such as ammonium acetate enhance ionization via the formation of $[M+NH_4]^+$ or $[M+H]^+$ ions. Analyses of modified DNA have been performed recently with the intact nucleosides and the hydrolyzed bases (98–101). All of the above examples involved *in vitro* reactions to generate suitable quantities of the analytes. At least some of the reports indicate considerable fragmentation, which resulted when ions formed due to the loss of the deoxyribose or the base moiety. These research articles usually provided no quantitative information or dealt with μg amounts of modifications. The only exception was in Ref. 99, where the spectrum for 10 ng of 7-methylguanine was shown. Callery's group at the University of Maryland investigated a number of anticancer drugs and reported thermospray LC-MS analyses with quadrupole MS (100). Figure 3.16 presents a representative nucleoside adduct (mitomycin *C*–deoxyguanosine) and shows the ~2 %RI $[M+H]^+$ ion and the fragmentation indicative of not only the guanine base, but also the mitomycin *C* moiety. In the same report, small base modifications were shown to give rise to spectra with 100 %RI $[M+H]^+$ ions and much-less-abundant fragment ions that indicated the nature of the base.

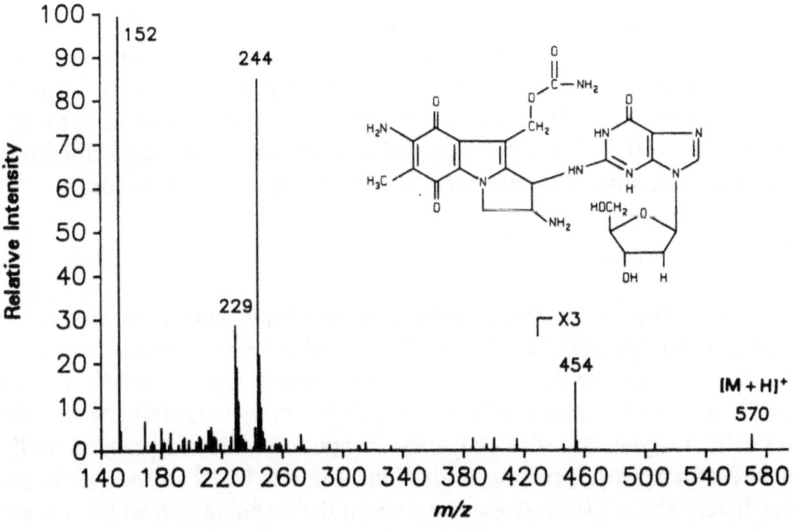


*Figure 3.16.* LC-MS analysis using a thermospray interface to a quadrupole MS. Structure and positive-ion full-scan spectrum of the mitomycin C–guanine nucleoside adduct. Modified from Ref. 100; used with permission.

## *10.3. Particle Beam Interface*

In the particle beam interface, the HPLC mobile phase is dispersed with an inert gas through a nebulizer. The aerosol beam travels through a heated desolvation chamber, where the more volatile solvent is preferentially dispersed and evaporated. The analyte-enriched beam is passed through a momentum separator, which skims off the solvent and retains the analyte-rich central portion of the beam. EI and CI spectra may be determined with this mode of LC-MS coupling. This interface was employed in the analysis of an exocyclic propano-deoxyguanosine (72). EI conditions resulted in a spectrum that revealed considerable fragmentation and a 5 %RI molecular ion. Standard probe EI analysis had failed to give a molecular ion of >1 %RI. FAB and probe CI results both had shown an intense $[M+H]^+$ ion, but had failed to show any fragmentation except for the aglycone. Quantitative results were not given.

## *10.4. Atmospheric Pressure Ionization*

With atmospheric pressure ionization, the analytes are ionized under CI conditions at or near atmospheric pressure. A number of different modes of ionization is possible, depending on whether or not heat is applied in the desolvation chamber or whether a discharge electrode is used. Flow rates of up to 2 ml/min can be accommodated without splitting. Atmospheric pressure ionization was used to help identify the mechanism of *in vitro* alkylation (addition of fluoroethyl) of dG by the anticancer drug 1,3-*bis*(2-fluoroethyl)-1-nitrosourea (102, 103). CID, TQ/MS, and deuterium labeling helped formulate exact structures and thus indicated modes of action of the drug.

### *10.5. Electrospray*

The atmospheric pressure ionization technique known as electrospray depends on the application of a ~3-kV potential difference between the end of the HPLC column and the entrance to the ion-source area of the MS. The flow regimes are usually much smaller (1–10 μl/min), and, in general, the technique favors the analysis of polar and ionic compounds. Electrospray, as well as standard atmospheric pressure ionization, was employed in the article cited immediately above (103). A modification of the technique, in which a sheathing gas coaxially envelops the nebulized spray of HPLC eluate, is known as ion spray, and accommodates even higher eluate flow rates. This analysis mode has been employed recently by Quilliam's group to investigate a number of amino-PAH bases, nucleosides, and nucleotides (104). Various HPLC columns with 0.32 to 4.6 mm inner diameter (i.d.) were used. Splitting of the eluate was necessary with the wider-bore columns. Figure 3.17a gives the background-subtracted full-scan positive-ion mass spectrum of synthetic dG(C8)AAF acquired by injection of 10 ng of adduct (unpublished data, used with permission; personal communication). (The structure for this component was given in Figure 3.15.) The entire 50 μl/min eluate of the 1 mm i.d. reverse-phase C-18 column was directed into the MS. The 0.1% trifluoroacetic acid, water-based mobile phase was increased from 50 to 90% MeOH. The MS analysis mode involved only the first stage of the TQ/MS and a technique sometimes referred to as "up-front" CID, because the collisions occurred in the high-pressure expansion region prior to the first quadrupole. The degree of dissociation of $[M+H]^+$ to $[AH_2]^+$ is directly related to the voltage on the sampling orifice leading to the first stage of the MS.

Similar results can be obtained with CID of the $[M+H]^+$ ion in the central stage of the TQ/MS. This latter technique should allow an improved signal-to-noise ratio, but Quilliam has determined that improved LODs are obtained with "up-front" CID and by using only the first stage of the TQ/MS. His analytical results (personal communication) depicted in Fig. 3.17a indicate that the LOD is much better than the 10 ng injected. Figure 3.17b details an extracted ion-current profile for the sum of two ions monitored in a similar SIM "up-front" CID experiment. However, in that latter case, the 5-μl injection

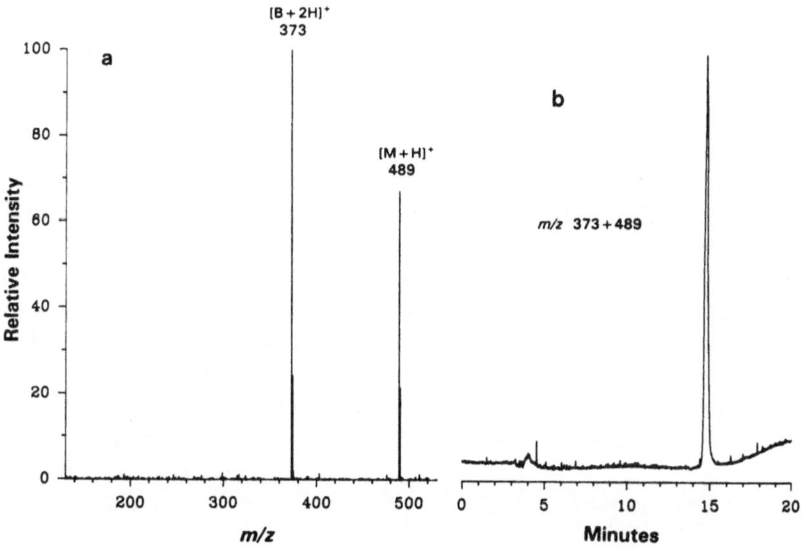


*Figure 3.17.* LC-MS analysis using an ion-spray interface and the first stage of a triple quadrupole. "Up-front" collision-induced dissociation of the $[M+H]^+$ ion, *m/z* 489, of the dG(C8)acetylaminofluorene nucleoside adduct (structure given in Fig. 3.15). (a) Positive-ion full-scan spectrum. (b) Extracted ion-current profile for the sum of the spectrum's two major ions, collected with selected ion monitoring. Unpublished data from M. A. Quilliam.

volume contained 2.5 ng dG(C8)AAF and unmodified nucleosides ranging in amount from 20 to 250 μg. The HPLC conditions ensure that only the adducted nucleosides will be initially retained at the head of the column. A significant advantage of this method is that no sample preparation was used to separate unmodified components from adducted components prior to the actual analysis. Calculations based on the individual extracted ion-current profiles give an LOD of 30 pg and suggest that adducts could be investigated at levels as low as 1 in $10^7$ unmodified nucleosides. Use of pre-LC separation technologies should allow investigations involving even lower adduct levels. The spikes observed in the extracted ion current profile in Fig. 3.17b are in part due to the unmodified nucleosides, which are present at near-saturation concentrations. This spiking may be eliminated by an effluent switching system under consideration by Quilliam.

## 10.6. Flow-FAB

HPLC effluents have also been coupled with Flow-FAB to allow separation of compound mixtures and to obtain improved LODs, in part through background subtraction techniques. Three common interfaces coaxially admit the effluent to the MS Flow-FAB source, use a liquid tee junction to mix FAB matrix with the HPLC effluent stream, or mix the FAB matrix with the HPLC solvent(s). The last-named type, with a 1% glycerol matrix, was recently used by Wolf and Vouros (105) to investigate various AAF–DNA adducts produced *in vitro* and separated from nonadducted nucleosides by butanol extraction. A packed fused silica C-18 HPLC column (320-μm i.d. × 30 cm) was interfaced with a transfer line of 50-μm i.d. × 65 cm to a TQ/MS. Injection volumes of 0.5 μl were subjected to isocratic HPLC analysis conditions (54% MeOH). LODs of less than 25 pg for nucleoside adducts were obtained with Flow-FAB MS/MS techniques. CNL techniques revealed minor deoxyguanosine adducts that eluted at ~14 minutes, whereas the common dG(C8)AAF adduct eluted at ~25.5 minutes, as shown in Fig. 3.18 (Susan Wolf, personal communication). CNL analysis successfully detected these adducts within a mixture of other

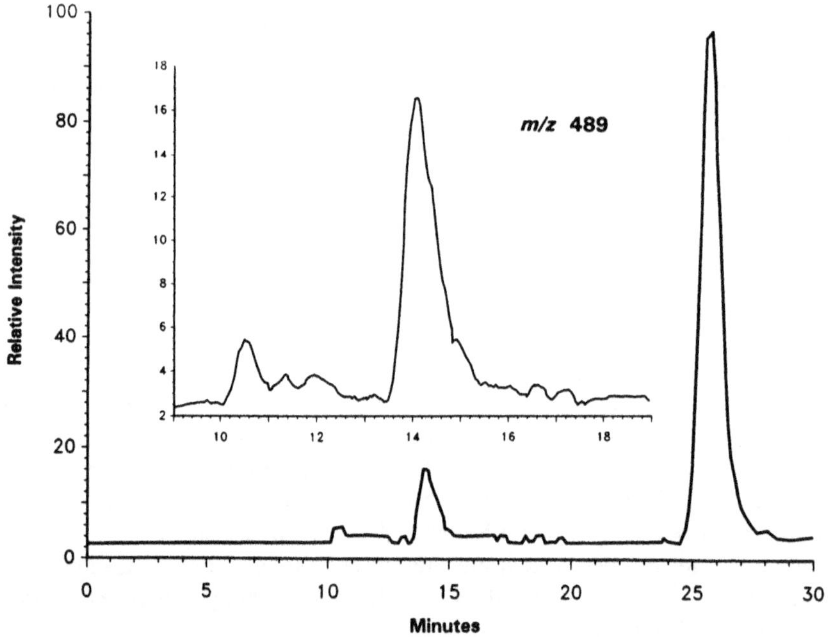


*Figure 3.18.* LC-MS analysis using a flow-fast atom bombardment interface and MS/MS with a triple quadrupole MS. Collision-induced dissociation and 116-Da constant neutral loss analysis resulted in the extracted ion-current profile for the precursor $[M+H]^+$ ion, *m/z* 489. In addition to the major dG(C8)acetylaminofluorene nucleoside adduct (structure given in Fig. 3.15) at 25.5 min, unidentified minor guanosine acetylaminofluorene adducts were revealed (inset). Unpublished data from Susan Wolf.

unidentified components revealed by UV detection. The structure and full-scan spectrum for dG(C8)AAF was given in Fig. 3.15.

## *11. COMBINED CAPILLARY ELECTROPHORESIS-MASS SPECTROMETRY*

A variety of different capillary electrophoresis techniques is possible, but to date only the linkage of capillary zone electrophoresis (CZE) with Flow-FAB has been reported for the analysis of DNA adducts. Recently, a collaboration of Northeastern University and EPA researchers used CZE-MS techniques to investigate four PAH and amino-PAH nucleoside adducts (106). CZE accomplished separations based on the charge and the size of the constituents involved—entailing a separation process different from HPLC. CZE conditions can be selected to give a flow regime on the order of 1 μl/min and peak widths of approximately 20 sec, although great changes in these values can be accomplished through manipulation of the CZE conditions. A modified liquid tee interface permitted addition of FAB matrix solution (1% glycerol, 5% acetonitrile, and 19% methanol in water) so that the total flow into the TQ/MS was ~2 μl/min. *In vitro*-produced dG(C8)AAF (structure shown in Fig. 3.15) gave full spectrum information with 1.3 ng, as seen in Fig. 3.19. Note that, in addition to the molecular ion, a number of fragments are visible and permit identification of the base, the adduct, and the point of attachment. MS/MS monitoring of the loss of sugar moiety was similarly successful for a mixture of four adducts.

CZE routinely involves injection of much smaller (nl) volumes of sample, so that only with extreme reduction of the reconstituted volume containing the analytes can a reasonable portion of the sample be injected. However, new sample-stacking techniques (107) permit μl-range injections, which should obviate the requirement for extreme volume reduction.

## *12. CONCLUSIONS*

The fundamental challenge in modified DNA research involves the identification of unknown modifications present at low levels in a complex mixture of DNA modifications. When μg quantities of relatively pure adduct are present, nearly any MS technique can provide additional structural information. However, for smaller amounts, the selection of a particular MS technique depends on a number of factors: the sensitivity of the technique, the level of identification required, the form of the DNA, the requisite sample processing, and the complexity of a mixture of components. Because of the low level of

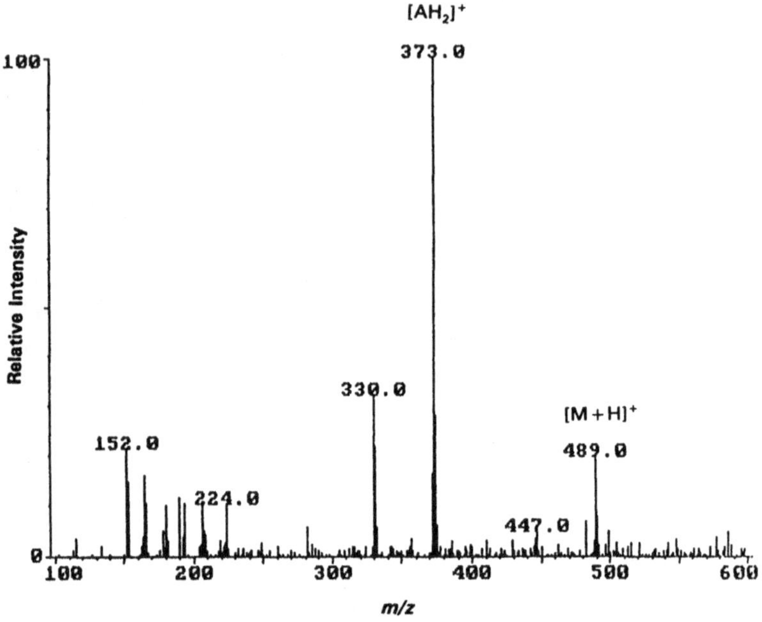


*Figure 3.19.* CZE-MS analysis using a flow-fast atom bombardment interface and the first stage of a triple quadrupole MS. The spectrum resulted from the positive-ion full-scan analysis of 1.3 ng of the dG(C8)acetylaminofluorene nucleoside adduct (structure given in Fig. 3.15). Modified from Ref. 106; used with permission.

modification present in many biological samples, primary significance must be placed on sensitivity. Sensitivity encompasses not only how much material (measured in, e.g., picograms) is required for a particular MS analysis, but also what portion of the total sample is analyzed, wasted, or recoverable. The potential for multiple analyses with one sample loading must also be considered. For most techniques (e.g., probe analyses or coupled GC- or LC-) 10–20% of the sample can be utilized per analysis.

The degree and type of information required may impart further constraints to the selection of a MS methodology. For example, AMS would suffice in dosimetry experiments, where the nature of the modification has already been established, and only a determination of the extent of modification is required. In fact, AMS can detect lower levels of modification than any non-MS technique. If further specificity is required, then other MS techniques offer more identification power, based on high resolution, fragmentation, and determination of the molecular species. High-resolution MS often permits the elimination of alterative structural possibilities. Degree of fragmentation is strongly dependent on the ionization mode and can be augmented by chemical derivatization (e.g., TMS derivatization). EI generally gives more fragmentation, whereas CI, FD, and PD tend to produce few fragments. The other modes generally yield intermediate amounts of fragmentation. The ease of identification of the molecular species is often inversely proportional to the degree of fragmentation and thus is subject to many of the same ionization considerations mentioned above. Some forms of derivatization (e.g., PFB derivatization) greatly augment the abundance of the molecular ion or another specific ion. CID with MS/MS can help induce and control the amount of fragmentation and determine the relationships among the observed fragments. Thus, MS/MS aids greatly in the identification process, as well as lending considerable ability to discriminate among components present in a mixture. None of these approaches to greater identification power should be lightly discarded if unknowns are subject to investigation. The combined GC (with TMS derivatization), CZE, and HPLC techniques offer LODs in the mid-picogram, high-picogram and low-nanogram ranges, respectively, when full-scan data are collected. LD offers mid- to high-picogram LODs, whereas AMS and GC-MS with PFB derivatization promise low-femtogram LODs. SIM and MS/MS techniques can give LODs two orders of magnitude better than those for full-scan acquisition. However, the utilization of these techniques, except for CNL MS/MS analyses, requires *a priori* knowledge of the nature of the modifications. These LODs were not always obtained for modified DNA from biological matrices. Thus, the matrix interferences have not always been assessed.

The selection of an MS technique involves a further practical consideration, which deals with the actual handling and preparation of the minute quantities of sample destined for a specific MS analysis. This sample "processing" can be quite moderate or very rigorous, depending on the chosen MS technique, the form of the DNA, and the level of adduction. In general, techniques are more problematic when they require more off-line separation, hydrolysis to a smaller form of the DNA, lower levels of modification, or derivatization. The DNA may be required to be of a particular form (bases, oligonucleotides, etc.). For example, genome-specific details may be available only when oligonucleotides are subjected to MS analysis, although specific nucleases and molecular-biology techniques might obviate even this requirement. Similarly, when adducts are cleaved from their bases prior to analysis, there follows a necessary loss of site-specific information and of some assurance that the source of the adduct was truly DNA, not RNA or proteins. AMS is also subject to this latter limitation.

The more challenging MS identification problems are confronted when a complex mixture of unknown adducts is contained in a biological matrix. In addition to requisite sample-preparation steps, this situation requires that many different components be resolved from one another and identified. MS/MS techniques alone are probably not sufficient to accomplish this separation. Off-line separation schemes necessitate additional collection of fractions and subsequent tedious multiple analyses. These factors in part explain the generally perceived trend toward on-line couplings of MS and MS/MS with an additional separation scheme. Such couplings are not compatible with many MS techniques such as FD, LD, and PD. However, GC-MS has long functioned as a convenient way to accomplish such a separation. Unfortunately, few if any bases, nucleosides, nucleotides, and certainly oligonucleotides have the prerequisite volatility, thermal stability, and low polarity to be compatible with the GC process, unless the sample is first derivatized. Derivatization does offer advantages in these and other areas, but incurs other debts with respect to sample processing and data reduction. Nevertheless, GC techniques are powerful and rightfully used for many research applications.

HPLC and CZE are the two other promising separation schemes that have been coupled with MS for DNA modification research (although supercritical fluid chromatography may also merit future attention). Numerous interfaces have been applied to the research of modified DNA, as discussed in this chapter. Notably, these separation technologies can accommodate many different forms of DNA without prior derivatization. A more powerful methodology is promised when capillary HPLC or CZE separations are coupled on-line to MS/MS, which can remove, or at least minimize, interferences or other chemical noise associated with the sample. Ultimately, the selection of an MS method should be guided by the overall goals of the research project. The current MS methods have just begun to approach the levels and expectations of environmental and biological scientists. However, as the MS instrumentation and data-handling techniques continue to evolve, new systems of higher sensitivity (e.g., ion traps) are likely to mature. The techniques and integrated methodology for sample handling, trace-level separations, and use with MS/MS will similarly continue to be improved. Combinations of these developments will undoubtedly allow applications of MS techniques to identify and quantify DNA modifications in complex environmental and biological samples.

ACKNOWLEDGMENTS. The authors would like to thank Dr. Eugene Jackim for his careful review of this chapter. Portions of this work were supported by the Environmental Protection Agency. Mention of trade names does not constitute endorsement by the U.S. EPA. Papyrus® reference software (Research Software Design, Portland, Oregon) was of tremendous aid in compiling this chapter.

## REFERENCES


1. Santella, R., 1991, DNA adducts in humans as biomarkers of exposure to environmental and occupational carcinogens, *Environ. Carcino. Ecotox. Rev.* **C9:**57–81.
2. McCloskey, J. A., and Crain, P. F., 1992, Progress in mass spectrometry of nucleic acid constituents: analysis of xenobiotic modifications and measurements at high mass, *Int. J. Mass Spectrom. Ion Proc.* **118/119:**593–615.
3. Lay, J. O., Jr., 1992, Mass spectrometry for the analysis of carcinogen DNA adducts, *Mass. Spectrom. Rev.* **6:**447–493.
4. Crain, P. F., 1990, Mass spectrometric techniques in nucleic acid research, *Mass Spectrom. Rev.* **9:**505–554.
5. Schram, K. H., 1990, Mass spectrometry of nucleic acid components, in *Methods of Biochemical Analysis, Biomedical Applications of Mass Spectrometry,* Vol. 34 (C. H. Suelter and J. T. Watson, eds.), pp. 203–287.
6. Mallis, L. M., Raushel, F. M., and Russell, D. H., 1987, Differentiation of isotopically labeled nucleotides using fast atom bombardment tandem mass spectrometry, *Anal. Chem.* **59:**980–984.
7. Shaulsky, G., Johnson, R. L., Shockcor, J. P., Taylor, L. C., and Stark, A. A., 1990, Properties of aflatoxin–DNA adducts formed by photoactivation and characterization of the major photoadduct as aflatoxin-N7-guanine, *Carcinogenesis* **11:**519–527.
8. Okuda, H., Nojima, H., Miwa, K., Watanabe, N., and Watabe, T., 1989, Selective covalent binding of the active sulfate ester of the carcinogen 5-(hydroxymethyl)chrysene to the adenine residue of calf thymus DNA, *Chem. Res. Toxicol.* **2:**15–22.
9. Chung, F. L., Wang, M. Y., and Hecht, S. S., 1989, Detection of exocyclic guanine adducts in hydrolysates of hepatic DNA of rats treated with N-nitrosopyrrolidine and in calf thymus DNA reacted with alpha-acetoxy-N-nitrosopyrrolidine, *Cancer Res.* **49:**2034–2041.
10. Cushnir, J. R., Naylor, S., Lamb, J. H., and Farmer, P. B., 1990, Deuterium exchange studies in the identification of alkylated DNA bases found in urine, by tandem mass spectrometry, *Rapid Commun. Mass Spectrom.* **4:**426–431.
11. Li, F., Solomon, J. J., Mukai, F., and Segal, A., 1990, *In vitro* reactions of isopropyl methanesulfonate with DNA and with 2′-deoxyribonucleosides, *Cancer Biochem. Biophys.* **11:**253–264.
12. Bhanot, O. S., Grevatt, P. C., Donahue, J. M., Gabrielides, C. N., and Solomon, J. J., 1992, *In vitro* DNA replication implicates $O^2$-ethyldeoxythymidine in transversion mutagenesis by ethylating agents, *Nucleic Acids Res.* **20:**587–594.
13. Li, F., Segal, A., and Solomon, J. J., 1992, *In vitro* reaction of ethylene oxide with DNA and characterization of DNA adducts, *Chem.–Biol. Interact.* **83:**35–54.
14. Roy, A. K., Upadhyaya, P., Evans, F. E., and El-Boyoumy, K., 1991, Structural characterization of the major adducts formed by reaction of 4,5-epoxy-4,5-dihydro-1-nitropyrene with DNA, *Carcinogenesis* **12:**577–581.
15. Spratt, T. E., Trushin, N., Lin, D., and Hecht, S., 1989, Analysis for $N^2$-(pyridyloxobutyl)deoxyguanosine adducts in DNA of tissues exposed to tritium-labeled 4-(methylnitrosamino)-1-(3-pyridyl)-1-butanone and N′-nitrosonornicotine, *Chem. Res. Toxicol.* **2:**169–173.
16. Sodum, R. S., and Chung, F. L., 1988, 1,$N^2$-ethenodeoxyguanosine as a potential marker for DNA adduct formation by *trans*-4-hydroxyt-2-nonenal, *Cancer Res.* **48:**320–323.
17. Van den Eeckhout, E., De Bruyn, A., Pepermans, H., Esmans, E. L., Vryens, I., Claereboudt, J., Claeys, M., and Sinsheimer, J. E., 1990, Adduct formation identification between phenyl glycidyl ether and 2′-deoxyadenosine and thymidine by chromatography, mass spectrometry and nuclear magnetic resonance spectroscopy, *J. Chromatogr.* **504:**113–128.

18. Savela, K., Hesso, A., and Hemminki, K., 1986, Characterization of reaction products between styrene oxide and deoxynucleosides and DNA, *Chem.–Biol. Interact.* **60:**235–246.
19. Pongcracz, K., and Bodell, W. J., 1991, Detection of 3′-hydroxyl-1,$N^6$-benzetheno-2′-deoxyadenosine 3′-phosphate by $^{32}P$ postlabeling of DNA reacted with *p*-benzoquinone, *Chem. Res. Toxicol.* **4:**199–202.
20. Wood, J. M., Hoke, S. H., II, Cooks, R. G., Chae, W. -G., and Chang, C. -J., 1991, Chemical modification of deoxyribonucleic acids: quantitation of 3-methylthymidine and O4-methylthymidine by tandem mass spectrometry, *Int. J. Mass Spectrom. Ion Proc.* **111:**381–394.
21. Reardon, D. B., Prakash, A. S., Hilton, B. D., Roman, J. M., Pataki, J., Harvey, R. G., and Dipple, A., 1987, Characterization of 5-methylchrysene-1,2-dihydrodiol-3,4-epoxide–DNA adducts, *Carcinogenesis* **8:**1317–1322.
22. Wang, M., Chung, F. L., and Hecht, S. S., 1989, Formation of acyclic and cyclic guanine adducts in DNA reacted with alpha-acetoxy-N-nitrosopyrrolidine, *Chem. Res. Toxicol.* **2:**423–428.
23. Jowa, L., Witz, G., Snyder, R., Winkle, S., and Kalf, G. F., 1990, Synthesis and characterization of deoxyguanosine–benzoguinone adducts, *J. Appl. Toxicol.* **10:**47–54.
24. Pongracz, K., Kaur, S., Burlingame, A. L., and Bodell, W. J., 1990, Detection of (3′-hydroxy)-3,$N^4$-benzetheno-2′-deoxycytidine-3′-phosphate by $^{32}P$-postlabeling of DNA reacted with *p*-benzoguinone, *Carcinogenesis* **11:**1469–1472.
25. Silk, N. A., Lay, J. O., Jr., and Martin, C. N., 1989, Covalent binding of 4,4′-methylene-*bis*-(2-chloroaniline) to rat liver DNA *in vivo* and of its N-hydroxylated derivative to DNA *in vitro, Biochem. Pharmacol.* **38:**279–287.
26. Chadha, A., Sayer, J. M., Yeh, H. J. C., Yagi, H., Cheh, A. M., Pannell, L. K., and Jerina, D. M., 1989, Structures of covalent nucleoside adducts formed from adenine, guanine and cytosine bases of DNA and the optically active bay-region 3,4-diol 1,2-epoxides of dibenz[*a,j*]anthracene, *J. Am. Chem. Soc.* **111:**5456–5463.
27. Frenkel, K., Zhong, Z., Wei, H., Karkoszka, J., Patel, U., Rashid, K., Georgescu, M., and Solomon, J., 1991, Quantitative high-performance liquid chromatography analysis of DNA oxidized *in vitro* and *in vivo, Anal. Biochem.* **196:**126–136.
28. Weng, Q. -M., Hammargren, W. M., Slowikowskim, D., Schram, K. H., Borysko, K. Z., Wotring, L. L., and Townsend, L. B., 1989, Low nanogram detection of nucleotides using fast atom bombardment-mass spectrometry, *Anal. Biochem.* **178:**102–106.
29. Trainor, T. M., Giese, R. W., and Vouros, P., 1988, Mass spectrometry of electrophore-labeled nucleosides—pentafluorobenzyl and cinnamoyl derivatives, *J. Chromatogr.* **452:**369–376.
30. Saha, M., Kresbach, G. M., Giese, R. W., Annan, R. S., and Vouros, P., 1989, Preparation and mass spectral characterization of pentafluorobenzyl derivatives of alkyl and hydroxyalkyl-nucleobase DNA adducts, *Biomed. Environ. Mass Spectrom.* **18:**958–972.
31. Freeman, J. A., Johnson, J. V., Hail, M. E., Yost, R. A., and Kuehl, D. W., 1990, Estimation of mutagenic/carcinogenic potential of environmental contaminants by ion–molecule reactions and tandem mass spectrometry, *J. Am. Soc. Mass Spectrom.* **1:**110–115.
32. Kappen, L. S., Lee, T. R., Yang, C. C., and Goldberg, I. H., 1989, Oxygen transfer from the nitro group of a nitroaromatic radiosensitizer to a DNA sugar damage product, *Biochemistry* **28:**4540–4542.
33. Stillwell, W. G., Glogowski, J., Xu, H. -X., Wishnok, J. S., Zavala, D., Montes, G., Correa, P., and Tannenbaum, S. R., 1991, Urinary excretion of nitrate, N-nitrosoproline, 3-methyladenine, and 7-methylguanine in a Colombian population at high risk for stomach cancer, *Cancer Res.* **51:**190–194.
34. Stillwell, W. G., Xu, H. -X., Adkins, J. A., Wishnok, J. S., and Tannenbaum, S. R., 1989,

Analysis of methylated and oxidized purines in urine by capillary gas chromatography-mass spectrometry, *Chem. Res. Toxicol.* **2**:94–99.

35. Weston, A., Rowe, M. L., Manchester, D. K., Farmer, P. B., Mann, D. L., and Harris, C. C., 1989, Fluorescence and mass spectral evidence for the formation of benzo(*a*)pyrene anti-diol-epoxide–DNA and –hemoglobin adducts in humans, *Carcinogenesis* **10**:251–257.
36. Dizdaroglu, M., 1991, Chemical determination of free radical–induced damage to DNA, *Free Radical Biol. Med.* **10**:225–242.
37. Malins, D. C., and Haimanot, R., 1991, The etiology of cancer: hydroxyl radical–induced DNA lesions in histologically normal livers of fish from a population with liver tumors, *Aquat. Toxic.* **20**:123–130.
38. Friesen, M. D., Garren, L., Prevost, V., and Shuker, D. E. G., 1991, Isolation of urinary 3-methyladenine using immunoaffinity columns prior to determination by low-resolution gas chromatography-mass spectrometry, *Chem. Res. Toxicol.* **4**:102–106.
39. Gansewendt, B., Foest, U., Xu, D., Hallier, E., Bolt, H. M., and Peter, H., 1991, Formation of DNA adducts in F-344 rats after oral administration or inhalation of $^{14}C$ methyl bromide, *Food Chem. Toxicol.* **29**:557–563.
40. Krewet, E., Muller, G., and Norpoth, K., 1989, The excretion of chlorophenylmercapturic acid, chlorophenols and a guanine adduct in the urine of chlorobenzene-treated rats after phenobarbital pretreatment, *Toxicology* **59**:67–79.
41. Musser, S. M., Pan, S. -S., Egorin, M. J., Kyle, D. J., and Callery, P. S., 1992, Alkylation of DNA with aziridine produced during the hydrolysis of N,N′,N″-triethylenethiophosphoramide, *Chem. Res. Toxicol.* **5**:95–99.
42. Hammargren, W. M., Luffer, D. R., Schram, K. H., Reimer, M. L. J., Nakano, K., Yasaka, T., and Moorman, A., 1992, The identification of 5′-deoxy-5′methylthioguanosine in human urine by gas chromatography/mass spectrometry, *Nucleosides and Nucleotides* **11**:1275–1292.
43. Foiles, P. G., Akerkar, S. A., Carmella, S. G., Kagan, M., Stoner, G. D., Resau, J. H., and Hecht, S. S., 1991, Mass spectrometric analysis of tobacco-specific nitrosamine–DNA adducts in smokers and nonsmokers, *Chem. Res. Toxicol.* **4**:364–368.
44. Fedtke, N., Boucheron, J. A., Turner, M. J., Jr., and Swenberg, J. A., 1990, Vinyl chloride–induced DNA adducts. I: Quantitative determination of $N^2$,3-ethenoguanine based on electrophore labeling, *Carcinogenesis* **11**:1279–1285.
45. Bonfanti, M., Magagnotti, C., Galli, A., Bagnati, R., Moret, M., Gariboldi, P., Fanelli, R., and Airoldi, L., 1990, Determination of $O^6$-butylguanine in DNA by immunoaffinity extraction/gas chromatography-mass spectrometry, *Cancer Res.* **50**:6870–6875.
46. Bakthavachalam, J., Abdel-Baky, S., and Giese, R. W., 1991, Release of 2-aminofluorene from N-(deoxyguanosin-8-yl)-2-aminofluorene by hydrazinolysis, *J. Chromatogr.* **538**:447–451.
47. Li, W., Sotiriou, C., Abdel-Baky, S., Fisher, D. H., and Giese, R. W., 1991, Superoxide chemical transformation of diolepoxide polyaromatic hydrocarbon DNA adducts. Determination of benzo[*a*]pyrene-*r*-7,*t*-8,9,*c*-10-tetrahydrotetrol by gas chromatography, *J. Chromatogr.* **588**:273–280.
48. Abdel-Baky, S., and Giese, R. W., 1991, Gas chromatography/electron capture negative-ion mass spectrometry at the zeptomole level, *Anal. Chem.* **63**:2986–2989.
49. Solomon, J. J., and Segal, A., 1985, Direct alkylation of calf thymus DNA by acrylonitrile. Isolation of cyanoethyl adducts of guanine and thymine, and carboxyethyl adducts of adenine and cytosine, *Environ. Health Perspect.* **62**:227–230.
50. Solomon, J. J., Fedyk, J., Mukai, F., and Segal, A., 1985, Direct alkylation of 2′-deoxynucleosides and DNA following *in vitro* reaction with acrylamide, *Cancer Res.* **45**:3465–3470.
51. Viari, A., Ballini, J. P., Vigny, P., Voituriez, L., and Cadet, J., 1989, Plasma desorption mass

spectrometric study of UV-induced lesions within DNA model compounds, *Biomed. Environ. Mass Spectrom.* **18:**547–552.

52. Voituriez, L., Ulrich, J., Gaboriau, F., Viari, A., Vigny, P., and Cadet, J., 1990, Identification of the two *cis-syn*[2+2] cycloadducts resulting from the photoreaction of 3-carbethoxypsoralen with 2′-deoxycytidine and 2′-deoxyuridine, *Int. J. Radiat. Biol.* **57:**903–918.
53. Moysan, A., Viari, A., Vigny, P., Voituriez, L., Cadet, J., Moustacchi, E., and Sage, E., 1991, Formation of cyclobutane thymine dimers photosensitized by pyridopsoralens: quantitative and qualitative distribution within DNA, *Biochemistry* **30:**7080–7088.
54. Negra, S. D., Ginot, Y. M., Le Beyec, Y., Spiro, M., and Vigny, P., 1982, Fast heavy ion induced desorption mass spectrometry: future applications in the area of molecular modifications of nucleic acids induced by physical or chemical agents, *Nucl. Instr. Meth.* **198:**159–163.
55. Hercules, D. M., Day, R. J., Baulesanmugan, K., Deng, T. A., and Li, C. P., 1982, Laser microprobe mass spectrometry 2. Applications to structural analysis, *Anal. Chem.* **54:**280A.
56. Lay, J. O., Jr., Chiarelli, M. P., Bryant, M. S., and Nelson, R. W., 1993, Detection and characterization of DNA adducts at the femtomole level by desorption ionization mass spectrometry, *Environ. Health Perspect.* **99:**191–193.
57. Giam, C. S., Holliday, T. L., Zheng, Y., Williams, J. L., and Ahmed, M. S., 1991, DNA adducts as early bioindicators of chemical exposure, *Environ. Monit. Assess.* **19:**335–340.
58. Giam, C. S., Ahmed, M. S., McAdoo, D. J., Zheng, Y., and Holliday, T. L., 1991, Characterization of ethylguanine isomers and structurally related compounds by laser-desorbed cationization in a Fourier transform mass spectrometer, *Anal. Chim. Acta* **247:**235–241.
59. Hettich, R. L., 1989, The differentiation of methyl guanosine isomers by laser ionization Fourier transform mass spectrometry, *Biomed. Environ. Mass Spectrom.* **18:**265–277.
60. Hettich, R. L., Buchanan, M. V., and Ho, C. -H., 1989, Characterization of photo-induced pyrimidine cyclobutane dimers by laser desorption Fourier transform mass spectrometry, *Biomed. Environ. Mass Spectrom.* **19:**55–62.
61. Hettich, R. L., and Buchanan, M. V., 1991, Matrix-assisted laser desorption Fourier transform mass spectrometry for the structural examinationof modified nucleic acid constituents, *Int. J. Mass Spectrom. Ion Proc.* **111:**365–380.
62. Hettich, R., and Buchanan, M., 1991, Structural characterization of normal and modified oligonucleotides by matrix-assisted laser desorption Fourier transform mass spectrometry, *J. Am. Soc. Mass Spectrom.* **2:**402–412.
63. Felton, J. S., Knize, M. G., Roper, M., Fultz, E., Shen, N. H., and Turteltaub, K. W., 1992, Chemical analysis, prevention, and low-level dosimetry of heterocyclic amines from cooked food, *Cancer Res.* **52**(7 Suppl):2103s–2107s.
64. Turteltaub, K. W., Felton, J. S., Gledhill, B. L., Vogel, J. S., Southon, J. R., Caffee, M. W., Finkel, R. C., Nelson, D. E., Proctor, I. D., and Davis, J. C., 1990, Accelerator mass spectrometry in biomedical dosimetry: relationship between low-level exposure and covalent binding of heterocyclic amine carcinogens to DNA, *Proc. Natl. Acad. Sci. U.S.A.* **87:**5288–5292.
65. Lyon, P., ed., 1985, *Desorption Mass Spectrometry. Are SIMS and FAB the Same?* American Chemical Society, Washington, D.C.
66. Martin, L. B., III, Schreiner, A. F., and van Breemen, R. B., 1991, Characterization of cisplatin adducts of oligonucleotides by fast atom bombardment mass spectrometry, *Anal. Biochem.* **193:**6–15.
67. Pei, G. K., and Moschel, R. C., 1990, Aralkylation of 2′-deoxyguanosine: medium effects on sites of reaction with 7-(bromomethyl)benz[*a*]anthracene, *Chem. Res. Toxicol.* **3:**292–295.
68. Koepke, S. R., Kroeger-Koepke, M. B., and Michejda, C. J., 1990, Evidence for an unstable DNA adduct from N-nitroso-N-methylaniline, *Chem. Res. Toxicol.* **3:**17–20.

69. Lasko, D. D., Basu, A. K., Kadlubar, F. F., Evans, F. E., Lay, J. O., Jr., and Essigmann, J. M., 1987, A probe for the mutagenic activity of the carcinogen 4-aminobiphenyl: synthesis and characterization of an M13mp10 genome containing the major carcinogen–DNA adduct at a unique site, *Biochemistry* **26:**3072–3081.
70. Delclos, K. B., Miller, D. W., Lay, J. O., Jr., Casciano, D. A., Walker, R. P., Fu, P. P., and Kadlubar, F. F., 1987, Identification of C8-modified deoxyinosine and $N^2$- and C8-modified deoxyguanosine as major products of the *in vitro* reaction of N-hydroxy-6-aminochrysene with DNA and the formation of these adducts in isolated rat hepatocytes treated with 6-nitrochrysene and 6-aminochrysene, *Carcinogenesis* **8:**1703–1709.
71. Fu, P. P., Miller, D. W., Von Tungeln, L. S., Bryant, M. S., Lay, J. O., Jr., Huang, K., Jones, L., and Evans, F. E., 1991, Formation of C8-modified deoxyguanosine and C8-modified deoxyadenosine as major DNA adducts from 2-nitropyrene metabolism mediated by rat and mouse liver microsomes and cytosols, *Carcinogenesis* **12:**609–616.
72. Marinelli, E. R., Johnson, F., Iden, C. R., and Yu, P. L., 1990, Synthesis of 1,$N^2$-(1,3-propano)-2′-deoxyguanosine and incorporation into oligodeoxynucleotides: a model for exocyclic acrolein–DNA adducts, *Chem. Res. Toxicol.* **3:**49–58.
73. Shibutani, S., Gentles, R., Johnson, F., and Grollman, A. P., 1991, Isolation and characterization of oligodeoxynucleotides containing dG-$N^2$-AAF and oxidation products of dG-C8-AF, *Carcinogenesis* **12:**813–818.
74. Wood, M. L., Smith, J. R., and Garner, R. C., 1988, Structural characterization of the major adducts obtained after reaction of an ultimate carcinogen aflatoxin $B_1$-dichloride with calf thymus DNA *in vitro, Cancer Res.* **48:**5391–5396.
75. Tilby, M. J., Lawley, P. D., and Farmer, P. B., 1990, Alkylation of DNA by melphalan in relation to immunoassay of melphalan–DNA adducts: characterizationof mono-alkylated and cross-linked products from reaction of melphalan with dGMP and GMP, *Chem.–Biol. Interact.* **73:**183–194.
76. Pongracz, K., Kaur, S., Burlingame, A. L., and Bodell, W. J., 1989, $O^6$-substituted-2′-deoxyguanosine-3′-phosphate adducts detected by $^{32}P$ postlabeling of styrene oxide treated DNA, *Carcinogenesis* **10:**1009–1013.
77. Kim, D. H., Humphreys, W. G., and Guengerich, F. P., 1990, Characterization of *S*-[2-($N^1$-adenyl)ethyl]glutathione as an adduct formed in RNA and DNA from 1,2-dibromoethane, *Chem. Res. Toxicol.* **3:**587–594.
78. Mouret, J. -F., Polverelli, M., Sarrazini, F., and Cadet, J., 1991, Ionic and radical oxidation of DNA by hydrogen peroxide, *Chem.–Biol. Interact.* **77:**187–201.
79. Nukaya, H., Iwami, T., Ishida, H., Tsuji, K., Suwa, Y., Wakabayashi, K., Nagao, M., Sugimura, T., and Kosuge, T., 1990, $N^2$ acetylation of 2′-deoxyguanosine by coffee mutagens, methyl glyoxal and hydrogen peroxide, *Mutat. Res.* **245:**251–257.
80. Vasseur, J. -J., Rayner, B., Imbach, J. -L., Verma, S., McCloskey, J. A., Lee, M., Chang, D. -K., and Lown, J. W., 1987, Structure of the adduct formed between 3-aminocarbazole and the apurinic site oligonucleotide model *d*[Tp(Ap)pT], *J. Org. Chem.* **52:**4994–4998.
81. Sharma, M., Jain, R., and Isac, T. V., 1991, A novel technique to assay adducts of DNA induced by anticancer agent *cis*-diamminedichloroplatinum(II), *Bioconjugate Chem.* **2:**403–406.
82. Urata, H., Yamamoto, K., Akagi, M., Hiroaki, H., and Uesugi, S., 1989, A 2-deoxyribonolactone-containing nucleotide: isolation and characterization of the alkali-sensitive photoproduct of the trideoxyribonucleotide *d*(ApCpA), *Biochemistry* **28:**9566–9569.
83. Comess, K. M., Costello, C. E., and Lippard, S. J., 1990, Identification and characterization of a novel linkage isomerization in the reaction of *trans*-diamminedichloroplatinum(II) with 5′-*d*(TCTACGCGTTCT), *Biochemistry* **29:**2102–2110.
84. Tomer, K. B., Guenat, C., Dino, J. J., Jr., and Deterding, L. J., 1988, Applications of fast atom

bombardment and tandem mass spectrometry, *Biomed. Environ. Mass Spectrom.* **16**:473–476.

85. RamaKrishna, N. V. S., Cavalieri, E. L., Rogan, E. G., Dolnikowski, G., Cerny, R. L., Gross, M. L., Jeong, H., Jankowiak, R., and Small, G. J., 1992, Synthesis and structure determination of the adducts of the potent carcinogen 7,12-dimethylbenz[*a*]anthracene and deoxyribonucleosides formed by electrochemical oxidation: models for metabolic activation by one-electron oxidation, *J. Am. Chem. Soc.* **114**:1863–1874.
86. RamaKrishna, N. V. S., Gao, F., Padmavathi, N. S., Cavalieri, E. L., Rogan, E. G., Cerny, R. L., and Gross, M. L., 1992, Model adducts of benzo[*a*]pyrene and nucleosides formed from its radical cation and diol epoxide, *Chem. Res. Toxicol.* **5**:293–302.
87. Shieh, T. L., Hoyos, P., Kolodziej, E., Stowell, J. G., Baird, W. M., and Byrn, S. R., 1990, Properties of the nucleic acid photoaffinity labeling agent 3-azidoamsacrine, *J. Med. Chem.* **33**:1225–1230.
88. Bartczak, A. W., Sangaiah, R., Kelman, D. J., Toney, G. E., Deterding, L. J., Charles, J., Marbury, G. D., and Gold, A., 1989, Synthesis of $N^6$-adenosine adducts expected from cyclopenta-ring activation of acenaphthylene and aceanthrylene, *Tetrahedron Lett.* **30**:3251–3254.
89. Van den Eeckhout, E., Coene, J., Claereboudt, J., Borremans, F., Claeys, M., Esmans, E., and Sinsheimer, J. E., 1991, Comparison of the isolation of adducts of 2′-deoxycytide and 2′-deoxyguanosine with phenylglycidyl ether by high-performance liquid chromatography on a reversed-phase column and a polystyrene–divinylbenzene column, *J. Chromatogr.* **541**:317–331.
90. Cushnir, J. R., Naylor, S., Lamb, J. H., Farmer, P. B., Brown, N. A., and Mirkes, P. E., 1990, Identification of phosphoramide mustard/DNA adducts using tandem mass spectrometry, *Rapid Commun. Mass Spectrom.* **4**:410–414.
91. Mirkes, P. E., Brown, N. A., Kajbaf, M., Lamb, J. H., Farmer, P. B., and Naylor, S., 1992, Identification of cyclophosphamide–DNA adducts in rat embryos exposed *in vitro* to 4-hydroperoxycyclophosphamide, *Chem. Res. Toxicol.* **5**:382–385.
92. Cushnir, J. R., Naylor, S., Lamb, J. H., and Farmer, P. B., 1992, Tandem mass spectrometric approaches for the analysis of alkylguanines in human urine, *Org. Mass Spectrom.* **28**:552–558.
93. Annan, R. S., Giese, R. W., and Vouros, P., 1990, Detection and structural characterization of amino polyaromatic hydrocarbon–deoxynucleoside adducts using fast atom bombardment and tandem mass spectrometry, *Anal. Biochem.* **191**:86–95.
94. Bryant, M. S., Lay, J. O., Jr., and Chiarelli, M. P., 1992, Development of fast atom bombardment mass spectral methods for the identification of carcinogen–nucleoside adducts, *J. Am. Soc. Mass Spectrom.* **3**:360–371.
95. Maccubbin, A. E., Caballes, L., Riordan, J. M., Huang, D. H., and Gurtoo, H. L., 1991, A cyclophosphamide/DNA phosphoester adduct formed *in vitro* and *in vivo, Cancer Res.* **51**:886–892.
96. Wolf, S. M., Annan, R. S., Giese, R. W., and Vouros, P., 1992, Characterization of amino polyaromatic hydrocarbon–DNA adducts using continuous flow fast atom bombardment and collision induced dissociation. Positive and negative ion spectra, *Biol. Mass Spectrom.* **21**:647–654.
97. Kresbach, G. M., Itani, M., Saha, M., Rogers, E. J., Vouros, P., and Giese, R. W., 1989, Ester and related derivatives of ring N-pentafluorobenzylated 5-hydroxymethyluracil. Hydrolytic stability, mass spectral properties, and trace detection by gas chromatography-electron-capture detection, gas chromatography-electron-capture negative ion mass spectrometry, and moving-belt liquid chromatography-electron-capture negative ion mass spectrometry, *J. Chromatogr.* **476**:423–438.

98. Franz, R., and Neumann, H. -G., 1988, Reaction of *trans*-4-N-acetoxy-N-acetylaminostilbene with guanosine and deoxyguanosine *in vitro*: the primary reaction product at $N^2$ of guanine yields different final adducts, *Chem.–Biol. Interact.* **67**:105–116.
99. Liberato, D. J., Saavedra, J. E., Farnsworth, D. W., and Liijinsky, W., 1989, Thermospray liquid chromatography/mass spectrometry studies on mechanisms of nucleic acid alkylation by some deuterated carcinogens, *Chem. Res. Toxicol.* **2**:307.
100. Musser, S. M., Pan, S. S., and Callery, P. S., 1989, Liquid chromatography-thermospray mass spectrometry of DNA adducts formed with mitomycin C, porfiromycin and thiotepa, *J. Chromatogr.* **474**:197–207.
101. Van de Poll, M. L., Venizelos, V., Niessen, W. M., and Meerman, J. H., 1991, An unusual dearomatized adduct formed by reaction of 4′-fluoro-4-(acetylamino)biphenyl N-sulfate with deoxyadenosine, *Chem. Res. Toxicol.* **4**:318–323.
102. Ikounomou, M. G., Naghipur, A., Lown, W. J., and Kebarle, P., 1990, Characterization of the reaction products of deoxyguansine with the anticancer agent BFNU and BFNU-1,1,1′,1′-$d_4$ in different buffers by high-performance liquid chromatography/atmospheric pressure ionization tandem mass spectrometry, *Biomed. Environ. Mass Spectrom.* **19**:434–446.
103. Naghipur, A., Ikonomou, M. G., Kebarle, P., and Lown, J. W., 1990, Mechanism of action of (2-haloethyl)nitrosoureas on DNA: discrimination between alterntive pathways of DNA base modification by 1,3-*bis*(2-fluoroethyl)-1-nitrosurea by using specific deuterium labeling and identification of reaction products by HPLC/tandem mass spectrometry, *J. Am. Chem. Soc.* **112**:3178–3187.
104. Quilliam, M. A., 1992, DNA–carcinogen adduct analysis by ion-spray liquid chromatography-mass spectrometry, Proceedings of the 40th ASMS Conference on Mass Spectrometry and Allied Topics (A. Marshall, ed.), Port City Press, Baltimore, MD. Washington, DC, May 1992, pp. 430–431.
105. Wolf, S. M., and Vouros, P., Application of capillary liquid chromatography coupled with tandem mass spectrometric methods to the rapid screening of adducts formed by the reaction of N-acetoxy-2-acetylaminofluorene with calf thymus DNA, *Chem. Res. Toxicol.* (submitted).
106. Wolf, S. M., Vouros, P., Norwood, C., and Jackim, E., 1992, Identification of deoxynucleoside–polyaromatic hydrocarbon adducts by capillary zone electrophoresis-continuous flow fast atom bombardment mass spectrometry, *J. Am. Soc. Mass Spectrom.* **3**:757–761.
107. Chien, R. -L., and Burgi, D. S., 1992, Sample stacking of an extremely large injection volume in high-performance capillary electrophoresis, *Anal. Chem.* **64**:1046–1050.

# 4

# Mass Spectrometry in the Characterization of Variant Hemoglobins

*Cedric H. L. Shackleton and H. Ewa Witkowska*

## 1. INTRODUCTION

About 150 million people throughout the world carry abnormal hemoglobins, with consequences ranging from trivial to lethal. More than one in 800 people have hemoglobins that differ from normal by amino acid substitutions, globin chain elongations, contractions, and fusions. In determining this incidence, the most common ones—variants Hb S, Hb E, and Hb C—were not included (1).

Human hemoglobin (Hb) is a tetramer of two α chains (141 amino acids) and two non-α chains, β, δ, or γ (146 amino acids), each of them associated with a prosthetic heme group. Fetal and adult hemoglobin types are both built of α chains (coded by two gene copies on chromosome 16) and different non-α chains: $^{G}\gamma$ and $^{A}\gamma$ in fetal Hb F, β in adult Hb A, and δ in Hb $A_2$ (a minor component (2%) of adult hemoglobin. All non-α globin chains are coded by single-copy genes forming a β-type globin gene cluster on chromosome 11 (1).

Almost 600 globin variants of known structure have been described so far, and presumably many more remain undetected. Hemoglobin variants with amino acid substitutions in the protein chain generally arise as a consequence of single-base substitutions within globin-coding gene loci. Considering the α

---

*Cedric H. L. Shackleton and H. Ewa Witkowska* • Children's Hospital Oakland Research Institute, Oakland, California 94609.

*Mass Spectrometry: Clinical and Biomedical Applications, Volume 2,* edited by Dominic M. Desiderio. Plenum Press, New York, 1994.

and β chains alone, 2583 single-base substitutions are possible, but of these, only 1690 would result in amino acid replacement (1). About 33% of these variants have been characterized to date, but many others are yet to be identified in blood; a proportion of the possible amino acid substitutions may never be found because they may be incompatible with life. Hemoglobins with more than one point mutation in the same polypeptide chain are very uncommon and are likely to be derived from common variants. Seven hemoglobins that contain sickle cell mutation (β6Glu→Val) accompanied by another amino acid substitution have been described to date. However, more unusual combinations have also been encountered, and some of them include electrophoretically silent amino acid substitutions that have not been previously found as single mutation site variants (Hb Grenoble, Poissy, and Masuda). Apart from single-base mutations, there are other types of gene alterations (frameshifts, deletions, extensions, and nonhomologous crossing-over) that lead to hemoglobinopathy by producing a globin of variant sequence. Many of these mutants are present in peripheral circulation in relatively low abundance due to their low expression or impaired stability. In extreme cases, the amount of variant globins in blood is so low that phenotypically the condition is equivalent to thalassemia.* Those thalassemic and hyperunstable globins have to be characterized on the DNA level.

Although many known variants are not associated with any apparent clinical manifestations, the severity of the conditions produced by the clinically relevant variants, especially when present in the homozygous state (e.g., sickle cell hemoglobin, Hb S/S), or in compound heterozygosity with either noncompatible counterparts (e.g., Hb S/C, Hb S/D-Los Angeles, and Hb S/O-Arab) or thalassemia (e.g., Hb E/β thalassemia), justifies the growing interest in variant hemoglobin detection and identification.

A comment should be made on the nomenclature introduced above, which can be confusing. In the earliest studies, new hemoglobins identified by altered electrophoretic properties were given single initial names, e.g., S or C, but it was soon recognized that not only was the alphabet about to run out, but many of the hemoglobins so named in fact contained globin chains with different structure. Thus, compound names evolved, e.g., D-Los Angeles, but most recently the prefix letters have been dropped and the name given usually is that of a place, region, or institution. A comprehensive review of variant hemoglobins is given by Bunn and Forget (1), and a list of all variants is annually updated by the International Hemoglobin Information Center and published in *Hemoglobin.*

*The thalassemias constitute a heterogeneous group of naturally occurring inherited mutations characterized by abnormal globin gene function resulting in total absence or quantitative reduction of α and β globin chain synthesis in human erythroid cells (1).

Techniques routinely used nowadays to screen for abnormal hemoglobins are based on a difference in net charge between normal and variant protein: cellulose acetate electrophoresis (2), citrate agar electrophoresis (3), isoelectric focusing (4), and the recently adopted rapid-ion exchange HPLC (5). Charge-based detection of new variants is becoming increasingly difficult because only 575 of the 1690 possible α- or β-chain variants have amino acid replacements likely to result in a change of net charge large enough to be detected by routine techniques, and about 80% of these have already been identified. So, even restricting consideration to the α and β chains, more than 1000 possible variants will have essentially similar electrophoretic properties. An obvious need for more sensitive and more versatile methods for variant hemoglobin detection has been met by use of globin gel electrophoresis (6), isoelectric focusing in immobilized narrow pH gradients in nondenaturing (7,8) and denaturing conditions (9), reversed-phase HPLC (10), and electrospray mass spectrometry (11). However, more technological development is required before these methods can be offered to hemoglobinopathy laboratories on a routine basis.

When an unknown variant is found, attempts at isolation and characterization are made. To localize and characterize the mutation, it is invariably necessary to cleave the individual protein chains to provide fragments small enough to have their amino acid composition readily determined or to be sequenced. Proteolysis by trypsin, the most common technique, was used for characterization of the first variant, Hb S, in 1956 (12,13). Trypsin cleaves at the carboxyl side of the basic amino acids, arginine and lysine, so 15 β chain fragments (βT-1–βT-15), 14 α-chain fragments (αT-1–αT-14), 16 γ-chain fragments (γT-1–γT-16), and 16 δ fragments (δT-1–δT-16) can theoretically be expected (see Fig. 4.11 and 4.12).

Although trypsin is the most common proteolytic agent used, other enzymatic and chemical methods have also been used as secondary techniques in variant hemoglobin identification, often to further reduce the size of tryptic fragments. Endoproteinase Glu-C (EC 3.4.21.19) from *Staphylococcus aureus* V8 (SV-8), for example, cleaves peptides at the carboxylic acid side of glutamyl residues; α-chymotrypsin (EC 3.4.21.1) cleaves peptides at the carboxyl side of tryptophan, tyrosine, phenylalanine, leucine, and methionine residues; and CNBr attacks the peptide bond C-terminal of methionine residues.

In initial studies, the separation of tryptic peptides was achieved by two-dimensional paper chromatography combined with paper electrophoresis, so that in the preparative mode, abnormal peptides could be isolated for sequencing (12,13). Later, ion-exchange chromatography as described by Schroeder *et al.* (14) proved a useful interim method for preparative separation of tryptic peptides, but it was the introduction of HPLC (utilizing reversed-phase columns) to this field in 1979 that provided the universal method that continues to this day (15). The elution pattern produced by reversed-phase HPLC for tryptic peptides of globin chains has become so well known and consistent in variant characterization that usually only one or two peptides with unusual HPLC mobility require identification. Until 1981, when mass spectrometry was introduced to the field (16), variant tryptic peptides could be characterized only by determining their amino acid composition and sequence.

## 2. *MASS SPECTROMETRY IN HEMOGLOBIN ANALYSIS: A CHRONOLOGY*

Most possible single-nucleotide mutations in coding triplets give rise to amino acid changes that are reflected in changes in the molecular weight of the proteolytic fragments. A table can be constructed of mass changes that result from such amino acid replacements (Table 4.1). In addition to mass differences in proteolytic peptides caused by amino acid replacement, specific substitutions or deletions (for instance lysine or arginine in the case of tryptic digestion) also affect molecular mass through the creation of peptides of abnormal length during proteolytic digestion. Because most variants have altered molecular weights ($M_r$), a technique such as mass spectrometry, which can readily measure mass differences, is ideally suited for analysis of proteolytic fragments.

In 1981, Matsuo, Wada, and co-workers first used this technique for hemoglobin analysis using field desorption (FD) (16). In that initial research, they produced spectra of nonseparated tryptic peptides from normal α, β, γ, and δ chains and recorded molecular cations representing many of these peptides. They readily demonstrated the $m/z$ 952 → $m/z$ 922 shift for the $MH^+$ peptide βT-1 of sickle-cell hemoglobin reflective of the 30-Da decrease caused by replacement of glutamic acid by valine at the sixth position of the β-globin chain. That research was the first demonstration of variant identification by mass spectrometry, albeit the most common one. It is fortunate that these early studies coincided with the introduction of the successful fast atom bombardment (FAB) technique for mass spectrometric peptide analysis because it is unlikely that the difficult FD method would have been widely used for clinical variant hemoglobin analysis. In the early 1980s, the Osaka group had the field to themselves and soon published the characterization of a new hemoglobin, subsequently called Hb Meilahti (β36 Pro→Thr), primarily by using the FAB technique (17). Use of the allied desorption method liquid secondary ion mass spectrometry (LSIMS) with a CsI source for hemoglobin characterization was first reported by them in 1983 (18). Hayashi and co-workers used mass spectrometry as part of a neonatal screening program in Osaka Prefecture (19). Five different γ-chain variants were detected among 140,000 newborns, one of which (Hb F-Yamaguchi) was found in 40 cases. These five variants all had

*Table 4.1*
*Mass Differences between Amino Acid Residues Alterable through a Single-Nucleotide Mutation*

| Nominal mass difference (Da) | Amino acid change | Nominal mass difference (Da) | Amino acid change | Nominal mass difference (Da) | Amino acid change |
|---|---|---|---|---|---|
| 0 | Ile→Leu | 18 | Leu→Met | 42 | Gly→Val |
| | Gln→Lys | | Ile→Met | 43 | Leu→Arg |
| 1 | Lys→Glu | 19 | His→Arg | | Ile→Arg |
| | Ile→Asn | 22 | Asp→His | 44 | Ala→Asp |
| | Asn→Asp | 23 | Asn→His | | Cys→Phe |
| | Gln→Glu | 24 | Leu→His | 46 | Gly→Cys |
| 3 | Lys→Met | 25 | Met→Asn | 48 | Val→Phe |
| 4 | Pro→Thr | 26 | His→Tyr | | Asp→Tyr |
| 9 | Gln→His | | Ala→Pro | 49 | Asn→Tyr |
| 10 | Ser→Pro | | Ser→Leu | 53 | Cys→Arg |
| 12 | Thr→Ile | 27 | Ser→Asp | 55 | Thr→Arg |
| | Thr→Leu | | Thr→Lys | 58 | Gly→Asp |
| 13 | Thr→Asn | 28 | Lys→Arg | | Ala→Glu |
| 14 | Gly→Ala | | Ala→Val | 59 | Pro→Arg |
| | Ser→Thr | | Gln→Arg | 60 | Ser→Phe |
| | Val→Leu | 30 | Val→Glu | | Cys→Tyr |
| | Asp→Glu | | Arg→Trp | 69 | Ser→Arg |
| | Asn→Lys | | Thr→Met | 72 | Gly→Glu |
| 15 | Leu→Gln | | Gly→Ser | 73 | Leu→Trp |
| | Ile→Lys | | Ala→Thr | 76 | Ser→Tyr |
| 16 | Val→Asp | 31 | Pro→Gln | 83 | Cys→Trp |
| | Ser→Cys | 32 | Val→Met | 99 | Ser→Trp |
| | Ala→Ser | 34 | Leu→Phe | | Gly→Arg |
| | Phe→Tyr | | Ile→Phe | 129 | Gly→Trp |
| | Pro→Leu | 40 | Pro→His | | |

mutations resulting in a change of net charge and, because the primary screening technique used involved the isoelectric focusing (IEF) of globins, it was not surprising that no new hemoglobins with the neutral amino acid substitutions were detected. It is to be anticipated that many more variant hemoglobins would be present in such a large sample of newborns.

The next research groups to enter the field were those of Pucci in Naples, Rahbar in Duarte, California, and Castagnola in Rome. In 1985, Pucci and co-workers (20) described the identification by a combination of FAB and Edman degradation a variant in which lysine replaced glutamine at position 121 of β globin, thus creating a new site for tryptic cleavage. The variant had been previously identified and was known as O-Arab. In this study, the abnormal β chain was purified by CM-52 cellulose ion-exchange chromatography, and no isolation of individual peptides was performed. Rahbar *et al.* (21) made a comprehensive study using HPLC for separation of both α- and β-globin chains and tryptic digests, finalizing the identification by FAB MS. They characterized four α-chain and six β-chain variants.

Castagnola *et al.* (9,22) characterized two known variants, Hb D-Los Angeles and Hb J-Sardegna, using FAB analysis of reversed-phase HPLC purified abnormal peptides combined with a limited acid hydrolysis. They later emphasized the use of mass spectrometry in addition to sophisticated electrophoretic methods for the characterization of variants that were "silent" in routine electrophoretic techniques (9). Their key to finding such variants was the use of second-generation isoelectric focusing procedures, i.e., immobilized pH gradients (IPGs), which have greater resolving power than do standard techniques. Having shown the presence of a variant, these workers characterized the hemoglobin by proteolysis of a mixture of globin chains and FAB MS.

Another important group that started to use mass spectrometry in the late 1980s was that of Promé and co-workers in Toulouse. They discussed optimization of MS and tandem MS techniques for characterization of hemoglobin tryptic fragments (23) and demonstrated the use of a collisionally activated dissociation mass-analyzed ion kinetic energy (CAD-MIKE) analysis for sequencing the variant peptide in Hb Grenoble (24). This group has been particularly active in characterizing normal but minor hemoglobin fractions produced by posttranslational modifications such as glycerated hemoglobin (25) and adult hemoglobin $A_{1b}$, characterized as pyruvyl hemoglobin (26).

In 1989, the introduction of electrospray (ES) allowed for the first time the measurement of the molecular masses of intact proteins (27,28). When applied to the identification of variant hemoglobins, this technique was also the first use of quadrupole instruments in this field; all previous studies had been carried out on single and tandem magnetic sector mass spectrometers. The first publication using ESMS for hemoglobin characterization was by Green *et al.* (11). In this paper we demonstrated the analysis of normal adult and fetal hemoglobin and several hemoglobin variants. It was shown that, in most cases, globin mixtures could be satisfactorily analyzed, although separation of the fetal globins $^{G}\gamma$ ($M_r$ 15,995) and $^{A}\gamma$ ($M_r$ 16,009) was marginal. Poor resolution was caused by peak broadening due to the presence of several major isotopic species for each globin species. This study demonstrated the high accuracy of molecular mass determination that was achievable. Measurement of the molecular mass of normal β globin gave a value of 15,866.6 ± 0.3 Da for three determinations, an accuracy greater than 0.01% (Table 4.2). Green *et al.* (11) also utilized FAB on a quadrupole instrument for the identification of Hb G-Philadelphia and tandem mass spectrometry (MS/MS) on a magnetic sector instrument for the sequencing of tryptic peptides produced from variant hemoglobins. The Kratos Con-

*Table 4.2*
*Molecular Weights Measured in Electrospray ESMS and Expected Molecular Weights of Globin Chains*

| Globin assignment | Measured MW (Da) | Expected MW (Da) | Error (Da) | *n* |
|---|---|---|---|---|
| $\alpha$ | 15,125.7 ± 0.7 | 15,126.4 | −0.7 | 3 |
| Glycated $\alpha$ | 15,288.0 ± 0.2 | 15,288.5 | −0.5 | 2 |
| $\beta^A$ | 15,866.5 ± 0.3 | 15,867.2 | −0.7 | 3 |
| Glycated $\beta^A$ | 16,028.5 ± 0.7 | 16,029.4 | −0.9 | 2 |
| $G_\gamma$ | 15,995.8 ± 0.6 | 15,995.3 | +0.5 | 3 |
| $A_\gamma$ | 16,009.6 | 16,009.3 | +0.3 | 1 |
| $\beta^S$ | 15,838.2 ± 0.5 | 15,837.2 | +1.0 | 3 |

cept instrument (with an array detector) gave unambiguous product ion spectra, confirming the amino acid sequence. Through use of the ESMS, FAB, and MS/MS instruments, hemoglobins D-Iran, G-Philadelphia, and Lepore Washington–Boston were characterized; we believe that this study was the first time a Lepore-type hemoglobin had been identified by mass spectrometry. Hemoglobin Lepore is a class of structurally abnormal hemoglobins composed of normal $\alpha$ chains and abnormal non-$\alpha$ chains, the latter being hybrid proteins that contain the N-terminal sequence of the $\delta$ globin and the C-terminal sequence of the $\beta$ globin. Different types of Lepore are known, depending on where the fusion between $\delta$ and $\beta$ sequences occurs. De Caterina *et al.* (29) have recently published an identification of Lepore-type hemoglobins (Lepore-Boston and Lepore-Baltimore) by ESMS of the intact globins and LSIMS mass spectrometry of tryptic peptides. Electrospray MS was also used in the characterization of Hb Thionville by Vasseur *et al.* (30). This novel $\alpha$-chain variant has glutamate substituted for valine at the $\alpha 1$ position and an acetylated methionine N-terminus.

We recently utilized ESMS and MS/MS in a study that illustrates several important features of hemoglobin variant characterization:

1. the clinical importance of these studies,
2. mass spectrometry in the detection of electrophoretically silent mutations,
3. the limits of resolution of ESMS for separation of globin mixtures, and
4. distinguishing between two possible mutations in the same peptide by a four-sector array detector instrument (31).

We were alerted to a problem when a patient with sickle cell disease gave an electrophoretic pattern indicative only of sickle cell trait, a heterozygous condition normally without clinical symptoms. It thus seemed probable that what we assumed was normal Hb A by electrophoresis was in fact a "silent" mutant. When we analyzed hemolysates of the patient and her mother by ESMS, we obtained the transformed spectra illustrated in Fig. 4.1. It was immediately evident that, in addition to $\beta^S$, an additional variant was present of molecular mass 15,879 Da, 12 Da more than expected for the normal β chain (Fig. 4.1). Following S-aminoethylation, tryptic digestion, and HPLC analysis, it was shown that the S-aminoethylated βT-10 had altered mobility and had increased in mass by 12 Da, indicating substitution of Thr→Leu or Ile (Table 1). Given the threonine codons used in this region of β globin, only a Thr→Ile conversion could occur (assuming a single base mutation), but the problem was complicated by βT-10 having two threonine residues. Sequencing the βT-10 peptide by MS/MS was described in detail in the publication of Falick *et al.* (32), which proved that it was the β87 threonine that had been replaced. The abnormal hemoglobin was named Hb Quebec-Chori after the Canadian prov-

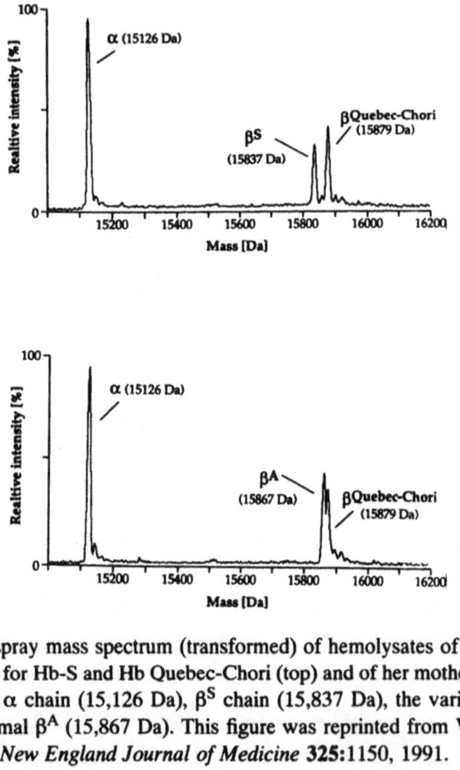


*Figure 4.1.* Electrospray mass spectrum (transformed) of hemolysates of a patient with a compound heterozygosity for Hb-S and Hb Quebec-Chori (top) and of her mother (bottom). Illustrated are peaks for normal α chain (15,126 Da), $\beta^S$ chain (15,837 Da), the variant Hb Quebec-Chori (15,879 Da), and normal $\beta^A$ (15,867 Da). This figure was reprinted from Witkowska *et al.* (31), by permission of the *New England Journal of Medicine* **325:**1150, 1991.

ince where the patient lived and the Children's Hospital Oakland Research Institute, where the characterization was achieved. Analysis of the heterozygote mother by ESMS (Fig. 4.1) illustrates well the "maximum" resolution of the technique. The mass differential of 12 Da between normal β and $\beta^{Quebec\text{-}Chori}$ allows only partial separation of the two compounds.

The successful conclusion of the Hb Quebec-Chori story illustrates the importance of hemoglobin variant identification and the role that mass spectrometry can play. From researchers in the field, there are many ancedotal descriptions of symptomatic sickle cell trait patients. In many cases, a second but unfound variant may be present containing a silent mutation, and only by exhaustive studies using the sophisticated methodologies described here will these disorders be satisfactorily interpreted.

Ferranti and co-workers (33) principally utilizing ESMS, have also recently characterized a silent variant. This variant was found during screening for hemoglobinopathies on Sardinia. It was an α-chain variant, so this chain was isolated and digested by trypsin. A major difficulty in analysis was due to the fact that the mutation was in αT-9, one of the largest α-chain peptides obtained by tryptic digestion and representing amino acids 62–90. The mass of the peptide had increased from 2995 to 3023 Da, indicating that only the amino acid changes Ala→Val, Lys→Arg, and Gin→Arg could have occurred. The αT-9 peptide does not contain glutamine, so the Gln→Arg conversion could be excluded. However, this peptide contains seven alanines available for conversion to valine, and the lysine β90 could have been changed to an arginine. Secondary hydrolysis with CNBr was used to cleave the peptide bond C-terminal to the methionine at position 76. This time, the anomalous peptide (as determined by FAB-MS) contained amino acids 62–76, but even this peptide still contained four alanines (at position 63, 65, 69, and 71), each of which could be replaced by a valine. Secondary hydrolysis of αT-9 (residues 62–90) with pepsin showed the presence of an anomalous peptide of mass 1209 Da, which was 28 Da more than peptide 70–80, thus proving that this peptide contains the mutation. Because this peptide has only one alanine (at position 71), the position of the mutation was confirmed. This identification is a textbook example of the use of sequential and selective proteolytic methods for narrowing down the mutation site. The mass of the primary tryptic digestion product (3023 Da) was too high for MS/MS sequencing, but had this technique been available, confirmatory sequencing could probably have been achieved after the CNBr hydrolysis through sequencing of the 62–76 peptide rather than by employing the third digestion with pepsin. This new hemoglobin was called Hb Ozieri (α71 Ala→Val).

Covey and co-workers (34,35) showed that ESMS was a viable technique for sequencing tryptic peptides of hemoglobin in addition to its more obvious use in intact protein analysis. In an early study, they demonstrated that these peptides tend to give, principally, doubly charged ions (Fig. 4.2), and when subjected to CID in triple quadrupole instruments, fragmented mostly to y-mode product ions. It is evident that, of all mass spectrometric techniques, ESMS shows the most promise for semi-routine and cost-effective hemoglobin analysis, particularly when it is combined with HPLC in a single HPLC/MS instrument. Molecular masses of on-line HPLC-separated globin chains as well as tryptic peptides have been determined this way (36). The limitations of ESMS in sequencing principally relate to the mass accuracy achievable because we have not been confident in confirming masses of product ions to greater than

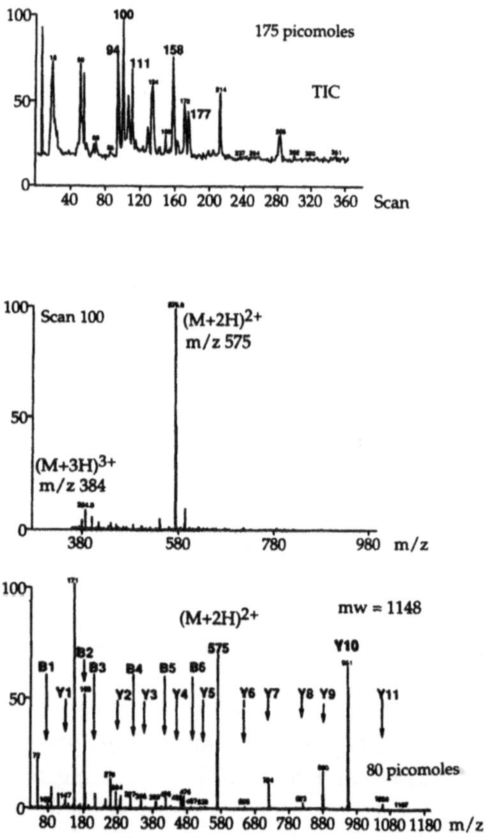


*Figure 4.2.* The original example of microbore HPLC-MS/MS used for the characterization of a hemoglobin tryptic peptide. The upper panel is the HPLC TIC chromatogram; the middle panel, the spectrum of βT-14, showing the dominant doubly charged molecular ion; and the bottom panel, the MS/MS spectrum, illustrating the sequence information obtained through production of y-mode and b-mode ions. Reproduced with permission of the American Society of Mass Spectrometry, from Covey *et al* (34).

± 1-amu accuracy, leaving some ambiguity in the sequence determined. At the present time, MS/MS on a sector instrument is still the only foolproof method of mass spectrometric sequencing. Details of our success with this technique have been described in the publications by Falick *et al.* (32,37).

Roepstorff and co-workers (38–40) have utilized plasma desorption (PD)/time of flight (TOF) instruments for analysis of proteolytic fragments. Optimized conditions for *S. aureus* V8 digestion of β globin with PDTOF gave a peptide map covering 100% of the protein structure in a single experiment (39,40). The mass measurement accuracy of PDTOF is inferior to that of LSIMS or ESMS; it is equivalent to 0.05% for peptides in the 300–2000 mass range, so the PDTOF technique is unlikely to be widely used.

The structure of recombinant hemoglobins has also been studied with the aid of mass spectrometry. We applied ESMS and LSIMS to the characterization of hemoglobins assembled in transgenic mice carrying human α and $\beta^{S\text{-Antilles}}$-genes (41), and the Japanese group used FAB to check the structure of recombinant artificial mutants of α and β globins expressed in *E. coli* (42,43). Recently, Coghlan *et al.* took advantage of ESMS to confirm the fidelity of expression of normal human hemoglobin in *Saccharomyces cerevisiae* (44).

We conclude this section with Table 4.3, summarizing the hemoglobin variants that have been characterized using mass spectrometry in our laboratory or by other workers. More than 10% of known hemoglobin variants have now been additionally characterized by mass spectrometry, although only 12 new variants have been identified exclusively by this technique. It is obvious, however, that this number will increase with an accelerating identification of electrophoretically silent variants. A more comprehensive review of mass spectrometry in hemoglobin characterization has been published by Wada *et al.* (45), and Oliver and Green (46) have reviewed the use of ESMS in this field.

## 3. *CURRENT STRATEGIES AND METHODS*

### 3.1. *Mass Spectrometry of Intact Hemoglobin*

#### 3.1.1. *Quadrupole ESMS*

Electrospray mass spectrometry of hemoglobin gives rise to a series of molecular ions of different charge states (11). Because the maximum number of charges corresponds approximately to the number of basic residues, the maximum charge state obtained is 24+ for the β chain (3 arginine, 11 lysine, 9 histidine, and the amino terminus), although under common instrument conditions the highest charge state seen is about 21 and lowest about 11 (Fig.

*Table 4.3*
*Hemoglobins Characterized by Mass Spectrometry*

| Hemoglobin | Mutation | Mass differential (Da) | Reference[a] |
|---|---|---|---|
| α-Chain variants | −1 +1 | | |
| Thionville[b] | α1 extension Ac-Met-Glu | 173 | 30 |
| J-Toronto | α5 Ala→Asp | 44 | 21 |
| I | α16 Lys→Glu | 1 | S |
| Fort Worth | α27 Glu→Gly | −72 | 21 |
| Torino | α43 Phe→Val | −48 | 9 |
| Hasharon | α47 Asp→His | 22 | 39, S |
| Montgomery | α48 Leu→Arg | 43 | S |
| J-Sardegna | α50 His→Asp | −22 | 22 |
| Shimonoseki | α54 Gln→Arg | 28 | 45 |
| M-Boston | α58 His→Tyr | 26 | 110 |
| G-Waimonalo | α64 Asp→Asn | −1 | P |
| G-Philadelphia | α68 Asn→Lys | 14 | 11, P |
| Ube II | α68 Asn→Asp | 1 | 111 |
| Ozieri[b] | α71 Ala→Val | 28 | 33 |
| Stanleyville II | α78 Asn→Lys | 14 | 78 |
| J-Broussais | α90 Lys→Asn | −14 | S |
| J-Cape Town | α92 Arg→Gln | −28 | 38 |
| Setif | α94 Asp→Tyr | 48 | 21 |
| Manitoba | α102 Ser→Arg | 69 | S |
| Tarrant | α126 Asp→Asn | −1 | 21 |
| Chicago | α136 Leu→Met | 18 | 112 |
| Hanamaki | α139 Lys→Glu | 1 | R |
| β-Chain variants | −1 +1 | | |
| South Florida[b] | β1 extension Ac-Met-Met | 205 | 113 |
| Deer Lodge | β2 His→Arg | 19 | S |
| S | β6 Glu→Val | −30 | 16,40,74,78 |
| C | β6 Glu→Lys | −1 | 40,74,110 |
| Siriraj | β7 Glu→Lys | −1 | 114 |
| G-San Jose | β7 Glu→Gly | −72 | 37,78,84,115 |
| J-Baltimore | β16 Gly→Asp | 58 | S |
| Olympia | β20 Val→Met | 32 | 21 |
| D-Iran | β22 Glu→Gln | −1 | 11,37,112 |
| G-Coushatta | β22 Glu→Ala | −58 | 45,111 |
| E | β26 Glu→Lys | −1 | 38,110 |
| Volga | β27 Ala→Asp | 44 | S |
| Meilahti[b] | β36 Pro→Thr | 4 | 17,81 |
| North Chicago[b] | β36 Pro→Ser | 10 | 58, S |
| Alabama | β39 Gln→Lys | 0 | R |
| Hammersmith | β42 Phe→Ser | −60 | 21 |
| Louisville | β42 Phe→Leu | −34 | M |
| Hoshida | β43 Glu→Gln | −1 | 111 |
| Grenoble[b] | β51 Pro→Ser<br>β52 Asp→Asn | −11 | 24,112 |

*Table 4.3. (Continued)*

| Hemoglobin | Mutation | Mass differential (Da) | Reference[a] |
|---|---|---|---|
| Willamette | β51 Pro→Arg | 59 | 37,78 |
| J-Bangkok | β56 Gly→Asp | 58 | S |
| Hikari | β61 Lys→Asn | −14 | 45 |
| M-Saskatoon | β63 His→Tyr | 26 | M, S |
| Bristol | β67 Val→Asp | 16 | M |
| Great Lakes | β68 Leu→His | 24 | 21 |
| City of Hope | β69 Gly→Ser | 30 | 21,116 |
| Korle-Bu | β73 Asp→Asn | −1 | 32 |
| Aalborg[b] | β74 Gly→Arg | 99 | 117 |
| J-Iran | β77 His→Asp | −22 | 112 |
| Miyazano[b] | β79 Asp→Glu | 14 | 111 |
| Baylor | β81 Leu→Arg | 43 | 74 |
| Providence | β82 Lys→Asn→Asp | −14 (−13) | 23,54,112 |
| Quebec-Chori[b] | β87 Thr→Ile | 12 | 31,32 |
| D-Ibadan | β87 Thr→Lys | 27 | 118 |
| Vanderbilt | β89 Ser→Arg | 69 | S |
| M-Hyde Park | β92 His→Tyr | 26 | 84 |
| Newcastle | β92 His→Pro | −40 | M |
| Moriguchi | β97 His→Tyr | 26 | S |
| Köln | β98 Val→Met | 32 | 21,112, P, S |
| Kempsey | β99 Asp→Asn | −1 | P |
| Heathrow | β103 Phe→Leu | −34 | 119, S |
| Camperdown | β104 Arg→Ser | −69 | 40 |
| San Diego | β109 Val→Met | 32 | 7 |
| Indianapolis | β112 Cys→Arg | 53 | 65,86 |
| New York | β113 Val→Glu | −30 | S |
| Hafnia | β116 His→Gln | 9 | 38 |
| P-Galveston | β117 His→Arg | 19 | 36 |
| Fannin–Lubock | β119 Gly→Asp | 58 | S |
| Riyadh | β120 Lys→Asn | −14 | 120 |
| O-Arab | β121 Glu→Lys | −1 | 20,65, S |
| D-Punjab | β121 Glu→Gln | −1 | 22,23,112 |
| Ty Gard | β124 Pro→Gln | 31 | 38,112 |
| Yamagata | β132 Lys→Asn | −14 | 50 |
| Hope | β136 Gly→Asp | 58 | S, R |
| Andrew–Minneapolis | β144 Lys→Asn | 14 | M |
| McKees Rocks | β145 Tyr→Termination | −300 | 21,24,112, S |
| Tak | β146 frameshift | 1299 | S |
| γ-Chain variants | | | |
| F-Izumi (Kotobuki) | $^{A}\gamma$6 Glu→Gly | −72 | 18 |
| F-Iwata | $^{A}\gamma$72 Gly→Arg | 99 | 19 |
| F-Yamaguchi | $^{A}\gamma$T80 Asp→Asn | −1 | 121 |

*(continued)*

*Table 4.3. (Continued)*

| Hemoglobin | Mutation | Mass differential (Da) | Reference[a] |
|---|---|---|---|
| γ-Chain variants | | | |
| F-Fuchu[b] | $^{G}\gamma$21 Glu→Gln | -1 | 19 |
| F-Minoo[b] | $^{G}\gamma$72 Gly→Arg | 99 | 121 |
| δ-Chain variants | | | |
| $A_2$ Indonesia | δ69 Gly→Arg | 99 | 45 |
| $A_2$ Honai[b] | δ90 Glu→Val | 30 | 45,122 |
| Fusion hemoglobins | | | |
| Lepore Baltimore | δ50 . . . β86 | -45 | 29 |
| Lepore Washington–Boston | δ87 . . . β116 | -2 | 11,29 |
| P-Nilotic | β22 . . . δ50 | 89 | S |

[a]Unpublished results from various laboratories are indicated by M (Molchanova *et al.*), P (Pucci *et al.*), R (Rahbar *et al.*), and S (Shackleton *et al.*).
[b]Indicates hemoglobin variant originally characterized with the aid of mass spectrometry.

4.3). In the molecular ion envelope formed by electrospray, adjacent peaks differ by only one charge. Therefore, the charge state of each of those ions can be easily calculated and hence the molecular mass of the species derived. Averaging the results of multiple *m/z* measurements of differently charged molecular ions contributes to the very high accuracy of ESMS. In early studies using a prototype of the Fisons VG BioQ instrument, we demonstrated the accuracy with which molecular masses could be determined for globin chains by the electrospray technique. These results, shown in Table 4.2, illustrate an error of less than 1 Da for, typically, three measurements. Data processing methods were soon developed to transform the authentic spectra consisting of a series of multiply charged ions to spectra representing the average molecular mass of species on a molecular mass scale, facilitating interpretation (as in Fig. 4.1). In our studies, the best resolution obtained between two globins is 12 Da, as demonstrated in the spectrum of the new hemoglobin Hb Quebec-Chori, whose β chain differs from normal by this value (31). With this doublet, only 25% of baseline separation was achieved for the two proteins present in approximately equal amounts. Nevertheless, the masses of the two β globins can be measured accurately from the peak tops. This separation is close to the minimum that can be practically analyzed without resorting to mathematical deconvolution.

In the transformed spectra of globins produced by quadrupole ESMS

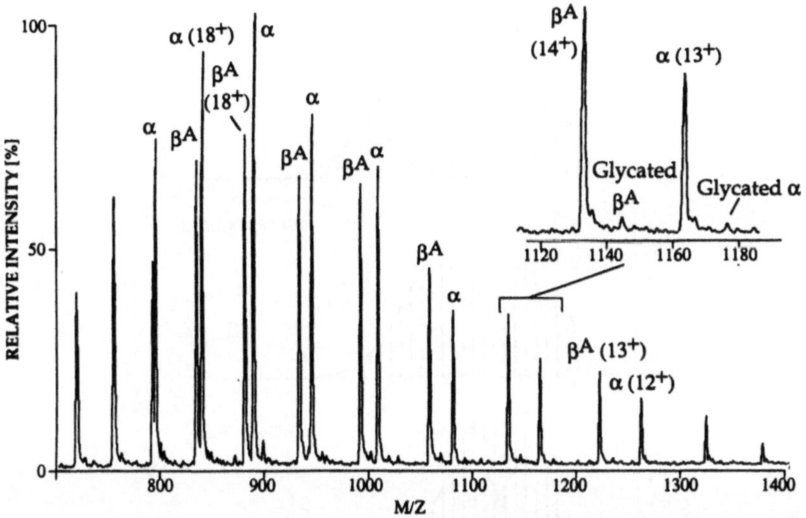


*Figure 4.3.* Electrospray mass spectrum of the hemolysate of an adult. This nontransformed spectrum shows multiply charged ions (from $11^+$ to $21^+$) of α and β globin. The $\alpha13^+$, $\beta14^+$ region has been expanded (inset) to illustrate the detection of the minor glycated hemoglobins. Reproduced with permission of the Elsevier Science Publishers BV, Academic Publishing Division, from Green *et al.* (11).

instruments, each molecular ion peak represents a composite of the various isotope species present, a composite that, in the absence of any broadening from the instrument itself, would have a theoretical width at half height of 8 Da (Fig. 4.4). It is this isotope effect that mainly limits the ability to separate two globins close together in mass. With state-of-the-art quadrupoles, the instrumental contribution is small for the globins and increases the isotopic width only from 8 Da to a little less than 10 Da, giving rise to the practical limit, without deconvolution, of 12 Da referred to above. Skilling and co-workers (47) have shown that a marked improvement in apparent resolution can be achieved by using the maximum entropy (MaxEnt®) software developed for the VG Bio-Q instrument. The MaxEnt® program creates and repeatedly processes a series of trial spectra and compares the results with the experimental data. The trial spectrum that best agrees with the data—i.e., that which is most probable—is presented as the final MaxEnt® spectrum. By also incorporating the isotopic and instrumental broadenings into the program, MaxEnt® is able to deconvolute those factors from the data, thus enhancing the resolution observed in the final spectrum. When this technique is used, the $\beta^{\text{Quebec-Chori}}$ peak is baseline-resolved from normal β (Fig. 4.5). The 9-Da differential between $\beta^{\text{Hafnia}}$ and normal β is almost baseline resolved, but the limit is approached at

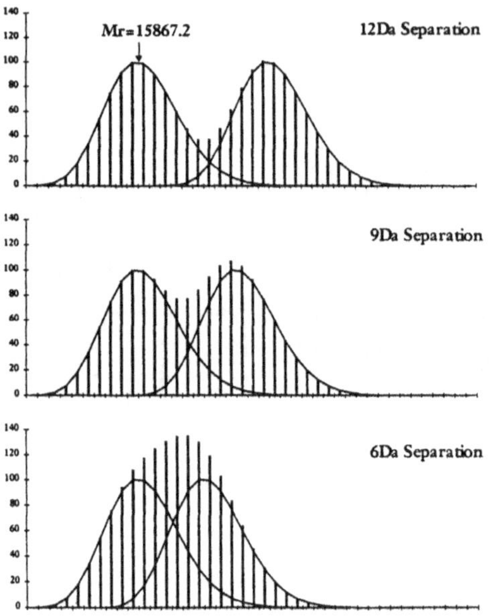


*Figure 4.4.* The limits of MS resolution of β globins. This figure, produced by Brian Green and colleagues (personal communication), shows the theoretical isotope distribution of normal β globin and $\beta^{\text{Quebec-Chori}}$ (upper panel), normal β and $\beta^{\text{Hafnia}}$ (middle panel), and normal β and a variant of 6 Da greater $M_r$ (lower panel). The vertical lines represent the isotopic intensities of the doublets as a whole, assuming no mass-spectrometric peak broadening. These examples illustrate the futility of expecting higher-resolution instrumentation to significantly improve the ability to resolve and measure the masses of two globins of closely related mass. However, computer deconvolution techniques (such as MaxEnt®) can greatly aid their resolution. It is of note that peak tops of globins of mass differential 9 Da (middle panel) are slightly pulled together (by about 1 Da). Problems in variant hemoglobin mass determination caused by this apparent convergence were noted in Ref. 78.

6 Da, as demonstrated by the 50%-to-baseline separation noted for the sheep and cow α-globin doublet (15,047 and 15,053 Da, respectively).

Using this type of instrumentation, it is clear that chromatographic or electrophoretic separation will remain essential in some cases for determining the molecular masses of closely mass-related proteins in mixtures.

*Method.* Hemolysates were prepared from blood by washing erythrocytes three times with normal saline, lysing packed red cells with 1–2 volumes of water, and removing stroma by centrifugation at 16,000 *g*. Globins were dehemed by treating 10–20 μl of hemolysates with 1.5 ml cold acetone/0.7% HCl, without mercaptoethanol (48). Precipitated protein was washed three times with cold acetone and dried at room temperature.

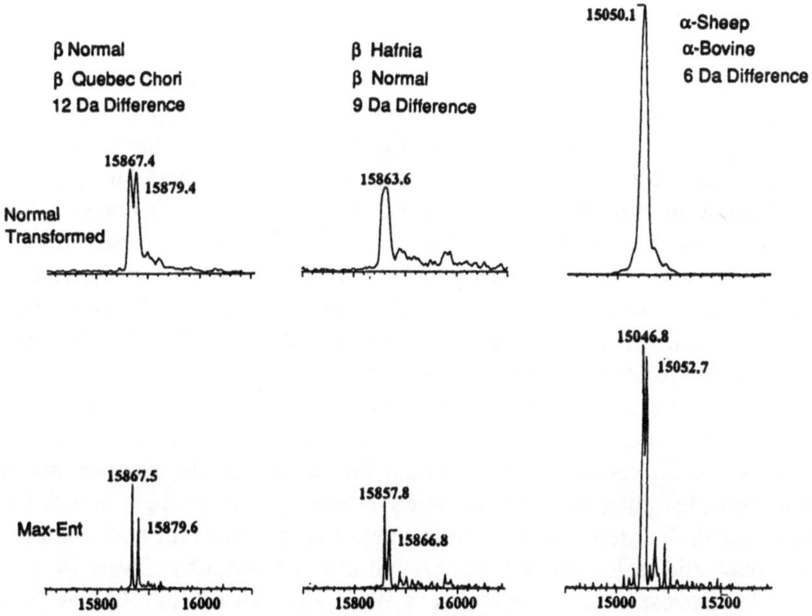


*Figure 4.5.* Improved apparent electrospray mass-spectral resolution through employment of the MaxEnt® program (47). In each set of panels, the upper panel contains the transformed data and the lower panel contains the data obtained by MaxEnt® processing. In the left-hand panel, baseline resolution of Quebec-Chori β globin and normal β (12 Da apart) is illustrated, although under normal conditions, these two are only slightly resolved. In the middle panel, the unresolved doublet normal $\beta/\beta^{Hafnia}$ (9 Da apart) is almost fully separated when the MaxEnt® program is used. In the right-hand panel, bovine and ovine α globins (6 Da apart) can be partially resolved and mass measured with MaxEnt®. Thus, 6 Da appears to be close to the maximum attainable resolution for proteins of this molecular mass. These spectra were obtained by B. Green of the Fisons VG Biotech Company, who granted permission for their reproduction.

Our mass spectrometry studies are carried out on a VG BioQ mass spectrometer with an ES ion source (VG BioTech/Fisons, Altrincham, England). A blood hemolysate taken up in 50% methanol:1% acetic acid is passed directly into the ion source at 3 μl/min. The mass spectrometer is scanned from *m/z* 970 to 1420 in 10 sec. For mass scale calibration, horse heart myoglobin (16,951 Da) is used, though α globin (15,126.4 Da) can alternatively serve as a useful internal calibrant in most cases (36).

### *3.1.2. Magnetic Sector Mass Spectrometers*

Recently, Larsen and McEwen (49) suggested advantages in resolution and mass accuracy when ESMS is carried out on double-focusing mass spectrometers. In theory, these advantages could be important when variant hemoglobins are being analyzed because globins of close molecular mass can be more easily distinguished in mixtures and more trust could be placed in the molecular mass determination of individual globins or tryptic peptides that differ by 1 Da. When Wada and co-workers (50) used ES magnetic sector mass spectrometry for analyzing Hb Yamagata, a β-chain variant with a 14-Da differential from normal, the separation was comparable to that achieved with quadrupoles. At an instrumental resolution of 2000, they were able to obtain 50% of baseline separation, equivalent to our 25% of baseline separation for Hb Quebec-Chori, which has a 12-amu differential from normal. In fact, although the isotope distribution of the molecular ion envelope of globins dictates a minimum separation of about 10 Da before deconvolution becomes necessary (Fig. 4.4), little improvement can be expected from magnetic sector instruments. Even with an instrument resolution of 16,000, some form of deconvolution would be necessary to determine the masses of the components of a doublet of less than about 9 Da separation. However, it should be stressed that, in practice, final resolution can be dramatically improved for both quadrupole and magnetic sector instruments through use of MaxEnt® software (47).

Of interest in the contribution of Wada and co-workers (50) was the use of the ES magnetic sector MS technique for measurement of the molecular masses of cyanogen bromide α-chain peptides. This analysis was carried out at a resolution of 5000, and CB1 and CB2 had masses of 3297.7 and 4757.3 Da, respectively. Under the conditions of analysis, it was possible to resolve isotopic peaks in the mass spectrum of each molecular ion, as shown in Fig. 4.6. Solely on the basis of those isotopic distributions, the assignment of molecular ion charge state could be made. The superior resolution of magnetic sector mass spectrometers might prove extremely useful when sequencing large peptides or intact proteins by collisionally induced dissociation in the ion source (51).

### *3.2. Separation of Intact Hemoglobins*

Cation- and anion-exchange chromatography were both used for separation of normal components of hemolysates and isolation of abnormal hemoglobins. Huisman and Wilson (52) discussed correlations between the site of mutation and the mobility of variant hemoglobins in cation-exchange HPLC. We have employed cation-exchange HPLC according to Wilson (53) on either analytical (Synchropak CM 300, SynChrom, Inc., Lafayette, IN) or semipreparative TSK CM-5 PW (Toyo Soda Manufacturing Co., Ltd., Tokyo, Japan) columns for isolation and further characterization (by reversed-phase HPLC, Triton urea gel electrophoresis, and mass spectrometry) of hybrid hemoglobins produced in transgenic mice expressing human α and $\beta^{S\text{-}Antilles}$globins (41). Wada *et al.* (54) used DEAE-52 (Whatman Scientific Ltd., Kent, England)

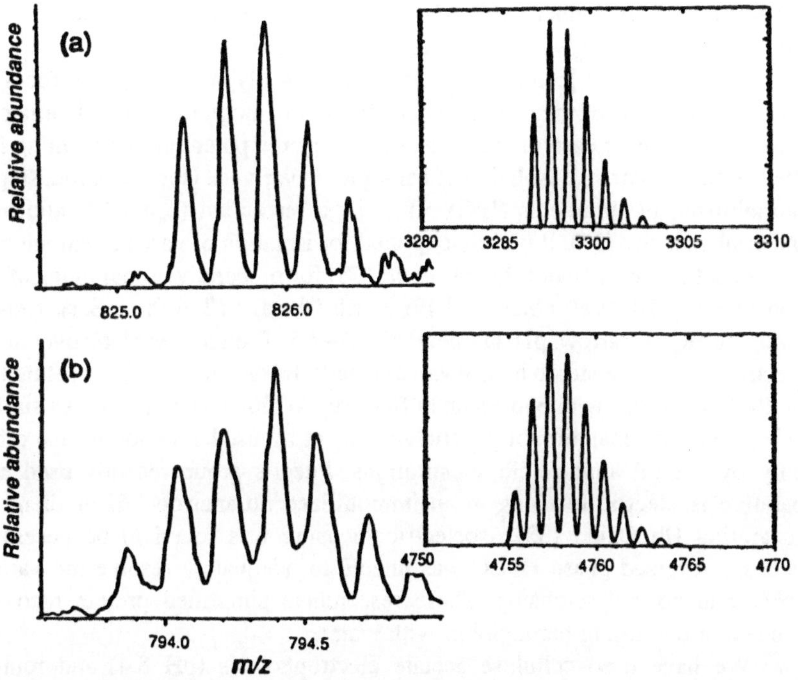


*Figure 4.6.* Electrospray mass spectra obtained on a magnetic sector instrument. ES mass spectra of cyanogen bromide peptides CB1 and CB2 of hemoglobin α chain at a resolution of 5000: (a) quadruply charged CB1 and (b) sextuply charged CB2. The insets to the right represent theoretical isotope distributions at the same resolution. Reproduced with permission of John Wiley and Sons, Ltd., from Wada *et al.* (50).

chromatography according to Abraham and co-workers' method (55) for isolation of Hb Providence variants, prior to globin-chain separation and LSIMS of tryptic fragments using a grand-scale double-focusing mass spectrometer constructed at Osaka University. Fast protein liquid chromatography (FPLC, with apparatus from Pharmacia, Sweden) was used by Pucci and co-workers for intact hemoglobin isolation (29). The method was applied for the separation of Lepore hemoglobins, as is illustrated in Fig. 4.7.

Although they are not strictly associated with variant hemoglobin identification, mention should be made of the problems of isolating the minor hemoglobin fractions (Hb $A_{1c}$, Hb $A_{1b}$, and $Hb_X$), which are present in blood at very low levels (e.g., $HbA_{1b}$<0.5%). Promé and co-workers (25, 26) used a combination of ion-exchange methods including Bio-Rex 70 chromatography (56) for separating minor components from Hb A, DEAE 52 chromatography (55) for separating $Hb_X$ (glycerated hemoglobin) from Hb $A_{1c}$ (glycated hemoglobin), and an Amberlite CG-50 column for purification of the Bio-Rex 70–isolated Hb $A_{1b}$ (pyruvylated hemoglobin) (26).

Electrophoresis [cellulose acetate of pH 8.4 (2)] and isoelectric focusing [IEF, (4)] are part of the routine repertoire of a hemoglobinopathy laboratory. Due to the relative ease of their use, they carry the potential of becoming fast methods of preparative isolation of hemoglobin variants. Isoelectric focusing in immobilized pH gradients (IPG) offers high resolution ($\Delta$pI 0.01) and high material capacity (57). It was first applied for isolation of variant hemoglobins by Rochette *et al.* (7) and Rahbar *et al.* (58) for preparative separations of Hb San Diego ($\beta$105 Val→Met) and Hb North Chicago ($\beta$36 Pro→Ser), respectively, using a narrow pH gradient of 7.1–7.5. Proteins were recovered by electrophoretic transfer to hydroxyapatite beds, from which they were eluted by multiple washings with phosphate buffer (59). Whitney *et al.* (8) used a similar gel system for separation of genetic variants of mouse hemoglobins that differ only by neutral amino acid substitutions. Pucci's group recently used preparative isoelectric focusing in an immobilized ultranarrow pH gradient for separating Hb Ozieri (33). Isoelectric focusing was found to be necessary because reversed-phase HPLC was unable to adequately resolve the variant $\alpha^{Ozieri}$ and normal $\alpha$ chains. These researchers simplified protein recovery procedure by eluting hemoglobin with water.

We have used cellulose acetate electrophoresis (pH 8.4) and routine isoelectric focusing (mobile pH gradient) for variant hemoglobin separation and preparative isolation. Cellulose acetate electrophoresis is the method of choice for the preparative isolation of a hemoglobin variant whenever good

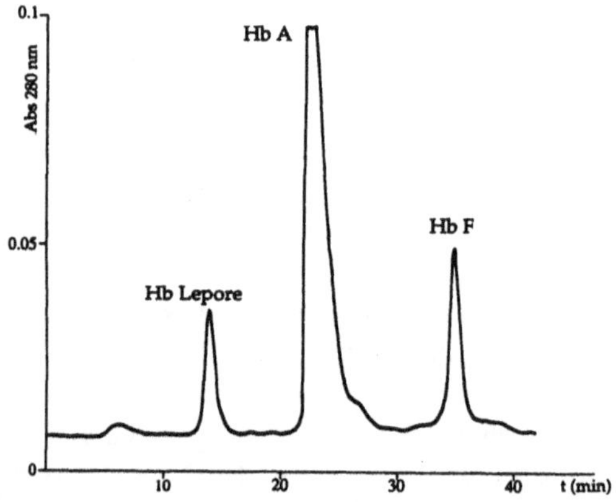


*Figure 4.7.* Fast protein liquid chromatography (FPLC) of intact hemoglobins from a patient with Lepore-type hemoglobin. This sample from an infant shows Hb Lepore, Hb A, and fetal hemoglobin. Reproduced with permission of the American Association for Clinical Chemistry, Inc., from De Caterina *et al.* (29).

separation from other hemolysate components can be achieved. Depending on circumstances, isolated hemoglobins are either subjected only to ESMS analysis for a final confirmation of their provisional identity or undergo full structural characterization. Normally, the amount of material isolated by a single electrophoretic step is enough to accomplish full structural analysis, including MS/MS sequencing, if required. Fig. 4.8 shows the example of an ES mass spectrum of Hb H isolated by cellulose acetate electrophoresis. The presence exclusively of β chains in this sample confirms the identity of Hb H and is diagnostic for α-thalassemia. Mass spectrometric analysis of hemoglobins isolated by isoelectric focusing allows for a rapid confirmation of the identity of abnormal bands detected in routine screening. We have demonstrated the feasibility of the method by performing ESMS of Hb A and Hb S isolated from agarose isoelectric focusing gels. Even though full removal of ampholines was not achieved in case of Hb S, the useful *m/z* range of over 950 Da is devoid of any interference (compare transformed spectra of Hb A and Hb S, Fig. 4.8 insets). Occasionally, trifluoroethanol or its mixtures with 50% MeOH/1% AcOH (1:1) were found to provide higher sensitivity in ESMS analysis of gel-isolated hemoglobins. Hemoglobins are eluted from gels with water, am-

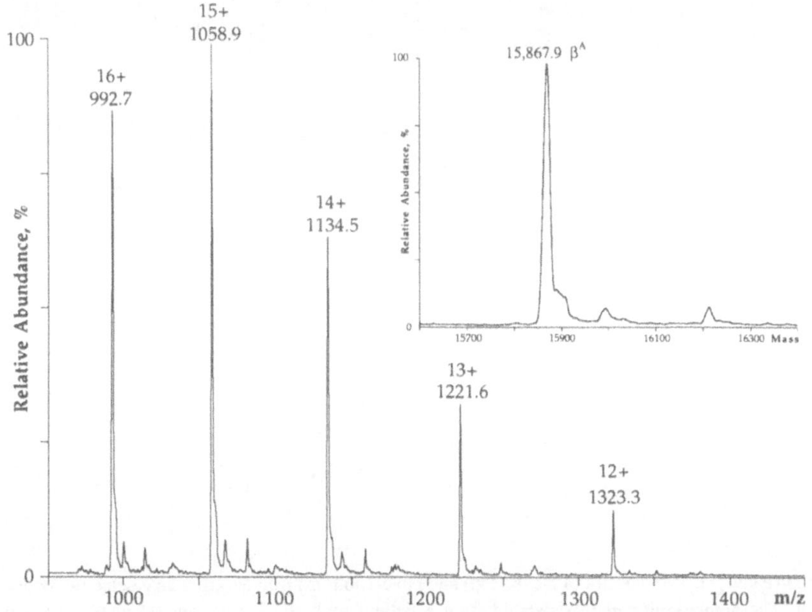


*Figure 4.8.* **ESMS spectrum of Hb H extracted from an electrophoresis strip, multiply charged species and (inset) transformed. This hemoglobin is a tetramer of four β-globin chains, and as expected, no α globins ($M_r$ 15,126) were detected in the spectrum.**

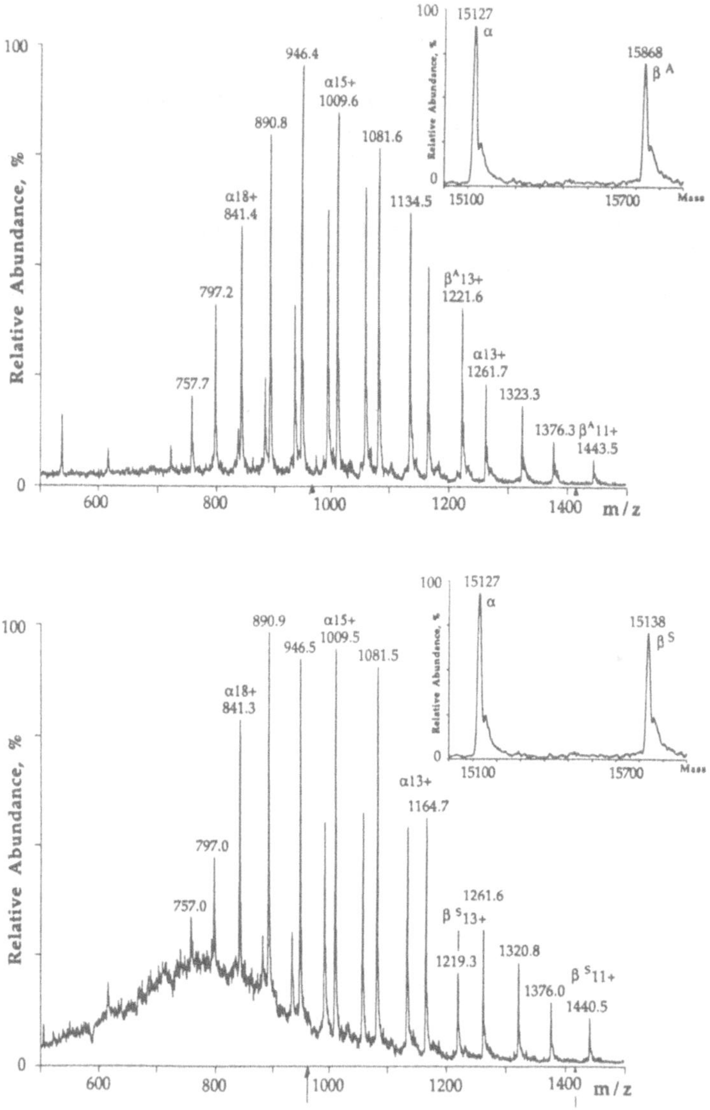


*Figure 4.9.* ESMS spectra of Hb-A (top) and Hb-S (bottom) isolated from an isoelectric focusing gel. The hemoglobins were separated from a patient with sickle cell trait (Hb-A/S); therefore, normal hemoglobin was also present in the sample. No evidence of cross-contamination was seen, illustrating the specificity of band elution. Transformation was performed within the *m/z* range indicated by the arrows (970–1420 Da) to give transformed spectra (insets).

pholines present in IEF preparations are removed by ultrafiltration through an Amicon Centricon 10, and if necessary hemoglobin solutions are concentrated on a SpeedVac® centrifuge before ESMS analysis. This method allows for an efficient recovery of major and minor components resolved by cellulose acetate electrophoresis and major components separated by isoelectric focusing. Recovery of minor bands isolated from IEF gels still requires methodological improvement.

*Method I: Fast Protein Liquid Chromatography.* The hemolysate is diluted to an Hb concentration of ~4 mg/ml with 50-mM Tris-HCl buffer (pH 8.5) containing KCN (0.1 g/l), and 500 μl is loaded onto a Mono Q column (Pharmacia, Sweden) previously equilibrated with the same buffer. The protein samples are eluted over 40 min using a linear gradient from 100% buffer A (50 mM Tris-HCl, pH 8.5) to 75% buffer B (50 mM Tris-HCl, pH 6.5), each containing KCN (0.1 g/l). Individual hemoglobins were collected manually (29).

*Method II: Cellulose Acetate Electrophoresis Preparative Method.* In our protocol (60), 1–4 μl of hemolysate (60–100 mg/ml) is applied to a cellulose acetate gel and electrophoresis is performed as described by Kim *et al.* (2). Hemoglobin bands are excised, placed in 1.5 ml Eppendorf tubes, soaked in water at room temperature for 1 hr, and spinned down in an Eppendorf centrifuge. The resulting supernatant is removed and filtered through a Nylon Acrodisc®4 (4 mm, 0.45 μm) filter (Gelman Sciences) if it appears turbid. A supernatant that still shows orange-reddish color can be analyzed directly by ESMS after a 1:1 dilution with MeOH/2% AcOH or trifluoroethanol. When the hemoglobin concentration is low, it is better to concentrate the solution on a SpeedVac® centrifuge prior to ESMS analysis. For further structural studies, globins are dehemed using an acetone/0.7% HCl treatment of supernatants concentrated to 20–50 μl volumes (48). Occasionally, the presence of a polymer with a monomer subunit of 44 Da is detected in the *m/z* range up to ~1000 Da. However, because we normally use a mass range of 970–1420 Da for variant hemoglobin mass measurement, the presence of a few low-intensity signals derived from this impurity does not interfere with the analysis.

*Method III: Isoelectric Focusing Preparative Method.* We perform isoelectric focusing of 1–3 μl hemolysate (diluted to 15–25 mg/ml) using a pH gradient of 6–8 (Isolab ampholines) on Isolab agarose plates according to the manufacturer's protocol. Hemoglobin elution from excised bands is achieved with water, as described for cellulose acetate isolated samples. Supernatants must be filtered through a Nylon Acrodisc®4 (4 mm, 0.45 μm) filter (Gelman Sciences) and are applied to an Amicon Centricon 10. Samples are concentrated to a 50-μl volume, washed with 0.7–1 ml water, and concentrated again; the process of washing and concentration is repeated twice. Preparation of samples for ESMS is performed as described above (60).

Pucci's method (33) is as follows: Isoelectric focusing in an ultranarrow immobilized pH gradient of 7.1–7.5 is performed on a lysate (60–80 mg/ml). The lysate (20 μl) was pipetted into 0.3 × 15 × 5 mm wells. After focusing, the portion of the gel containing the Hb tetramer was recovered with a spatula and eluted from the gel in a microcentrifuge tube with 200 μl of distilled water in a few hours. Following centrifugation, the hemoglobin solution was dried in a Savant SpeedVac® centrifuge (Savant Instruments, Farmingdale, NY) and subjected to reversed-phase HPLC analysis for separation of the variant α chain.

## *3.3. Preparative Globin Chain Separation*

Preparative globin chain separations have been carried out by chromatography using ion-exchange or reversed-phase columns. Given the limited material requirements of mass spectrometry, it is likely that existing *analytical* procedures—cellulose acetate electrophoresis (36, 61, 62), Triton acid urea polyacrylamide gels (6), or isoelectric focusing in the presence of Nonidet P-40 (63) or in immobilized pH gradients 9—can and will be adapted for isolation purposes, if they happen to offer any advantages over the abovementioned established chromatographic methods.

Carboxymethylcellulose (CM) chromatography as described by Clegg *et al.* (64) was a primary technique used in early studies and still remains a valuable method in many laboratories. The method separates denatured globins and employs a linear gradient of phosphate buffer in 8-M urea/50-mM mercaptoethanol on CM-cellulose. Samples are desalted prior to further analysis, either by dialysis or by using a Sep-Pak C18 cartridge (65). FPLC on a Mono S column was recently introduced by Pucci for globin-chain separation (65).

Early applications of reversed-phase HPLC for human globin separation employed a variety of supports and developing systems (10, 66, 67). With the introduction of large-pore reversed-phase columns, almost all current methods of reversed-phase HPLC of globins are adaptions of those described by Shelton and co-workers (68), who employed a Vydac C-4 protein column (pore size 300 Å). This method offers excellent resolution from a narrow-bore to a semipreparative column scale. In a comparative study, Rahbar and Asmerom (69) found it superior to other ion-exchange or reversed-phase alternatives as applied to the quantification of labeled globins in globin biosynthesis studies. For variant globin separation, we use this method exclusively. A chromatogram produced by us using chromatography on a semipreparative C-4 column is shown in Fig. 4.10. The sample whose separation is illustrated was obtained from cord blood, and in addition to normal constituents, the elution positions of a few variants are also indicated.

*Method.* For analytical purposes (36), we use a C-4 Vydac column (0.46 × 25 cm, 5 μm, 300 Å); the Separations Group, Hesperia, CA). Automated analysis is carried out at room temperature using the solvent system and elution gradient according to Shelton *et al.* (68). Hemolysates are diluted with water to provide a protein concentration at 0.25–0.5 mg/ml (typically 200-fold), and 100 μl samples are injected every 70 min. No more than 12 samples are kept in the autosampler at one time, to diminish any deteriorating effect of ambient temperature on hemolysate storage. For preparative purposes, an analytical size column can be used for up to ~0.3 mg of material or a semipreparative column (11) (Vydac 1 × 25 cm) can be employed to allow for separations of up to 2–3 mg of hemoglobin. In our hands, certain β globin variants co-eluted with heme on this column, and therefore we prefer to apply dehemed globin mixtures rather than hemolysates. The Shelton solvent system is used: [A, acetonitrile:water (20:80 v/v)/0.1% TFA; B acetonitrile:water (60:40 v/v)/0.09% TFA]; flow rate 1.7 ml/min; linear gradient 44% B to 65% B in 60 min, then to 60% B in 16 min; detection at 220 nm. For conditions of narrow-bore reversed-phase chromatography of globin chains, refer to Section 3.6.

## 3.4. Digestion

Proteolytic digestions can be undertaken either on globin mixtures or on separated globin chains. Increasingly, with an improvement of resolution of HPLC columns and the introduction of on-line HPLC/MS techniques, separation of the globins is less necessary. An extensive discussion of different

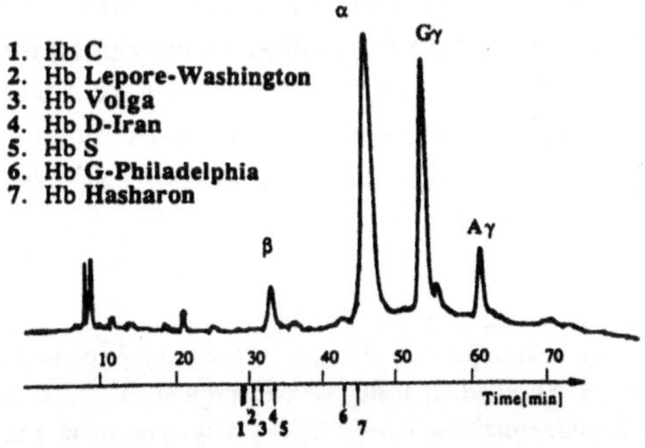


*Figure 4.10.* Preparative RP HPLC separation of precipitated globin from cord blood (250 μg). On the time axis, arrows with numbers indicate the elution time of globin chains of several variants. The column was a C4 Vydac (1 × 25 cm), flow was 1.7 ml/min, detection was at 220 nm, and the linear gradient was from 44% B to 56.6% B in 60 min, then to 60% B in 16 min. A: acetonitrile:water (20:80)/0.1% TFA, B: acetonitrile:water (60:40)/0.09% TFA. Reproduced with permission of the Elsevier Science Publishers BV, Academic Publishing Division, from Green *et al.* (11).

digestion methodologies as applied to hemoglobin can be found in the review by Wada *et al.* (45), and fragments obtained from α, β, γ, and δ chains by the most specific proteolytic methods are shown in Fig. 4.11 and 4.12.

### *3.4.1. S-Cysteine Derivatization and Recovery of Cysteine-Containing Peptides*

For effective recovery and analysis of cysteine-containing peptides, it is advisable to derivatize the thiol group before proteolytic digestion. Oxidation of cysteine to cysteic acid as well as alkylation are used, and aminoethylation has been applied most frequently. Aminoethylation of cysteines was introduced to produce smaller, less hydrophobic peptides upon trypsinization. S-aminoethylated cysteine is a close structural analog of lysine, and it is recognized as such by trypsin to deliver an additional proteolytic site at its C-terminal peptide bond. αT-12 and βT-12 peptides are both cleaved, producing αT-12.1 ($MH^+$, 615.8 Da), αT-12.2 ($MH^+$, 2413.9 Da), βT1-2.1 ($MH^+$, 872.1 Da), and βT-12.2 ($MH^+$, 907.1 Da). There are mixed reports about detection of aminoethylated αT-12.2, though its presence was reported in FAB maps of nonfractionated digests (70) and in reversed-phase HPLC–separated fractions (39). In our hands, αT-12.2 was recovered from a C-18 HPLC column, albeit with rather low yield, emerging right before the αT-8.9 peptide. Aminoethylated βT-10 ($MH^+$ 1464.7 Da) peptide does not normally undergo tryptic digestion, probably because the intrinsically low activity of trypsin toward modified cysteine is additionally undermined by the presence of the C-terminal aspartic acid residue. However, βT-10.2 ($MH^+$, 1222 Da) was once detected in the mixture of nonfractionated tryptic digest derived from the $\beta^{\text{North Chicago}}$ globin (58). We follow the method of Schwartz *et al.* (71) for aminoethylation of globins, using N-[β-iodoethyl) trifluoroacetamide] instead of ethyleneimine, which is banned in the United States. Wada and co-workers (45) give a detailed procedure for carboxymethylation and comment that this derivatization, when combined with a tandem tryptic/endoproteinase Lys-C digestion, allows for detection of a

*Figure 4.11.* The amino acid sequences of human α (top) and β (bottom) globins, illustrating the peptide fragments to be expected in using several proteolytic techniques (T, trypsin; L, endoproteinase Lys-C; S, endoproteinase Glu-C; CB, cyanogen bromide; and C, cyanylation/hydrolysis). Average molecular masses of most of the peptides are given, and the masses of cysteine-containing peptides assume the conversion to carboxyamidomethylated (CAM) derivatives (+57 Da), except for the C-2 peptide which contains 2-iminothiazolidine-4-carboxylic acid at the N-terminus. Masses of cyanogen bromide–generated peptides were calculated assuming generation of the homeserine residue. For α and β globin, the combined peptides shown for trypsin are those commonly found in our studies. The combined SV8 peptides are those reported by Jensen *et al.* (32,40).

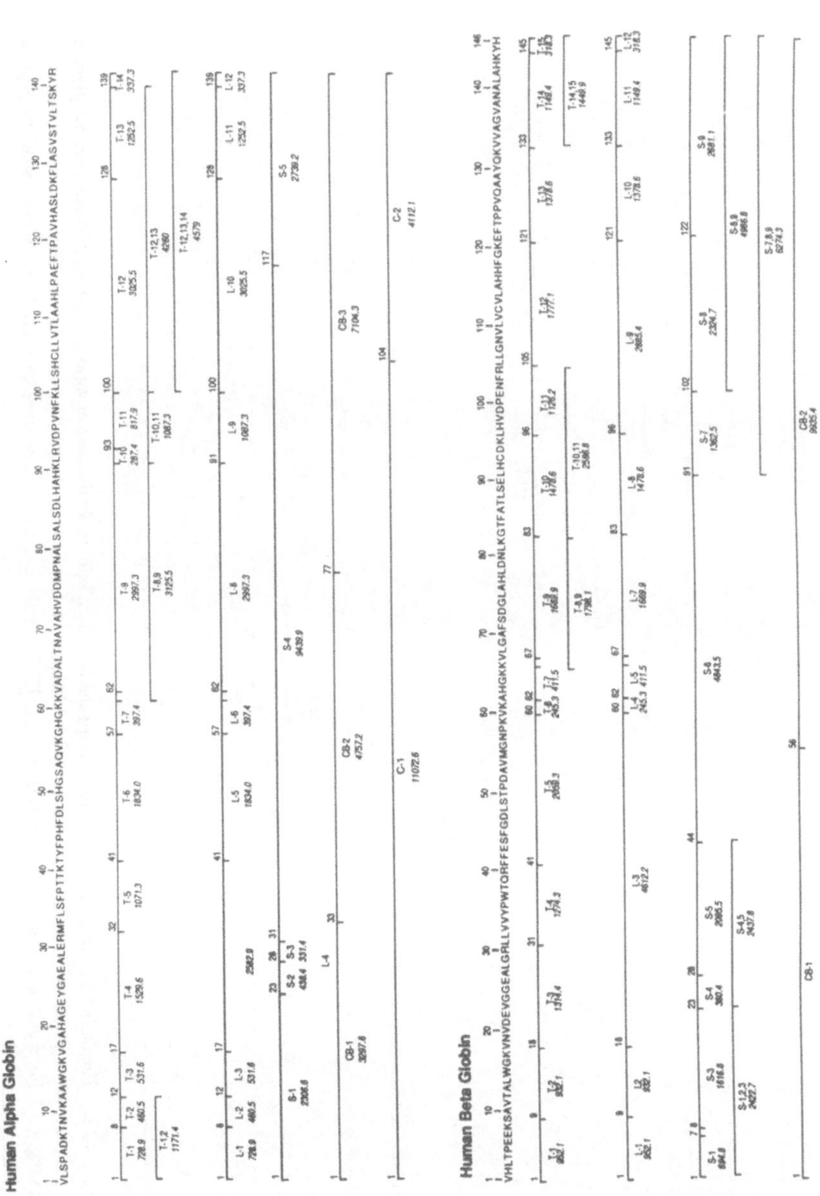
Human Alpha Globin
Human Beta Globin

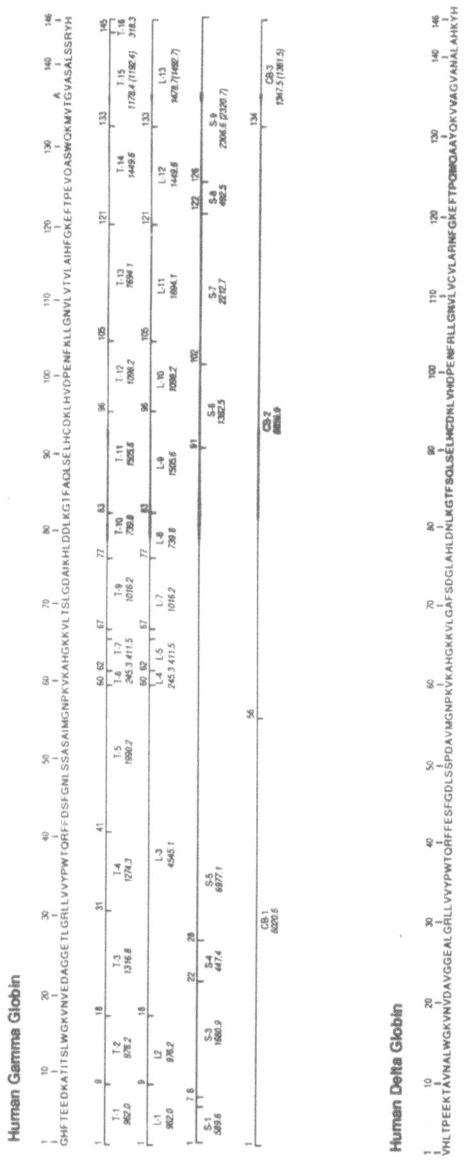


*Figure 4.12.* The amino acid sequences of human γ (top) and δ (bottom) globin. In the case of γ globin, residue 136 can be either glycine ($G_\gamma$) or alanine ($A_\gamma$); both are illustrated. The proteolytic techniques, molecular mass calculation, and cysteine derivatization are the same as in Fig. 4.11. For γ globin, two masses for peptides T-15, L-13, S-9, and CB-3 are given because the mass depends on whether the peptide originates from $^{G}\gamma$ or $^{A}\gamma$.

"core" αT-12 peptide by FAB of an unfractionated mixture of proteolytic fragments.

Jensen and Roepstorff (40) applied pyridylethylation to the analysis of β globin. Oxidation of cysteines to cysteic acid is applied to enhance the solubility of core peptides. This operation can be performed on globin before tryptic digestion (21), allowing the recovery of an oxidized αT-12 peptide from a reversed-phase column, or it can be applied to an insoluble fraction of tryptic peptide mixture before secondary proteolytic digestion (15). Until recently, not only was insolubility a problem in analysing the α-core region, but the high molecular mass (~3000) of the peptides also was at the limit for LSIMS analysis. This latter obstacle has essentially been alleviated by the introduction of ESMS, although it is still necessary to further fragment the peptides to produce smaller ones suitable for sequencing. Recently, we have adapted carboxyamidomethylation of cysteines for the analysis of β-chain variants, and we follow the fast alkylation/trypsinization procedure of Stone *et al.* (72). Recovery of carboxyamidomethylated peptides of the β chain is good, and there is evidence that carboxyamidomethylated αT-12 peptide can be recovered from a reversed-phase C-18 column as a hybrid peptide αT-12,13 ($MH^+$, 4258.3 Da) and αT-12,13,14 ($MH^+$, 4577.5 Da) (see Section 3.6.2 and Fig. 4.19).

*Method: Carboxyamidomethylation of Globins (36).* The procedure described by Stone *et al.* (72) is followed with slight modifications, i.e., ammonium bicarbonate (0.4 M) in 6M guanidine hydrochloride rather than in 8-M urea is used. Initial reduction/alkylation reaction volumes are: 0.4-M ammonium bicarbonate buffer 7.5 μl or 20 μl; 45-mM DTT 1 μl or 2 μl; 100-mM iodoacetamide 1 μl or 2 μl, for 0.5–2 nmol and 2–10 nmol protein, respectively. Buffers are degassed with nitrogen before use.

### *3.4.2. Trypsin (EC 3.4.21.4)*

Trypsin [general reference: Kostka and Carpenter (73)] is most universally used for hemoglobin digestion because it produces few very small or very large fragments, most being of a size suitable for the determination of molecular mass by LSIMS and many able to be sequenced using the tandem mass spectrometric technique (Figs. 4.11 and 4.12). Trypsin attacks the peptide bonds at the carboxyl side of the basic amino acids lysine and arginine. Certain bonds are poorly cleaved, particularly the lysine 126 in the α chain, resulting in the production of a very large and relative insoluble core peptide αT-12,13. In our hands, lysine 95 in the β chain is also not efficiently cleaved, resulting in a high proportion of the βT-10,11 peptide, which we analyze in its S-carboxyamidomethylated form. A novel approach to the analysis of tryptic digests has been proposed by Stachowiak and Dyckes (74), who performed a digestion with trypsin that was immobilized to glycophase-controlled pore-glass beads (Pierce Chemical, Rockford, IL) followed by HPLC and on-line thermospray mass spectrometry in the negative-ion mode. All α-and β-chain peptides were detected, including the αT-12 core, and recovery of cysteine-containing peptides was facilitated by globin pretreatment with performic acid. Immobilized trypsin was found to cleave preferentially arginine in a peptide containing both lysine and arginine amino acids, a result consistent with findings of Craik *et al.* on the higher selectivity of trypsin toward the former amino acid residue (75).

Residual chymotryptic activity of trypsin preparations very often results in the appearance of a small amount of chymotryptic peptides. Cleavages at β37 Trp and β130 Tyr were reported (58,76). We very often see a peptide α81-90 due to a cleavage at α80 Leu (unpublished observation). The abundance of chymotryptic fragments in tryptic digests seems to be very dependent upon the enzyme batch and reaction conditions: Rahbar *et al.* (58) saw exclusively chymotryptic cleavage at β35 Tyr in Hb North Chicago β36 Pro→Ser, whereas in our hands the same variant produced the expected tryptic fragment (unpublished results). Incidentally, the mutation in $\beta^{\text{North Chicago}}$ occupies the site whose chymotryptic sensitivity was exploited by Wada in his identification of Hb Meilahti (β36 Pro→Thr) (see Section 3.4.4). Apart from chymotryptic fragments, we very often observe an unexplained appearance of the deamidated βT-14 peptide, which contains Asp in place of normally present Asn at position β139 (77).

*Method.* We utilize the following proteolysis procedures: Nonderivatized globins are trypsinized with L-1-tosylamide-2-phenylethyl chloromethylketone–treated (TPCK) trypsin (Worthington) (6 μg per 150 μg protein) in 50-mM ammonium bicarbonate (pH 8.0–8.1) overnight at 37 °C with stirring (78). Half of the trypsin is added initially, followed by the second half within 30–60 min from the start of digestion. Digestion is stopped by acidifying to pH 3.0 with glacial acetic acid, and mixtures are stored at −20 °C. Trypsinization of S-carboxyamidomethylated globins is carried out according to Stone and co-workers (72) in the presence of an excess of alkylation reagents and denaturing buffer, upon 3.3-fold dilution with water using the enzyme-to-protein ratio and time and temperature conditions as described above (36).

### *3.4.3. Endoproteinase Lys-C (EC 3.4.99.30)*

Endoproteinase Lys-C [Wako Pure Chemicals, Richmond, VA; general reference: Jekel *et al.* (79)] specifically attacks lysyl but not arginyl bonds and has a higher specific activity than trypsin. Its use can therefore improve the recovery of those peptides that are notoriously resistant to trypsin. It proved useful in the analysis of the αT-12 core peptide and is generally superior to trypsin in the characterization of the α chain (45). It is also of value when tryptic digestion produces very small peptides, which are not easily amenable to LSIMS analysis. A good example of the advantage of its higher degree of specificity is its application to the characterization of $HbA_2$ Indonesia (δ69 Gly→Arg) (45). In this study, the electrophoretic pattern suggested that the mutant hemoglobin had a more basic nature than normal, implying the incorporation of either an arginine or a lysine residue. Mass spectrometry showed that the δT9 tryptic peptide (67–82) lost 269 amu, an amount too large for a single amino acid substitution, thus suggesting the incorporation of a new hydrolytic site. The replacement peptide had a mass of 1400 Da, equivalent to the mass of residues 70–82, proving that this sequence was normal. A mutation in residues 67–69 would have a mass of 358 if the substituent were lysine and 386 Da if it were arginine, but fragments of this small size could not be detected by FAB. Therefore, a method was utilized that gave greater selectivity in digestion, resulting in the production of larger peptides; i.e., endoproteinase Lys-C. In this case, the abnormal peptide found by FAB had a mass consistent with the substitution of arginine and not lysine at position 69.

The use of lysyl endopeptidase will surely become more common with the introduction of ESMS, a technique well able to analyze the larger peptides produced by this enzyme.

*Method.* The digestion with endoproteinase Lys-C (Wako Pure Chemicals, Richmond, VA) is carried out in 50-mM ammonium bicarbonate (pH 8.4) or in 50-mM Tris-HCl (pH 9.0). The globin chains were S-carboxymethylated prior to proteolysis (45).

### 3.4.4. *Chymotrypsin (EC 3.4.21.1)*

As previously mentioned, chymotrypsin [general reference: Hummel (80)] is used to further fragment the large peptides (particularly from α globin) produced during tryptic digestion. Chymotrypsin cleaves at the carboxyl side of tryptophan, tyrosine, phenylalanine, leucine, and methionine residues. In Wilson's studies (15), the insoluble αT-12,13 core peptide is oxidized by performic acid prior to proteolysis. HPLC analysis of the chymotryptic digest of αT-12,13 showed that seven peptides were produced, each containing between three and seven residues (see Section 3.3.11 and Fig. 4.14).

Wada *et al.* (81) utilized chymotrypsin in the characterization of the new hemoglobin Hb Meilahti (β36 Pro→Thr). Initial FAB results on tryptic peptides suggested substitution of the proline at position 36 by threonine. They proved this substitution by the particular selectivity of α-chymotrypsin. The residues that must be present adjacent to the cleavage site have been listed above, but if the following residue is a proline, then cleavage does not occur. Thus, normal β globin would not be cleaved at this point, but cleavage would occur if proline had been changed to threonine. Mass spectrometry of the chymotryptic digest showed the presence of peptide β 33–35 [$MH^+$, 380] and β 36(Thr)–37 [$MH^+$ 306], showing that cleavage had taken place, thus proving the mutation.

*Method.* In Wada's studies (81), chymotryptic digestion of globin was carried out in 0.05-M ammonium bicarbonate (pH 7.8) at 37 °C for 1 hr. The bovine pancreatic enzyme can be obtained from Boehringer, Fluka, Sigma, or Calbiochem.

The detailed experimental procedure of performic acid oxidation and chymotryptic digestion of the core α-globin peptide is given in Huisman and Jonxis (82). Wilson *et al.* (15) isolated the αT-12,13 core peptide as follows: following tryptic digestion of underivatized α globin at pH 8.5 to 8.0, the pH is adjusted with dilute HCl to 6.3–6.5, and the insoluble core peptide is isolated by centrifugation. The precipitate is washed with distilled water to remove traces of soluble peptides.

### *3.3.5. Endoproteinase Glu-C (*S. aureus *V8 EC 3.4.21.9)*

In 50-mM ammonium bicarbonate (pH 7.8), endoproteinase Glu-C [general reference: Houmard and Drapeau (83)] cleaves peptide bonds C-terminal to glutamyl residues, so theoretically five, nine, nine, and eight peptides can be expected for α, β, γ, and δ globin chains, respectively (Figs. 4.11 and 4.12). *S. aureus* V8 digestion is mostly used for β-globin variants. Wada *et al.* (45) showed a LSIMS mass spectrum of a nonfractionated endoproteinase Glu-C digestion mixture of S-carboxymethylated β globin where six of the nine expected peptides were detected. Roepstorff's group performed *S. aureus* V8 digestion followed by reversed-phase HPLC on S-pyridylethylated β globin and were able to measure by PDMS all of the expected fragments, thus mapping 100% of the β-globin sequence in the single step (38). They also described the application of the *S. aureus* V8 digestion/PDMS protocol for identification of Hb S, C, E, Hafnia, and Ty Gard. However, in most cases, further subdigestion of a *S. aureus* V8–created fragment was necessary to prove the sequence unequivocally. Pucci *et al.* (84) used the technique to facilitate the characterization of Hb Hyde Park, in which tryptic peptide βT-9 (83–95) indicated a 26-Da increase in mass, suggesting β86 Ala→Pro, β89 Ser→Ileu/Leu, β92 His→Tyr, or β93 Cys→Glu substitutions. *S. aureus* V8 proteolysis resulted in the production of peptides containing residues 80–90 and 91–101, each containing two of the possible mutation sites. The latter peptide showed an appropriate increase in mass, so that the mutation was localized to this region. Finally, distinction between the two mutation sites in this peptide was achieved by two steps of Edman degradation. Promé *et al.* (24) also used *S. aureus* V8 to prove the structure of Hb M$^c$Kees Rocks, in which the loss of the two C-terminal residues 145–146 was more clearly seen in the large βS-9 peptide than in the small βT1-5 peptide.

This enzyme is less versatile than trypsin because some peptides are much larger than can be analyzed by LSIMS or FAB, although the introduction of ESMS may alleviate this problem to some degree. One α-chain peptide would have a molecular mass of 9435, and four β-chain peptides have masses between 2000 and 5000 Da.

*Method.* The S-pyridylethylated β chain (30–40 μg) is dissolved in 20 μl of 8-M urea and diluted to a final concentration of 2.2-M urea with 50 μl 0.2-M ammonium bicarbonate (pH 7.7). *S. aureus* V8 (ICN Immunobiologies, Costa Mesa, CA) (4 μg) is added (enzyme:substrate mole ratio 1:10). The sample is incubated for 4 hr at 37 °C, and proteolysis is terminated by acidifying to pH 2 with TFA (38).

### *3.4.6. Carboxypeptidase B (EC 3.4.17.2)*

Mass spectrometry was used for the characterization of Hb Indianapolis [β112 Cys→Arg] by Pucci's group (85). Tryptic digestion of the whole hemoglobin showed that βT-11 (105–120) was missing and was replaced by two new peptides containing residues 105–112 and 113–120. It was thus likely that Cys112 had been replaced by arginine. The structure of these peptides was confirmed by an elegant procedure where one-step Edman degradation of the whole tryptic mixture resulted in the removal of Leu105 and Val113 from the aberrant peptides, and following carboxypeptidase B hydrolysis [general reference: Folk (86)], showed a 156-amu loss from the 106–112 peptide, indicating removal of arginine112.

*Method.* Carboxypeptidase B (Sigma, St. Louis, MO) digestion was performed by incubation of the tryptic digest in 0.4% ammonium bicarbonate buffer (pH 8) at 37 °C for 1 hr (86).

### *3.4.7. Carboxypeptidase Y (EC 3.4.16.1)*

Immobilized carboxypeptidase Y [general references: Hayashi *et al.* (87), Kuhn *et al.* (88)] was used for proteolysis of βT-1 peptide from sickle cell hemoglobin. The mixture of proteolytic fragments was passed directly into a mass spectrometer source through a thermospray interface and was analyzed in the negative-ion mode. The amino acid sequence of the aberrant peptide was derived unequivocally (74).

*Method.* Carboxypeptidase Y was immobilized to N-hydroxysuccinimide–derivatized glycophase-controlled glass pore beads and placed into the column equilibrated in 0.1-N ammonium acetate buffer, pH 5.5. An enzyme column was located between the outlet of the reversed-phase column and a mass spectrometer. Peptides of interest, whose separation on the reversed-phase column was monitored by TIC, were directed to the enzyme column, trapped for up to 20 min, and pumped to the mass spectrometer for continuous mass analysis (74).

### *3.4.8. Pepsin (EC 3.4.23.1)*

Contrary to the majority of proteolytic enzymes, pepsin [general reference: Ryle (89)] can be used at very low pH. Pepsin has relatively low specificity, though the cleavage of peptide bonds adjacent to aromatic and other hydrophobic residues, especially phenylalanine and leucine, is preferred. The use of pepsin as a secondary digestion method was recently described by Ferranti *et al.* (33) and was discussed in Section 2. Casey and Lang (90) used pepsin for the secondary digestion of the core peptide α 104–141 generated by the fragmentation of the S-cyanylated α chain. Two-dimensional electrophoresis/paper chromatography was subsequently used for the analysis of peptic peptides. Piot *et al.* (91,92) presented comprehensive studies on the optimization of separation (by gel-permeation and reversed-phase HPLC) and structure analysis (FAB and MS/MS) of large-scale peptic digests of bovine hemoglobin, produced for potential use in the food industry and biotechnology.

*Method.* The hydrolysis was carried out in 5% formic acid (33) or in 0.1-M HCl (90) at 37 °C for 2 hr. An alternative protocol that we used to ensure extensive peptic digestion of large peptides (albeit not derived from hemoglobin) involved digestion with pepsin (Worthington) for 2 d at 30 °C in ammonium acetate buffer, pH 3.5; enzyme was added in four portions to achieve a final enzyme-to-protein ratio of 1:10 (93).

### *3.4.9. Cyanogen Bromide*

Ferranti *et al.* (33) describe the use of secondary chemical cleavage of αT-9 peptide from Hb Ozieri with CNBr (Section 2), which occurs on the carboxyl side of methionine residues. Only two or three peptides are produced from globin chains by this technique (Figs. 4.11 and 4.12); consequently they have high molecular masses unsuitable for analysis by FAB or LSIMS. However, the advent of ESMS should permit greater use to be made of this technique, and Wada *et al.* (50) used the characterization of these fragments as an early example of the combinations of ES with magnetic sector mass spectrometers (Fig. 4.6).

*Method.* Wada *et al.* (50) describe the following method for CNBr hydrolysis: globin is dissolved to a concentration of about 10 mg/ml in 70% formic acid. An equal weight of CNBr to that of globin is added to the solution, and the mixture is incubated in the dark under an oxygen-free nitrogen atmosphere at ambient temperature for 24 hr. The mixture is diluted with water and freeze-dried.

### *3.4.10. Cyanylation*

Casey and Lang (90) describe a useful method that produces just two major fragments from α globin, one being the α-core region containing residues 104–141. Because ESMS can measure the molecular mass of such large fragments, this method should readily be able to determine in which portion of the protein a mutation has occurred. The core cysteine is S-cyanylated, and cleavage by exposure to mild alkali occurs at the amino-terminal side of the resulting β-thiocyanoalanine (Fig. 4.11). Casey and Lang's paper also reports the use of chymotrypsin and pepsin hydrolysis for producing peptide fingerprints of the α-core region fragments generated by their cyanylation procedure.

*Method.* The method was first described by Jacobsen *et al.* (94). Isolated α chains were reduced for 2 hr at 37 °C in 10 ml of 0.01-M dithiothreitol–0.2-M Tris–6-M guanidium chloride at pH 8.2. Cyanylation was achieved using a fivefold excess of 2-nitro-5-thiocyanobenzoic acid for 15 min at room temperature, and the reaction was stopped by addition of an equal volume of acetic acid. Excess reagents were removed by G25 Sephadex chromatography, and the remaining acetic acid was removed by freeze-drying. The S-cyanylated α chains were cleaved by incubation in 0.1-M sodium borate–6-M guanidinium chloride pH 9.0 for 10 hr at 37 °C. After adjusting the pH to 4 with acetic acid, excess borate was removed by centrifugation and the product was purified on Sephadex G75 columns (90).

### *3.4.11 HPLC Analysis of Proteolytic Digests*

Probably the most useful current method for tryptic fragment separation by HPLC is that of Rahbar *et al.* (21), and it is the one we use in our laboratory. The original separation published by this group is illustrated in Fig. 4.13 for the separation of tryptic peptides from whole globin, i.e., the α and β chains. For the α chain, the following peptides are illustrated: αT-1, αT-2, αT-3, αT-4, αT-5, αT-6, αT-9, αT-11, αT-13, and αT-14. The combination peptide αT-8,9 is also detected. For the β chain, βT-1, βT-2, βT-3, βT-4, βT-5, βT-6, βT-7, β-T9, β-T11, βT-13, βT-14, and βT-15 are detected. A peak representing the combination peptide βT-8,9 is also seen. In our studies, we generally do not find very small and hydrophilic peptides (βT-6, βT-7, βT-15, αT-7), but we do see additional peaks, e.g., a combination of αT-1,2 (eluting after αT-3), and αT-10,11 (eluting midway between αT-4 and αT-6), βT-14,15 (preceding αT-10,11 by less than 1 min) and very often though not consistently βT-12 (emerging a few minutes after αT-9). Providing S-derivatization had been carried out, we also detect βT-10 carboxyamidomethylated (CAM) or amino-ethylated (AE) (midway between βT-11 and βT-2), βT-12 CAM (eluting after αT-9 but earlier than its nonderivatized counterpart), βT-10,11 CAM (midway between βT-2 and βT-5), and in the case of aminoethylation αT-1 2.1 (before αT-1), αT-12.2 AE (before αT-8,9), βT-12.1 AE (co-eluting with βT-2), and βT-12.2AE (halfway between βT-1 and αT-4, αT-11). Although in principle all amino acid residues in the β chain and most in the α chain are included in the peptides listed, some cannot be effectively analyzed by mass spectrometry because of the high background level in the low-mass region. This limitation

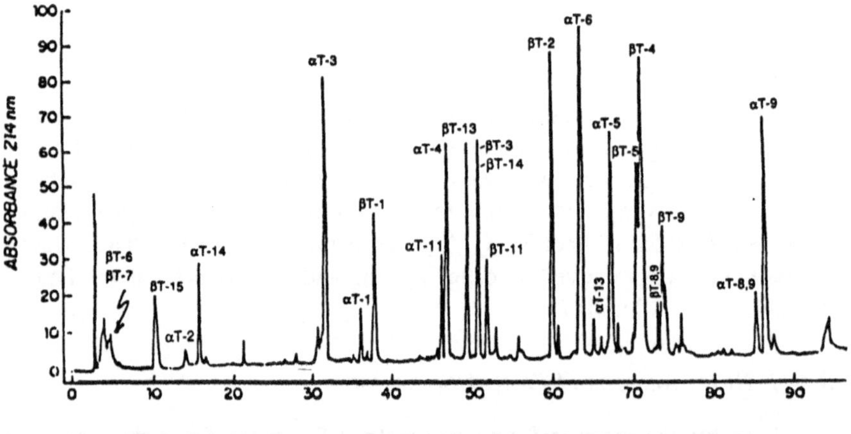


*Figure 4.13.* Separation on a Vydac C18 column of the tryptic peptides obtained from normal globin, according to the original method of Rahbar *et al.* (21). Reproduced from Rahbar *et al.* (21), by courtesy of Marcel Dekker, Inc.

applies particularly to the small peptides, βT-6, βT-7, βT-15, αT-2, αT-7, αT-10, and αT-14. Fortunately, αT-2 is seen in the combination αT-1,2, αT-10 in the combination αT-10,11, and βT-15 in the combination βT-14,15. Thus, the remaining difficult peptides are βT-6, βT-7, αT-7, αT-14, and the α globin core containing residues 100–127. The α-core peptides αT-12 (100–127) and αT-12,13 (100–139) produced by inefficient proteolysis at lysine 126 are very insoluble as well as being large and difficult to separate by HPLC or to characterize by mass spectrometry. Rahbar *et al.* (21) have obtained the αT-12 peak in chromatograms, providing they oxidized the globin with performic acid prior to proteolysis. This oxidation results in conversion of cysteine to cysteic acid. In our studies using HPLC/MS (see Section 3.5.2 and Figs. 4.19 and 4.20), a peak containing αT-12,13 CAM and αT-12,13,14 CAM is clearly seen and gives peptides of the appropriate mass.

Wilson and co-workers (15) precipitated and isolated the αT-12,13 peptide produced by the tryptic digestion of underivatized α globin so that they could further digest it with chymotrypsin. Separation of seven peptides produced by chymotrypsin is illustrated in Fig. 4.14. This type of sample has not yet been subjected to mass spectrometric analysis, and it is likely that the smaller peptides produced will present problems in detection.

Jensen and Roepstorff (40) digested S-pyridylethylated β chains with *S. aureus* V8, separated the peptide products by HPLC (Fig. 4.15), and determined their molecular masses with PDMS. Of the nine possible peptides (βS-1–βS-9), only a few were detected, viz., a combined βS-1,2,3, βS-4,5, βS-5, βS-6, βS-7, βS-8, βS-9, βS-7,8,9, and βS-8,9. The individual peptides βS-1, βS-3, and free glutamic acid βS-2 were not found, probably because endoproteinase Glu-C does not cleave, or cleaves only in low yield, after Glu residues in the sequence Glu-Glu-Lys (positions 6–8). The small polar peptide βS-4 is assumed to elute with the solvent front. Extensive utilization of *S. aureus* V8 for the proteolysis of the α chain was not undertaken in the above studies. The fact that such hydrolysis supposedly produces only five peptides, one of which covers more than half of the sequence, might hamper their easy analysis by HPLC and PDMS.

*Method.* We use the following technique for preparative tryptic peptide separation, essentially that of Rahbar *et al.* (21); Vydac C18 column (0.46 × 25 cm, 5 μm, 300 Å) (The Separations Groups, Hesperia, CA), flow rate 1 ml/min, detection 214 nm, isocratic development with A for 5 min followed by a linear gradient to 78% B in 90 min. A: water/0.125% TFA; B: acetonitrile:water (90:10)/0.125% TFA. Individual peptide fractions are collected manually (11).

Wilson *et al.* (15) describe the following system for separating chymotryptic digests of the α 100–139 core region: μBondapak C18 column (Waters); Solvent A: acetonitrile:0.01-M ammonium acetate (pH 5.7); B: 0.01 ammon-

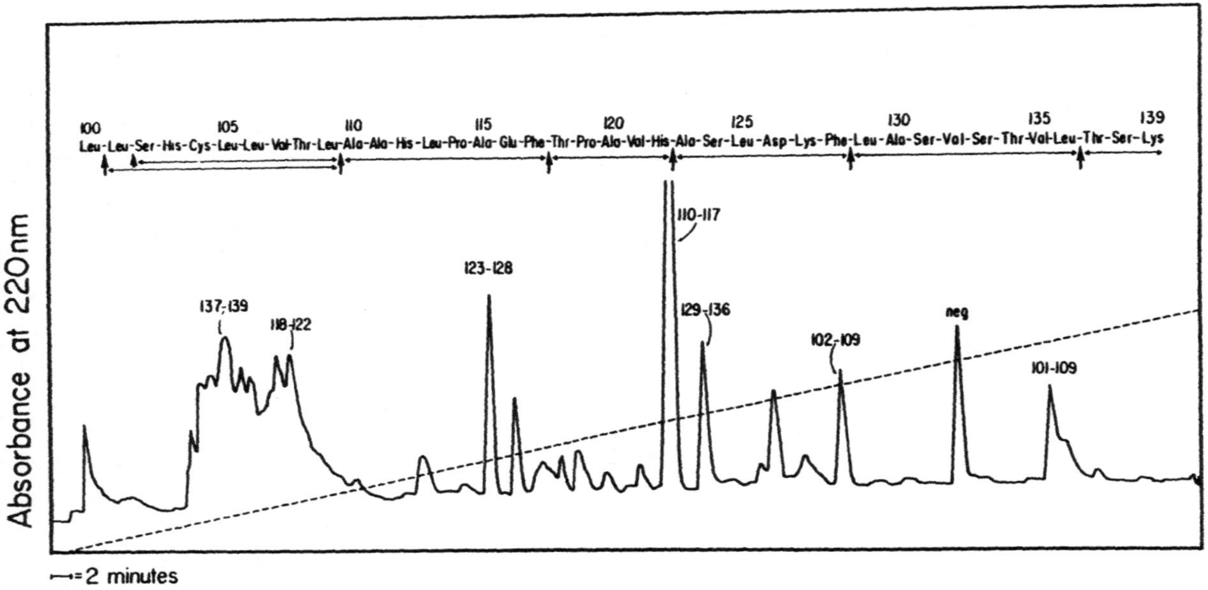


*Figure 4.14.* Subfractionation of the α-globin core region. HPLC separation of peptides from a chymotryptic digest of the oxidized core of a normal α chain. The amino acid sequence of the core peptide is listed at the top of the figure. The numbers identifying the various zones refer to the corresponding positions in the intact α globin. Dashed line indicates that a linear gradient of acetonitrile in ammonium acetate (pH 5.7) was used in the chromatographic separation. Reproduced from Wilson *et al.* (15), with permission of Elsevier Science Publishing BV, Academic Publishing Division.

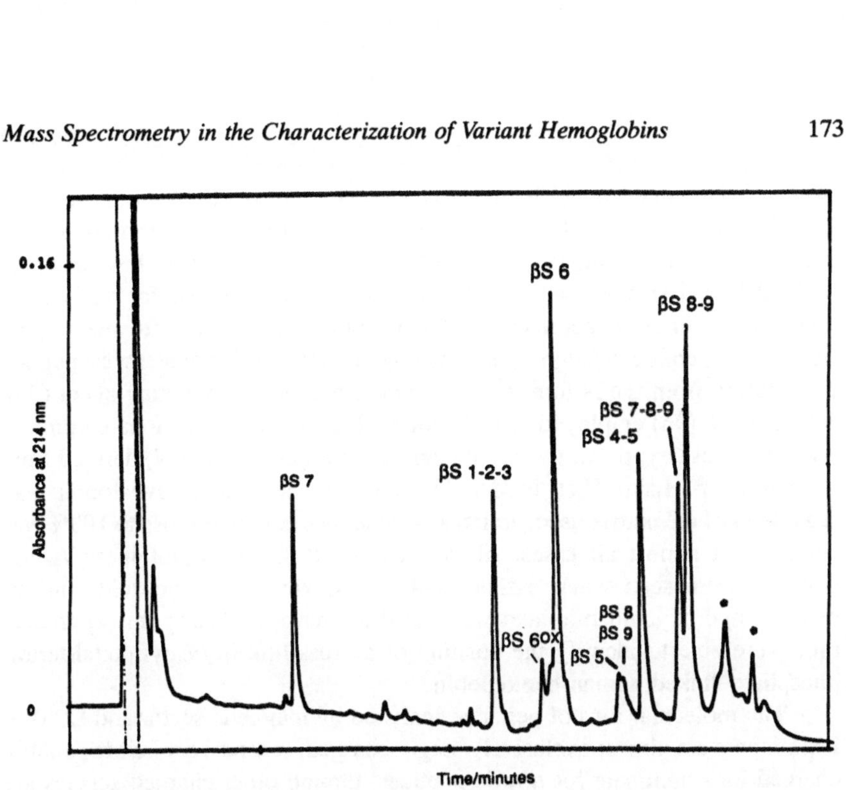


*Figure 4.15.* HPLC separation of normal β-globin peptides produced by *S. aureus* V8 proteolysis. Under the proteolytic conditions used (38), complete digestion was not achieved; thus, several combined peptides were isolated. Plasma desorption mass spectrometry (PDMS) was used for characterization, so there was a limited ability to characterize large peptide fragments such as those marked by asterisk. Reproduced from Jensen *et al.* (38), with permission of Academic Press.

ium acetate (pH 5.7); gradient: from 1% A to 60% A in 120 min; detection at 220 nm.

For separation of *S. aureus* V8–produced peptides, Jensen *et al.* (38) used the following system: C4 column (0.46 × 25 cm); solvent A: 0.1% TFA; solvent B: 0.08% TFA/60% acetonitrile. A flow rate of 1 ml/min was used, with a linear gradient from 10% to 85% B in 45 min; detection by UV at 214 nm.

### *3.5. Peptide Characterization*

The molecular masses of all peptides produced by tryptic digestion can be determined by LSIMS, ESMS, plasma desorption (PD-MS), and matrix-ionization-assisted laser desorption time-of-flight (MALDI-TOF) mass spectrometry. In practice, systems using LSIMS (including FAB) sources are limited to 4000–5000 Da for magnetic sector instruments and 2000–3000 Da for quadrupoles. When sequencing information is required, the limit for low- and high-energy CID is about 2500 Da, although it is preferable to analyze much smaller peptides. This mass range is increased dramatically if ESI sources are used. PD-MS and MALDI-TOF offer much higher mass limits. PD-MS measured *S. aureus* V8 peptides bigger than 5000 Da, albeit with rather low accuracy (0.05–0.1%). LD mass spectrometry offers a great potential for analysis of complex mixtures of peptides, as demonstrated by a successful mass spectrometry of peptide "ladders" generated by repeated cycles of stepwise peptide degradation from the N-terminus in the presence of a terminating agent (95). Juhasz *et al.* (96) employed LD-MS for evaluation of nonfractionated proteolytic mixtures (tryptic, *S. aureus* V8, and endoproteinase Asp-N) derived from the normal β chain. They used a 337-nm laser beam and have found that, regardless of the matrix used, amino acid sequence coverage close to 100% was achieved in almost all cases. All *S. aureus* V8 and endoproteinase Asp-N peptides were seen when caffeic acid, 2,5-dihydroxybenzoic acid, and α-cyano-4-hydroxycinnamic acid were used as matrices. Using this approach, they were able to identify the position of a cross-link in *bis*pyridoxal-tertraphosphate–linked human hemoglobin.

The molecular ions of peptides analyzed by magnetic sector and LDTOF instruments are almost exclusively singly charged, whereas for ESMS, doubly charged ions dominate for tryptic peptides, though other charged species are found if more basic residues are present. Identification of two charged species in an ESMS spectrum permits determination of the number of charges present and the molecular mass of the peptide.

In the CID analysis of peptides ionized by LSIMS, a fragmentation of the $[M+H]^+$ ion is usually carried out. In ESMS, CID is predominantly carried out on the doubly charged ions, although the fragments produced usually have only one charge each.

### *3.5.1. Sequencing on Quadrupole Instruments*

Hunt and co-workers (97) have pioneered the field of sequencing on quadrupole spectrometers and have brought the technology to the point of approaching in utility the mass spectrometry carried out on much more sophisticated instruments. When using a LSIMS source, the peptide molecular ion is encouraged to fragment in a low-energy CID cell (10–30 eV), and a series of fragment ions is produced. Because of the low resolution of quadrupoles and the requirement of high sensitivity, all isotope species of the molecular ion are passed to the collision cell, i.e., a spread of about 2–6 amu. This low resolution means that multiplets are formed for each ion type in the product ion spectra. Ions of types y and b dominate the spectra, although other ion types are sometimes seen. To help in identification, Hunt *et al.* (97) employ two derivatizations:

1. acetylation, in order to count free amino groups (N-terminus and lysines), and
2. methylation, in order to count carboxylic acid moieties.

The former gives a mass increase of 42 Da per active residue, and the latter increases mass by 14 Da per acid group.

The accuracy of sequence determination is lacking in quadrupole instruments for reasons such as the following:

1. A single isotopic species cannot be selected in the first quadrupole, so that product ion spectra contain fragment multiplets, rendering accurate mass determination very difficult.
2. The low collision energies employed in triple quadrupole instruments preclude the types of fragmentation needed for distinguishing the isomeric amino acids leucine and isoleucine.

The introduction of ESMS has greatly enlarged the sequencing possibilities of quadrupole instruments. It is much easier to interface to a separation system (HPLC) than is LSIMS, and it also offers the possibility of producing fragments in the source region as well as in the collision cell, thus permitting partial sequences to be obtained on single-stage instruments (98). When the electrostatic field in the high pressure region is low (low nozzle–skimmer bias), only molecular ions are seen, but increasing this electrostatic field causes an increase in energetic collisions of ionic species, leading to collisional activation and fragmentation. From tryptic peptides, mainly y- and b-mode series are formed.

Fortunately, peptides produced during the study of hemoglobin variants differ very little from normal ones, either by having an amino acid replaced or by being either longer or shorter. Thus, one has to have only enough information to recognize a single aberrant amino acid. An example from our work, shown in Fig. 4.16, is a source-generated-CID spectrum of βPG CAM of a variant subsequently characterized as $\beta^{\text{P-Galveston}}$. The y-mode ions produced in the higher nozzle–skimmer bias spectrum match completely those predicted in a MacProMass® program (99), thus confirming the sequence (36). Source fragmentation offers the exciting possibility of conducting MS/MS/MS experiments, because any of the fragment ions produced in the source could be further fragmented in the CID cell (Q2) and analyzed in Q3. In practice, this process is not easy because of sensitivity limitations. Unpublished data produced by Terry Lee of the Beckman Research Institute of the City of Hope on variant peptide sequencing by capillary HPLC/ESMS/MS is described in Section 3.6.2 and Fig. 4.21.

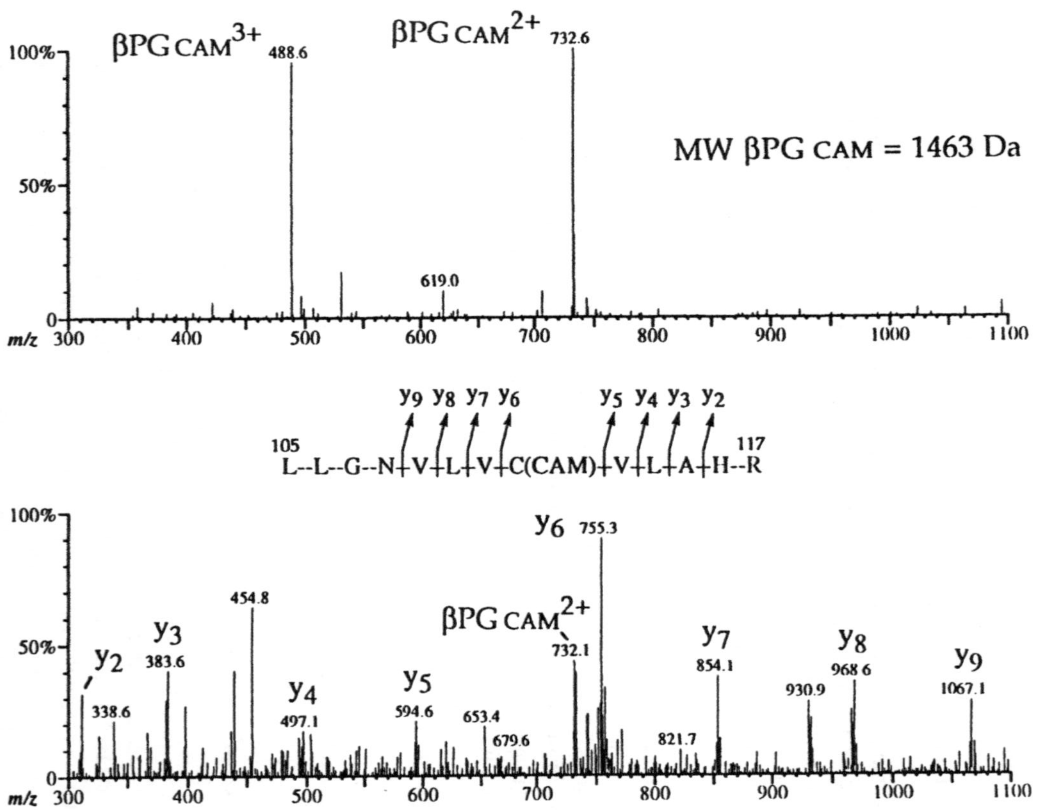


*Figure 4.16.* ES mass spectra of the carboxyamidomethylated $\beta^{P\text{-}Galveston}$-T12 (βPG CAM) fragment acquired at a low ($V_c$=50 V, top) and high ($V_c$=100 V, bottom), nozzle-to-skimmer bias. At $V_c$=100 V, fragmentation of doubly and triply charged ions of a precursor ion occurs, generating a series of y-mode ions. The fragmentation spectrum confirms that a peptide of mass 1463 Da is indeed derived from a sequence that spans the N-terminal portion of native βT-12 up to position 117, where a substitution of arginine for histidine occurs. This figure is reproduced from Witkowska *et al.* (36) by courtesy of Marcel Dekker, Inc.

### *3.5.2. Sequencing by Tandem Four-Sector Instruments*

In spite of the advances made in obtaining amino acid sequence information from data produced by modest single- and triple-stage quadrupole instruments, in some cases there is no substitute for analysis on a four-sector instrument, particularly if it is equipped with a highly sensitive array detector. The advantages of four-sector instruments are particularly obvious for peptides that show mass differences of ≤1 Da from normal (such as those arising from Ile↔Leu, Gln↔Lys, Lys↔Glu, Ile↔Asn, Asn↔Asp, and Gln↔Glu substitutions).

In association with Dr. Arnie Falick of the UCSF mass spectrometry facility (Director, Dr. A. L. Burlingame), we have extensively used four-sector mass spectrometry for characterizing tryptic peptides. The instrument is a Kratos Concept (of EBEB configuration) with an LSIMS source and array detector covering 4% of the mass range. In our hands, abnormal tryptic peptides have always been sequenced accurately by using this instrumentation and technique. However, it must be emphasized that these peptides do not usually present complex problems because, often, only one amino acid is inappropriate within a known structure.

A good example is the identification of Hb D-Iran (37). The variant was detected by isoelectric focusing, under which it showed a more positive charge than normal. An abnormal β chain was detected, isolated, and trypsin-hydrolyzed. Reversed-phase HPLC showed the presence of an abnormal peak eluting before βT-3. This peak was shown to give a protonated molecular ion at *m/z* 1313.7, a mass 1 Da less than that of normal βT-3. Possible mutations Asn→Ile, Glu→Gln, or Asp→Asn were therefore inferred, though the former was very unlikely because it did not explain the altered net charge of the abnormal Hb. This problem was difficult not only because the mass differential was very small, but also because the βT-3 peptide had three residues that could be changed to produce the required mass change (VNV$^{21}$D$^{22}$EVGG$^{26}$EALGR). It is unlikely that a peptide of this degree of difficulty could be readily sequenced by a triple-quadrupole instrument. Figure 4.17 shows a portion of the tandem CID spectrum of the variant peptide. Some of the key fragment ions are summarized in Table 4.4. The MS/MS spectrum permitted the rapid and conclusive identification of the variant as Hb D-Iran. This example demonstrated that unit resolution of precursor and CID-fragment ions, which is offered by a four-sector mass spectrometer, can be essential for the correct identification. Good precursor ion resolution is important because the normal βT-3 peptide differs in mass by only 1 Da from the abnormal one. Accurate selection of only the $^{12}$C species by MS1 remove all possible ambiguities caused by the presence of $^{13}$C-containing fragment ions.

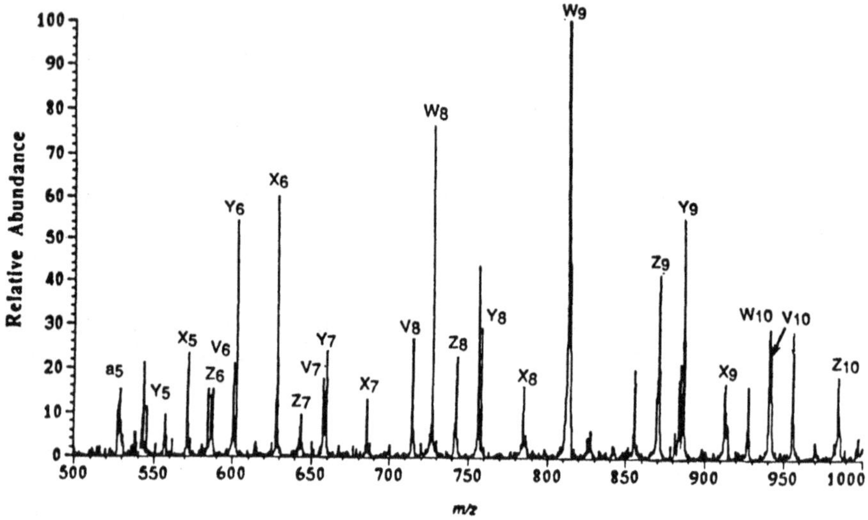


*Figure 4.17.* A portion of the CID spectrum of a variant βT-3 peptide, obtained on a tandem four-sector mass spectrometer of EBEB geometry equipped with an array detector. The precursor ion was the protonated molecule ($MH^+$) at 1313.7 Da, and the measured masses of the fragment ions observed are listed in Table 4.4. There were four possible amino acid changes in the peptide that could give rise to a peptide of the mass found ($\beta^{19}$N→I; $^{21}$D→N; $^{22}$E→Q; $^{26}$E→Q), but only one mutation could give a product ion spectrum of the type found. A $\beta^{19}$N→I mutation was unlikely, because it is inconsistent with the electrophoretic properties of the variant hemoglobin. Reproduced from Falick *et al.* (37), with permission of John Wiley & Sons, Ltd.

## 3.6. Combined HPLC/MS

### 3.6.1. Intact Globins

Introduction of on-line separation/MS analysis of intact globins extends the scope of ESMS in the evaluation of protein mixtures that contain species very close in mass because ESMS has difficulty in mass-analyzing proteins of close average molecular mass. It also helps to clarify any ambiguity that might arise if variants in low concentrations are obscured or confused by sodium and potassium adducts of major species. Whereas globin chain separation can be performed off-line on analytical or preparative columns, this step is wasteful in terms of time and solvent usage. For these reasons, we developed an on-line microbore HPLC/ESMS technique for fractionating and mass-analyzing globin mixtures (36, 100). In similar fashion to preparative technology, the method was based on using C4 reversed-phase columns. Figure 4.18 is the TIC chromatogram of the separation of globin chains from a patient with compound

*Table 4.4*
*Calculated and Experimental Fragment-Ion Masses from a Variant βT3 Peptide*[a]

| Fragment type[b] | Calculated CID fragment-ion masses (Da) from:[c] Cocody | β-26-EQ | D-Iran | Measured mass (Da) | Possible identification[d] |
|---|---|---|---|---|---|
| $Y_5$ | 545.3 | 544.3 | 545.3 | 545.1 | C,D |
| $X_5$ | 571.3 | 570.3 | 571.3 | 571.2 | C,D |
| Z6 | 586.3 | 585.3 | 586.3 | 586.1 | C,D |
| V6 | 600.3 | 599.3 | 600.3 | 600.3 | C,D |
| Z7 | 643.3 | 642.3 | 643.3 | 643.4 | C,D |
| V7 | 657.3 | 656.3 | 657.3 | 657.0 | C,D |
| V8 | 714.4 | 713.4 | 714.4 | 714.2 | C,D |
| X8 | 784.4 | 783.4 | 784.4 | 784.5 | C,D |
| W9 | 812.4 | 811.4 | 812.4 | 812.5 | C,D |
| Z9 | 871.4 | 870.5 | 870.5 | 870.6 | B,D |
| X9 | 913.4 | 912.5 | 912.5 | 912.5 | B,D |
| W10 | 941.5 | 940.5 | 940.5 | 940.5 | B,D |
| V10 | 942.5 | 941.5 | 941.5 | 941.5 | B,D |

[a] $[MH]^+$ = 1313.7.
[b] Nomenclature is described in Ref. 123.
[c] Calculated ion masses are for three possible variants: Cocody ($\beta^{21}D \rightarrow {}^{21}N$), VNV$^{21}$NEVGGEALGR; β-26-EQ ($\beta^{26}E \rightarrow {}^{26}Q$), VNVDEVGG$^{26}$QALGR; D-Iran ($\beta^{22}E \rightarrow {}^{22}Q$), VNVD$^{22}$QVGGEALGR.
[d] This column indicated which of the possible variants are consistent with each measurement: (C=Cocody, D=D-Iran, B=β-26-EQ.). D-Iran is the only variant consistent with all the measurements.

heterozygosity for Hb S and Hb Lepore Washington–Boston. The separated β globin peaks gave molecular masses of 15,837 and 15,865 Da, both measured within 2 Da error, appropriate for a $\beta^S$ and $\beta^{\text{Lepore Washington–Boston}}$ variant. This example illustrates the importance of the chromatographic technique because distinguishing the $\beta^{\text{Lepore Washington–Boston}}$ variant from normal β (15,867 Da) would always remain impossible within a mixed sample.

Under our conditions, a typical HPLC/MS run takes ~30 min, a considerable improvement over the off-line semipreparative technique (70 min). Also, because the flow rate is only 40 μl/min, solvent use is considerably lower compared to the off-line technique. Combined HPLC/MS of globin chains offers an additional advantage in that 90% of the sample is conserved and can be collected as individual peaks for further analysis.

*Method.* Separation of hemoglobin subunits is performed on a Michrom microbore HPLC system (Michrom BioResources, Pleasanton, CA) using a Vydac C-4 column (0.1 × 15 cm, 5 μm, 300 Å) (36, 100). The chromatography is carried out at 35 °C with a flow rate of 40 μl/minute. Solvent A is aceto-

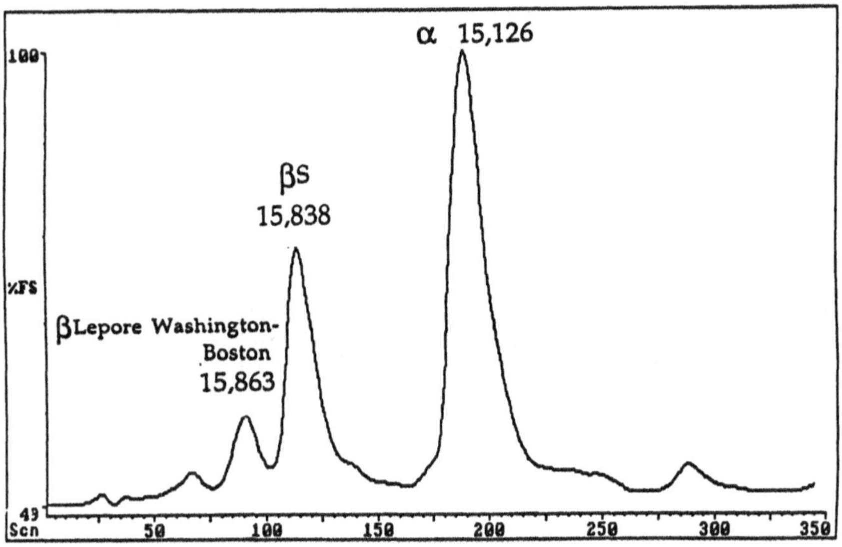


*Figure 4.18.* Combined microbore HPLC/ESMS for the analysis of intact globin chains. The patient had compound heterozygosity for Hb S and Hb Lepore Washington–Boston. After 100 pmol of hemoglobin was injected onto a $C_4$ Vydac protein microbore column (1 × 100 mm), it was eluted with a gradient of acetonitrile/TFA at a flow rate of 40 μl/min. The eluate was split 1:10 and mixed on the tip of probe with sheath liquid (MeOH/0.2% TFA, 2 μl/min) as it passed into a VG BioQ mass spectrometer.

nitrile:water (20:80)/0.1% TFA, and Solvent B is acetonitrile:water (60:40)/ 0.09% TFA. Globins were separated using a linear gradient from 39% to 47% B in 20 min, followed by a linear gradient to 60% B in 8 min; detection was achieved at 214 nm. Splitting (10:1) of the flow was achieved in an especially designed UV cell allowing 4 μl/min to be analyzed by the mass spectrometer. This HPLC system is interfaced to the VG BioQ mass spectrometer (VG BioTech/Fisons, Altrincham, England) through a modified VG electrospray probe. Typically, 2 μl of 0.2% TFA in methanol is added as sheath liquid prior to nebulization. The electrostatic source of the mass spectrometer is operated at 3.2 kV. The nozzle-to-skimmer bias value is typically 50 V. The mass spectrometric resolution was adjusted so that the *m/z* 998 peak from horse heart myoglobin was 1.5–2 Da wide at 10% height.

### *3.6.2. Tryptic Digests*

Stachowiak and Dyckes (74) applied thermospray LC/MS in the negative ionization mode for peptide mapping of normal and variant hemoglobins. The method involved a combination of on-line proteolysis with reversed-phase

HPLC separation followed by mass analysis of the peptides produced. The globin chains were passed through a column filled with trypsin immobilized to glycophase controlled-pore glass beads (Pierce Chemical, Rockford, IL). After a 20-min stop-flow incubation, proteolytic fragments were passed to a Hi-pore RP-304 column (Bio-Rad, Richmond, CA) interfaced with a Hewlett Packard (Palo Alto, CA, model 5988A) mass spectrometer fitted with a Vestec Thermospray LC/MS Interface (Vestec, Houston, TX). Nanomole quantities of peptides were applied, and the feasibility of the method was demonstrated by the analysis of variant hemoglobins S, C, and Baylor. This methodology allows for the addition of a secondary digestion step between the HPLC column and the mass spectrometer, creating a sequencing capability without the need for isolation of the peptide of interest. This sequencing capability was used to prove the presence of a sickle cell mutation in the βT1 peptide by digesting this peptide on line with carboxypeptidase Y.

The analysis of hemoglobin tryptic digests was one of the first applications of the combined HPLC/electrospray MS technique (Fig. 4.2). At the 1989 ASMS meeting, Covey, Conboy, and Henion (34) showed an HPLC/MS analysis of a tryptic peptide mixture of normal β globin. They used a C8 column ($0.1 \times 10$ cm) with a linear gradient of 5% $CH_3CN$ to 70% $CH_3CN$ (0.1% TFA). They demonstrated that many peptides could be resolved and produced predominantly doubly charged ions that were particularly amenable to collisionally induced fragmentation in the collision cell of a triple quadrupole instrument. A more detailed description of their HPLC/ESMS/MS technique was presented in a later publication (35).

We routinely employ LC/MS for analyzing tryptic peptides obtained from whole globin or from separated globin chains. Cysteines are alkylated with iodoacetamide prior to proteolysis. Figure 4.19 shows the chromatogram produced following proteolysis of whole hemoglobin. Almost all fragments are detectable, even the notoriously difficult to analyze α-core region, albeit occasionally with low yield. The final peak in the chromatogram is a mixture of αT-12,13 CAM and αT-12,13,14 CAM. The ES mass spectrum of these peptides is shown in Fig. 4.20, indicating the presence of components of $M_r$ 4260 Da (αT-12,13 CAM) and 4579 Da (αT-12,13,14 CAM). Obtaining sequence information of peptides during HPLC/ESMS is straightforward and has been described above. In fact, the spectrum illustrating product ions of βPG CAM previously illustrated (Fig. 4.16) was obtained during a similar run to that illustrated in Fig. 4.19.

Even though at present material limitation is hardly a problem in a hemoglobinopathy laboratory, the future might bring a growing need for hemoglobin diagnostics performed on scarce amounts of dried cord blood collected on filter paper. Packed-capillary LC can be coupled directly with MS without the need of postcolumn splitting, and this arrangement considerably lowers sample

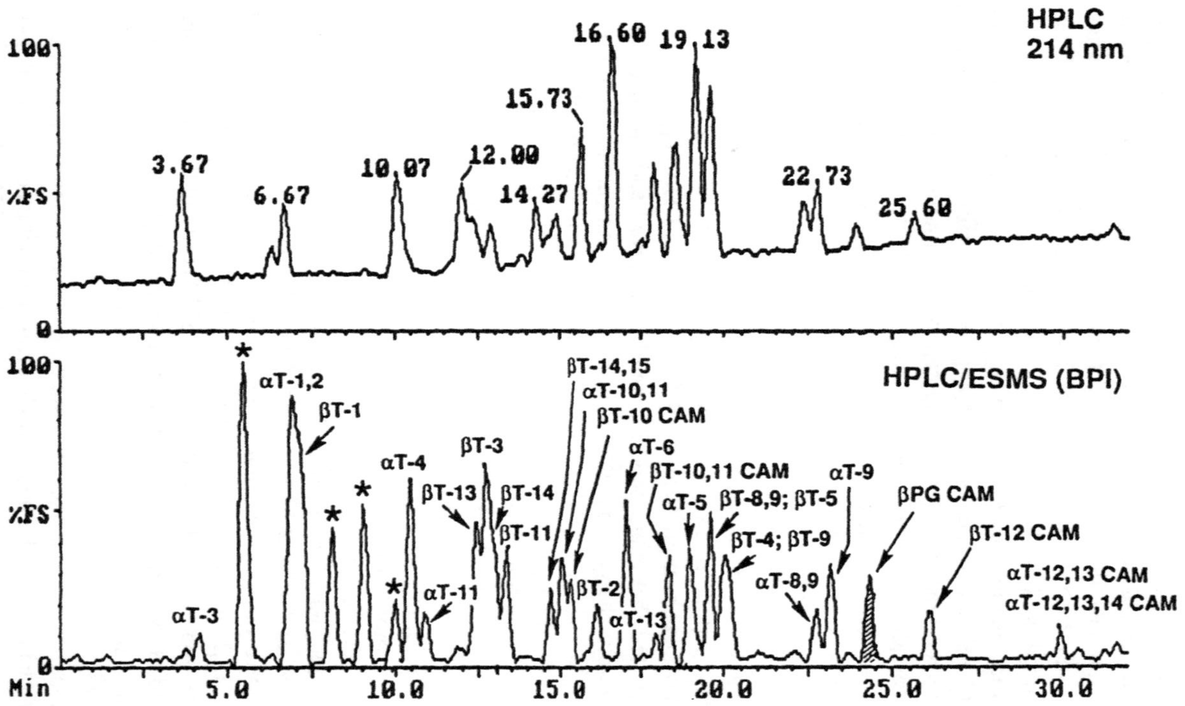


*Figure 4.19.* Microbore HPLC/ESMS of tryptic peptides from whole hemoglobin of the carrier of $Hb^{P\text{-}Galveston}$. The upper panel is the UV chromatogram at 214 nm and the lower is the HPLC/ESMS base peak intensity (BPI) profile. Almost all expected components are seen, including some combined peptides and the difficult α-core peptides. (CAM refers to S-carboxyamidomethylation.) A shaded peak contains the carboxyamidomethylated $B^{P\text{-}Galveston}$ T-12 (βPG CAM) peptide. Peaks annotated with stars contain unidentified material eluted from the filters used for sample preparation for HPLC; application of an alternative filtration device (Nylon Acrodisc®4) eliminated this contamination problem (not shown). Reproduced from Witkowska *et al.* (36) by courtesy of Marcel Dekker, Inc.

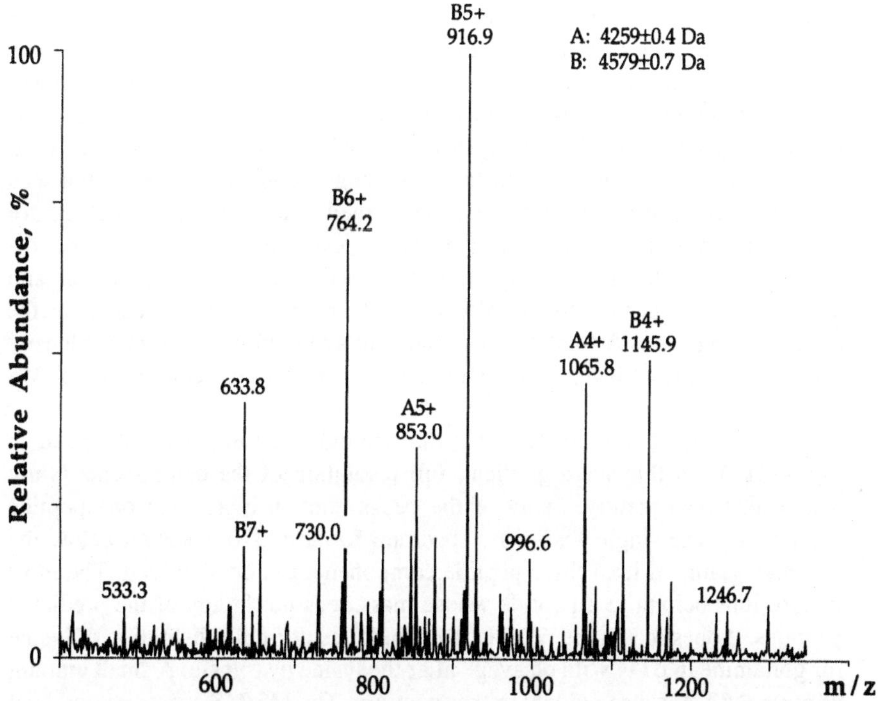


*Figure 4.20.* ES mass spectrum of α-globin core peptides. The spectrum was obtained during an HPLC/ESMS analysis. Two components are present: A represents αT12-13 CAM (Average $M_r$ 4260), and B represents αT12-13-14 CAM (Average $M_r$ 4579). The dominant charge states of both of these peptides are 4, 5, and 6.

consumption if coupled with FAB/MS (101) and ESMS (97, 102). There has been a notable development in the technique during recent years. High-quality packed-capillary columns are now commercially available along with a flow-rate controller and a low volume–long pathway flow cell, both of which allow for adaptation of standard HPLC systems to the requirements of low flow rates (0.5–50 μl/min) [LC Packings (USA), Inc., San Francisco, CA]. Davies and Lee (102) offer excellent practical guidelines regarding adjustments of the pumping system, construction of the flow cell, and packing of capillary columns to those willing or forced to rely on homemade instrumentation.

In a personal communication, Terry Lee of the Beckman Research Institute of the City of Hope illustrated the use of capillary HPLC-ESMS/MS for identification of variant Hb Alabama. The abnormal β chain was isolated by C4 reverse-phase HPLC and digested with trypsin. A portion of the tryptic digest mixture (approximately 20 pmole) was loaded onto a 0.25 mm × 150 mm C18 packed fused silica capillary column and eluted with an acetonitrile gradient (2–60% in 30 min) at a flow rate of 2 μl/min. The effluent from the column was directed through the capillary flow cell of a uv detector and then into the electrospray interface of a Finnigan MAT TSQ-700 mass spectrometer. Centroid spectra were collected over the mass range of *m/z* 500–2000 with a scan time of 3 sec. The instrument was programmed to analyze each spectrum for ions whose intensity was greater than a preset level. If such an ion is found, the instrument is set for the MS/MS mode. It takes 0.6 sec to analyze the scan and about 2.5 sec to change to MS/MS. Product ion spectra are collected for the selected parent ion. A total of four scans are acquired unless signal intensity falls below a preset level. The instrument is then reset to acquire normal MS scans.

The result of this analysis for the hemoglobin sample is illustrated in Fig. 4.21. With this steep gradient, full resolution of the components is not achieved. Consequently, many of the peaks contain more than one peptide component. The single-scan mass spectrum for scan 370 is shown below the chromatogram. At least three peptide components can be identified. The most intense ions belong to a peptide whose mass does not fit any of the predicted peptides. It does match the peptide that would result from substitution of lysine for glutamine in βT-4 with cleavage after the lysine by trypsin. A small amount of normal βT-4 is also evident in the spectrum. The MS/MS spectrum obtained from scans 371 and 372 is sufficient to confirm the sequence assignment.

This example illustrates the current limit of the technology for tryptic peptide detection and characterization. In practice, only time and experience will tell whether capillary HPLC is sufficiently robust, versatile, and easy to use routinely on HPLC/ESMS instruments. It may be that the intermediate technology of microbore HPLC is more generally applicable for a wide variety of analytes.

*Method.* [Microbore HPLC/MS technique, Witkowska *et al.* (36)]. A tryptic peptide mixture derived from 100–200 pmol total protein (25–50 pmol hemoglobin) is separated on the Michrom microbore HPLC system using a Vydac C-18 column (0.1 × 15 cm, 5 μm, 300 Å) at 40 °C, with a flow rate of 40 μl/min. Isocratic elution at 2% B for 5 min was followed by a linear gradient to 27% B in 20 min, and afterwards to 50% B in 15 minutes. Solvent A was acetonitrile:water (2:98)/0.1% TFA; solvent B was acetonitrile:water (98:2)/0.08% TFA; detection was at 214 nm. The probe was introduced to the mass spectrometer after the first 8 min of the run, and acquisition was started at 10 min. The flow-splitting, LC/MS interface, and mass spectrometric conditions are the same as described above for the LC/MS of proteins (36).

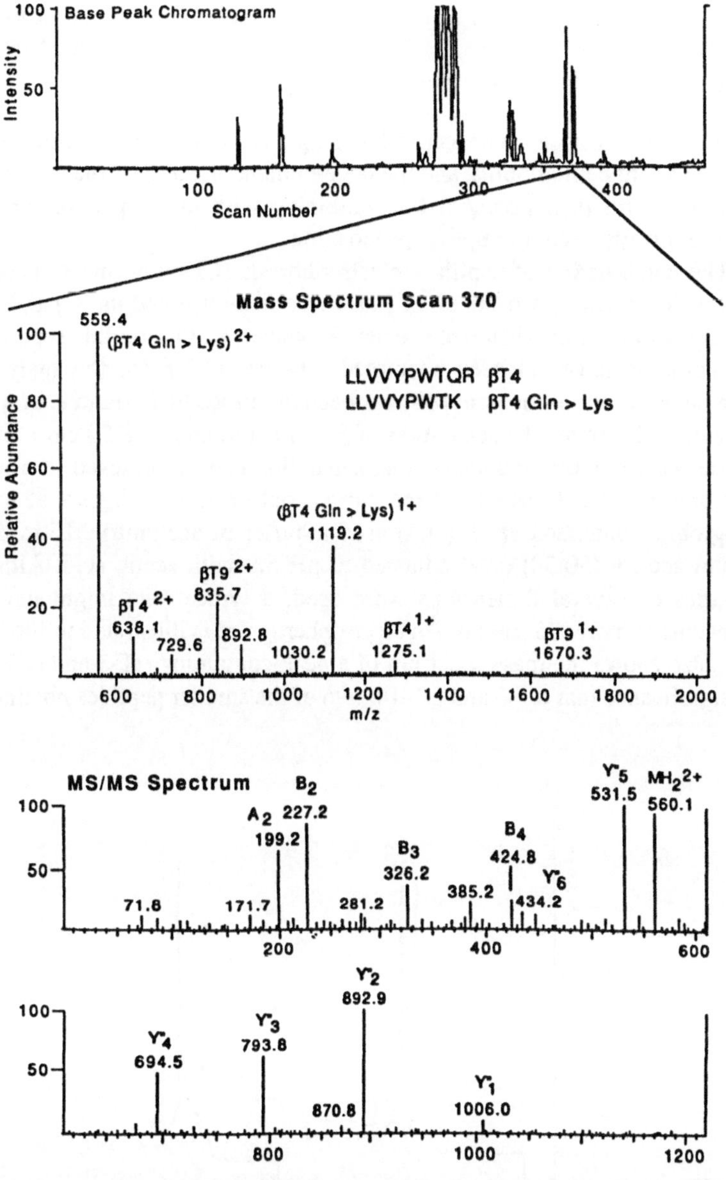


*Figure 4.21.* Capillary HPLC/ESMS/MS of β-globin tryptic peptides. These data, produced by Dr. Terry Lee of the Beckman Research Institute of the City of Hope, who granted permission for their reproduction, shows (top) the base peak intensity (BPI) chromatogram of the mixture and (middle) details of scan 370. This spectrum shows predominantly the doubly charged molecular ions of a variant βT-4 and normal βT-9. The MS/MS spectrum of the variant βT-4 (bottom) confirms the amino acid sequence primarily through interpretation of the y-mode and b-mode series ions and shows that the peptide is a clipped version of normal βT-4 formed by replacement of a glycine by lysine, thus introducing a new tryptic site. The product ion mass spectrum was obtained in the same HPLC run through the ability of the instrument to rapidly switch to the MS/MS mode after detection of the relevant peak. This variant hemoglobin is known as Hb Alabama.

## *3.7. Capillary Electrophoresis and Mass Spectrometry*

Electrophoresis and isoelectric focusing have always been central to the study of variant hemoglobins, and the development of capillary electrophoretic techniques, with their potential for combining with mass spectrometry, has therefore always been a tempting proposition.

The combination of capillary electrophoresis (CE) with mass spectrometry was described by Udseth *et al.* (103) who demonstrated the separation of three myoglobins from different species. However, similar on-line separations of hemoglobin have not been attempted. The use of CE for the analysis of globin chains (not on-line with MS) has been described by Ferranti *et al.* (104). Figure 4.22 illustrates the separation of β and α globins on a 50 cm × 50 μm (i.d.) coated capillary, and shows that separation can be achieved in 5 min.

Combined CE-ESMS has been carried out on tryptic digests of human hemoglobin. Johansson *et al.* (105) used a buffer of acetonitrile:15-mM ammonium acetate (50:50) (v/v) adjusted to pH 5.0 with acetic acid. Although capillaries of several dimensions were used, a typical one might have the dimensions 50 cm × 75 μm i.d. An electropherogram is illustrated in Fig. 4.23, which also shows the mass spectrum of a peak containing αT-3 and αT-14. It is of significance that αT-3 and αT-14, two of the smaller peptides obtained by

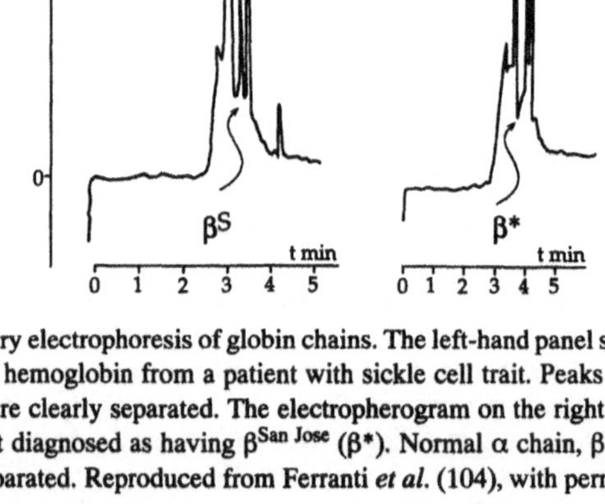


***Figure 4.22.*** **Capillary electrophoresis of globin chains. The left-hand panel shows the separation, achieved in 5 min, of hemoglobin from a patient with sickle cell trait. Peaks representing normal β, $\beta^S$, and normal α are clearly separated. The electropherogram on the right shows separation of globins from a patient diagnosed as having $\beta^{San\ Jose}$ (β*). Normal α chain, β globin, and $\beta^{San\ Jose}$ globins are clearly separated. Reproduced from Ferranti *et al.* (104), with permission of Academic Press.**

tryptic digestion, are normally quite difficult to determine by HPLC/ESMS. The technique also was able to isolate and determine the molecular masses of two of the largest tryptic fragments, αT-12 with a mass of 2965 Da and αT-9 with mass of 2996 Da. These two peptides primarily appeared as $[M+H]^{5+}$ ions of masses 594 and 600, respectively. Johansson and co-workers (105) also illustrated the CE-MS/MS of tryptic peptides, which gave sufficiently intense a-, b-, and y-mode ions to allow sequence confirmation for small peptides. Excellent preliminary data have been obtained using this technology, but the combined technique has been hindered by the inability to separate more than very small quantitative amounts of the analyte mixture. Even though ESMS instruments now have excellent sensitivity, ESMS is still only borderline when it comes to the small amounts that CE technique separates. However, recent results with CE-ESMS of proteins exploiting narrow-bore (5 μm) polyimide-coated fused-silica capillaries showed unexpectedly high analyte sensitivity

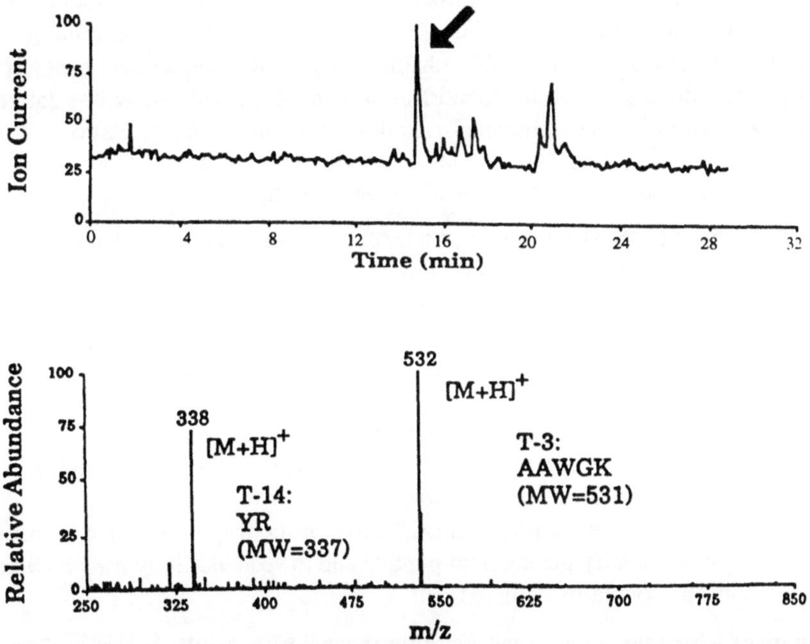


*Figure 4.23.* *Top*: Capillary electrophoresis ES mass spectrometric TIC profile (CE/ESMS-TIC) of an α-globin tryptic digest (10 pmol). ***Bottom***: The ES mass spectrum of a peak (annotated with an arrow) shows a co-eluting mixture of the peptides αT-3 and αT-14 of masses 552 and 339 Da, respectively. This figure is reproduced from Johansson *et al.* (105), with permission of Elsevier Science Publishers BV, Academic Publishing Division.

when compared with older systems that used larger-bore capillaries (106). Wahl *et al.* demonstrated CE-ESMS analysis of proteins at low attomole and subattomole levels. Although the reason for the improvement in sensitivity is not yet understood, a match between the CE and ES currents, improved ion transmission due to beam dynamic effects, and a reduction in space-charge expansion related to lower ES current appear to contribute to the phenomenon.

Aesthetically, the CE-ESMS technique is the most pleasing to the hemoglobinopathy laboratory workers because it combines the classical technique for hemoglobin separation (electrophoresis) with molecular mass determination by mass spectrometry.

## *4. SUMMARY AND FUTURE METHODOLOGICAL IMPROVEMENTS*

This chapter has shown that mass spectrometry has achieved a firm foothold among the techniques for characterizing variant hemoglobins. Acceptance of the technology has occurred principally following the recent introduction of the ESMS technique, which alone is capable of accurately determining molecular masses of variant globin chains and partially sequencing proteolytic peptides. Moreover, variant identification can be carried out with relatively inexpensive mass spectrometers of modest mass range and resolution.

Basic methods for characterizing variants have been described and in the usual circumstances can be reduced to the following:

1. An ESMS spectrum is obtained of a hemolysate to determine the molecular masses of any variant protein present, providing it has a mass differential of at least 12–14 Da from other mixture components (or 6–8 if MaxEnt® is used).
2. Combined HPLC/MS is used to separate globin chains and to determine their molecular masses if the mass differential is *less* than 12–14 Da (6–8 with MaxEnt®) from "normal."
3. Alkylation (preferably carboxyamidomethylation) of cysteine and tryptic hydrolysis are performed.
4. Microbore or capillary HPLC/ESMS of the tryptic peptides is carried out to identify the aberrant peptide and to sequence it by fragmentation in the source or collision cell.

In cases where the above analysis gives inconclusive results, a series of alternative or additional steps could be included, viz., preparative isoelectric focusing or electrophoresis to isolate the abnormal hemoglobin; isolation of a globin chain prior to proteolysis; altering the primary digestion technique or adding a secondary one; acetylation and/or methylation of peptides prior to MS sequencing; and, finally, amino acid sequencing using a four-sector tandem mass spectrometer (or Edman sequencer).

We may anticipate several developments of this procedure. One will be partial automation, whereby many hemolysate samples can be placed in a carousel and the molecular masses of components automatically determined by ESMS. These data could be automatically compared to molecular masses of known variants, or possible new variants, stored in a reference library so that listings of possible structures for an unknown variant could be given. Another advance will be the automated sequence determination of tryptic peptides. Yates and co-workers (107) have made great advances in this field and have produced a program that gives a high success rate for correctly sequencing peptides analyzed by triple quadrupole instruments. Capillary electrophoresis/ESMS will also be rendered more routine in coming years and will prove to be a useful complementary technique to microbore (or capillary) HPLC/MS. There will be improvements in wet chemistry methods for use prior to mass spectrometric analysis, particularly the refinement of techniques aimed at recovering proteins from separatory gels.

Finally, the instrumentation itself will improve. Greater sensitivity may be more widely achieved when ES ion-trap mass spectrometers become commercially available. These instruments are also capable of $MS^n$ (multiple MS/MS) experiments so that sequencing will be facilitated. As for magnetic sector machines, sensitivity in terms of sequencing capability should be improved when array detectors are able to collect ions over a greater section of mass range than is currently possible. Our studies (31,32,37) have been carried out on a Kratos Concept IIHH 4-sector instrument of EBEB geometry, where 4% of mass range is covered; the latest VG Autospec-FPD with focal plane (array) detector can cover 20%. The combination of ESMS and magnetic instruments may allow molecular masses to be determined with a greater degree of accuracy than is presently possible with quadrupoles, although this technology will be restricted to those able to afford the more expensive hardware. For intact proteins, however, the resolution of proteins of close molecular mass by magnetic sector instruments will probably not be significantly improved over that achievable with MaxEnt® software on quadrupole instruments.

Another important new technique, MALDI-TOF mass spectrometry, is likely to become a major player in hemoglobin identification. It has been shown to give reasonable accuracy in the determination of molecular masses of tryptic peptide mixtures (96), but at present has insufficient resolution to determine accurately the molecular masses of intact globins in mixtures with closely related variants. Its weak point is the lack of an interface to a separatory method, but a particular strength is the easy loading of multiple samples into the instrument and the speed at which they can be analyzed. Also, the potential use of MALDI-TOF mass spectrometry in peptide sequencing, as described by

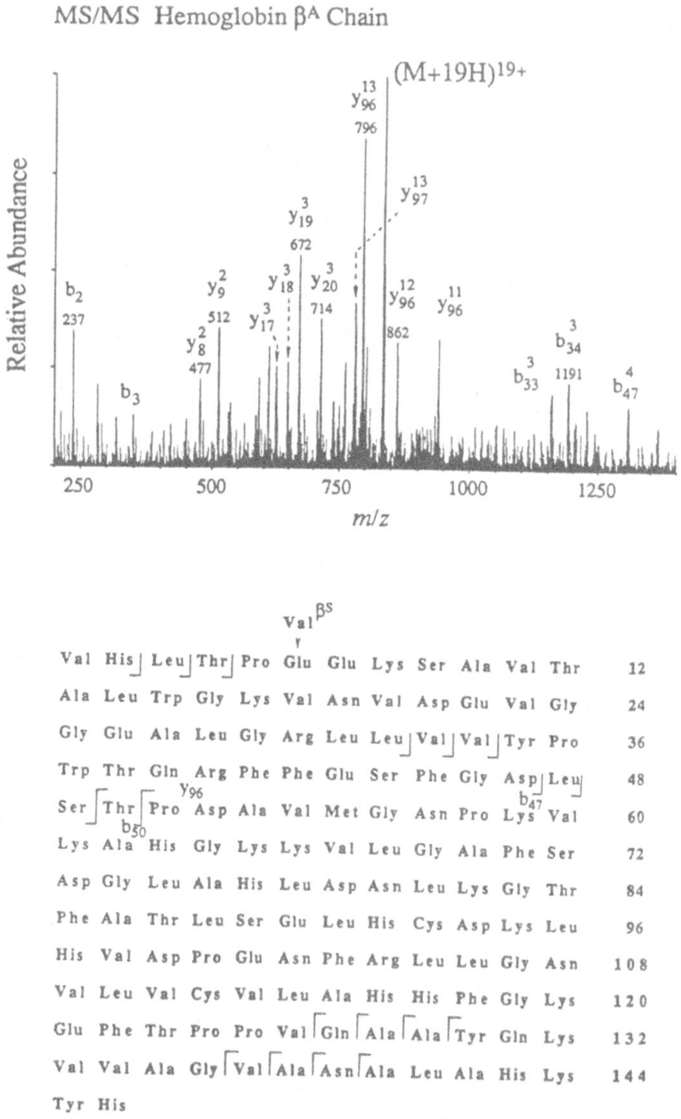


*Figure 4.24.* ES source CID of intact globin chains. The spectrum (top) shows the fragmentation of the $[M+19H]^{19+}$ ion to a series of y-mode and b-mode product ions. The amino acid sequence of the human hemoglobin $\beta^A$ chain (bottom) shows cleavage sites for which fragment ions were identified (breaks in the amino acid sequence). Shaded residues indicate the position of the amino acid substitutions in mutant β-globins that were studied. Reproduced from Light-Wahl *et al.* (109), with permission of John Wiley & Sons, Ltd.

Chait and Kent (95) might render laser desorption an invaluable addition to any structural laboratory in the near future.

A future direction of mass spectrometry in variant hemoglobin characterization may lie in MS/MS of intact globin chains analyzed by ESMS. Loo and co-workers (108) have described the interpretation of multiply charged product ion spectra of intact protein multiply charged ions, and recently, in collaboration with Richard Smith's group, we have obtained preliminary data on the CID fragmentation of multiply charged ions of intact normal and variant β globins (109) (Fig. 4.24). From the multiply charged b- and y-mode product ions found, we have been able to distinguish and confirm the sequence of the amino acids at both ends of the protein.

In the next decade, we can expect that current methodologies and these new developments will rapidly increase the number of variant hemoglobins known. With a mandatory requirement for newborn screening for hemoglobinopathies becoming common, the structural characterization of variants will be increasingly required.

ACKNOWLEDGMENTS. Some of the studies described were carried out by us in association with colleagues at Children's Hospital Oakland Research Institute (CHORI); the University of California, San Francisco; and the Batelle Laboratories, Richland, WA. We gratefully acknowledge their contributions to the studies. Thanks are due to Brian Green of the VG Biotech Company, who introduced us to the electrospray field, and to all the hemoglobin analysts who submitted their publications and latest data. We are indebted to Michelle Carlson of CHORI, who prepared the manuscript for submission.

The work was financially supported by NIH grants DK34400 and HL20925. Most of the mass spectrometry from this laboratory was carried out on an instrument provided by the NIH Shared Instrumentation Grant program (RR06505).

## *REFERENCES*

1. Bunn, H. F., and Forget, B. G., 1986, *Hemoglobin: Molecular, Genetic and Clinical Aspects*, W. B. Saunders Co., Philadelphia.

The figure for number of individuals worldwide with abnormal hemoglobins is approximate and was generated by Dr. William P. Winter.

2. Kim, H. C., Atwater, J., and Schwartz, E., 1983, Separation of hemoglobins, in *Hematology 3rd ed.* (W. J. Williams, E. Beutler, A. J. Ersley, and M. A. Lichtman, eds.), McGraw-Hill, New York, pp. 1611–1619.
3. Schneider, R. G., and Barwick, R. C., 1982, Hemoglobin mobility in citrate agar electrophoresis—its relationship to anion binding, *Hemoglobin* **6**:199–208.
4. Basset, P., Beuzard, Y., Garel, M. C., and Rosa, J., 1978, Isoelectric focusing of human hemoglobin: its application to screening, to the characterization of 70 variants, and to the study of modified fractions of normal hemoglobins, *Blood* **51**:971–982.

5. Loomis, S. J., Go, M., Kupeli, L., Bartling, D. J., and Binder, S. R., 1990, An automated system for sickle cell screening, *Am. Clin. Lab.* **Oct.**:33–40.
6. Alter, B. P., Goff, S. C., Efremov, G. D., Gravely, M., and Huisman, T. H. J., 1980, Globin chain electrophoresis: A new approach to the determination of the $^G\gamma/^A\gamma$ ratio in fetal haemoglobin and to studies of globin synthesis, *Br. J. Haematol.* **44**:527–534.
7. Rochette, J., Righetti, P. G., Bosisio, A. B., Vertongen, F., Schneck, G., Boissel, J. P., Labie, D., and Wajcman, H., 1984, Immobilized pH gradients and reversed-phase high-performance liquid chromatography: A strategy for characterization of haemoglobin variants with electrophoretic mobility identical to that of Hb A., *J. Chromatog.* **285**:143–152.
8. Whitney III, J. B., Cobb, R. R., Popp, R. A., and O'Rourke, T. W., 1985, Detection of neutral amino acid substitutions in proteins, *Proc. Natl. Acad. Sci. U.S.A.* **82**:7647–7650.
9. Castagnola, M., Dobosz, M., Landolfi, R., Pascali, V. L., De Angelis, F., Vettore, L., and Perona, G., 1988, Determination of neutral haemoglobin variants by immobilized pH gradient, reversed-phase high-performance liquid chromatography and fast-atom bombardment mass spectrometry: the case of a Hb Torino $\alpha$43 (CE1) Phe→Val, *Biol. Chem. Hoppe–Seyler* **369**:241–246.
10. Shelton, J. B., Shelton, J. R., and Schroeder, W. A., 1981, Further experiments in the separation of globin chains by high performance liquid chromatography, *J. Liquid Chromatog.* **4**:1381–1392.
11. Green, B. N., Oliver, R. W. A., Falick, A. M., Shackleton, C. H. L., Roitman, E., and Witkowska, H. E., 1991, Electrospray MS, LSIMS and MS/MS for the rapid detection and characterization of variant hemoglobins, in *Biological Mass Spectrometry* (A. L. Burlingame and J. A. McCloskey, eds.), Elsevier, Amsterdam, pp. 129–146.
12. Ingram, V. M., 1956, A specific chemical difference between the globins of normal human and sickle-cell anemia haemoglobin, *Nature* **178**:792–794.
13. Ingram, V. M., 1958, Abnormal human haemoglobins. I. The comparison of normal human and sickle-cell haemoglobins by "fingerprinting," *Biochim. Biophys. Acta* **28**:539–545.
14. Schroeder, W. A., Jones, R. T., Cormick, J., and McCalla, K., 1962, Chromatographic separation of peptides on ion exchange resins. Separation of peptides from enzymatic hydrolyzates of the $\alpha$, $\beta$, and $\gamma$ chains of human hemoglobins, *Anal. Chem.* **34**:1570–1575.
15. Wilson, J. B., Lam, H., Pravatmuang, P., and Huisman, T. H. J., 1979, Separation of tryptic peptides of normal and abnormal $\alpha$, $\beta$, $\gamma$, and $\delta$ hemoglobin chains by high-performance liquid chromatography, *J. Chromatog.* **179**:271–290.
16. Matsuo, T., Matsuda, H., Katakuse, I., Wada, Y., Fujita, T., and Hayashi, A., 1981, Field desorption mass spectra of tryptic peptides of human hemoglobin chains, *Biomed. Mass Spectrom.* **8**:25–30.
17. Wada, Y., Hayashi, A., Fujita, T., Matsuo, T., Katakuse, I., and Matsuda, H., 1983, Structural analysis of human hemoglobin variants by mass spectrometry, *Int. J. Mass Spectrom. Ion Phys.* **48**:209–212.
18. Wada, Y., Hayashi, A., Masanori, F., Katakuse, I., Ichihara, T., Nakabushi, H., Matsuo, T., Sakurai, T., and Matsuda, H., 1983, Characterization of a new fetal hemoglobin variant, Hb F Izumi $^A\gamma$6Glu→Gly, by molecular secondary ion mass spectrometry, *Biochim. Biophys. Acta* **749**:244–248.
19. Hayashi, A., Wada, Y., Matsuo, T., Katakuse, I., and Matsuda, H., 1987, Neonatal screening and mass-spectrometric analysis of hemoglobin variants in Japan, *Acta haemat.* **78**:114–118.
20. Pucci, P., Carestia, C., Fioretti, G., Mastrobuoni, A. M., and Pagano, L., 1985, Protein fingerprint by fast atom bombardment mass spectrometry: Characterization of normal and variant human haemoglobins, *Biochem. Biophys. Res. Commun.* **130**:84–90.
21. Rahbar, S., Lee, T. D., Baker, J. A., Rabinowitz, L. T., Asmerom, Y., Legesse, K., and Ranney, H. M., 1986, Reverse phase high-performance liquid chromatography and secondary

ion mass spectrometry. A strategy for identification of ten human hemoglobin variants, *Hemoglobin* **10:**379–400.

22. Castagnola, M., Landolfi, R., Rossetti, D. V., De Angelis, F., and Ceccarelli, S., 1986, Determination of abnormal hemoglobins by the combined use of reversed-phase high performance liquid chromatography and fast atom bombardment mass spectrometry, *Anal. Lett.* **19:**1793–1807.
23. Promé, D., Promé, J. -C., Blouquit, Y., Lacombe, C., Rosa, J., and Robinson, J. D., 1987, FAB mapping of proteins: Detection of mutation sites in abnormal human hemoglobins, *Spectros. Int. J.* **5:**157–170.
24. Promé, D., Promé, J. -C., Pratbernou, F., Blouquit, Y., Galacteros, F., Lacombe, C., Rosa, J., and Robinson, J. D., 1988, Identification of some abnormal haemoglobins by fast atom bombardment mass spectrometry and fast atom bombardment tandem mass spectrometry, *Biomed. Environ. Mass Spec.* **16:**41–44.
25. Blouquit, Y., Rhoda, M. -D., Delanoe-Garin, J., Rosa, R., Promé, J. -C., Poyart, C., Puzo, G., Bernassau, J. M., and Rosa, J., 1986, Glycerated hemoglobin, $\alpha_2{}^A\beta_2 82$ (EF6) $N^\varepsilon$-Glyceryllysine. A new post-translational modification occurring in erythrocyte bisphosphoglyceromutase deficiency. *J. Biol. Chem.* **261:**6758–6764.
26. Promé, D., Blouquit, Y., Ponthus, C., Promé, J. -C., and Rosa, J., 1991, Structure of the human adult hemoglobin minor fraction $A_{1b}$ by electrospray and secondary ion mass spectrometry. Pyruvic acid as amino-terminal blocking group, *J. Biol. Chem.* **266:**13050–13054.
27. Fenn, J. B., Mann, M., Meng, C. K., Wong, S. F., and Whitehouse, C. M., 1989, Electrospray ionization for mass spectrometry of large biomolecules, *Science* **246:**64–71.
28. Smith, R. D., Loo, J. A., Edmonds, C. G., Barinaga, C. J., and Udseth, H. R., 1990, New developments in biochemical mass spectrometry: Electrospray ionization. *Anal. Chem.* **62:**882–899.
29. De Caterina, M., Esposito, P., Grimaldi, E., Di Maro, G., Scopacasa, F., Ferranti, P., Parlapiano, A., Marlorni, A., Pucci, P., and Marino, G., 1992, Characterization of hemoglobin Lepore variants by advanced mass-spectrometric procedures, *Clin. Chem.* **38:**1444–1448.
30. Vasseur, C., Blouquit, Y., Kister, J., Promé, D., Kavanaugh, J. S., Rogers, P. H., Guillemin, C., Arnone, A., Galacteros, F., Poyart, C., Rosa, J., and Wajcman, H., 1992, Hemoglobin Thionville. An $\alpha$-chain variant with a substitution of glutamate for valine at NA-1 and having an acetylated methionine $NH_2$ terminus, *J. Biol. Chem.* **267:**12,682–12,691.
31. Witkowska, H. E., Lubin, B. H., Beuzard, Y., Baruchel, S., Esseltine, D. W., Vichinsky, E. P., Kleman, K. M., Bardakdjian-Michau, J., Pinkoski, L., Cahn, S., Roitman, E., Green, B. N., Falick, A. M., and Shackleton, C. H. L., 1991, Sickle cell disease in a patient with sickle cell trait and compound heterozygosity for hemoglobin S and hemoglobin Quebec–Chori, *N. Engl. J. Med.* **325:**1150–1154.
32. Falick, A. M., Witkowska, H. E., Lubin, B. H., Nagel, R. L., and Shackleton, C. H. L., 1991, Identification of variant hemoglobins by tandem mass spectrometry, in *Techniques in Protein Chemistry II* (J. J. Villafranca, ed.), Academic Press, San Diego, pp. 557–565.
33. Ferranti, P., Parlapiano, A., Malorni, A., Pucci, P., Marino, G., Cossu, G., Manca, L., and Masala, B., 1993, Hemoglobin Ozieri: a new $\alpha$-chain variant [$\alpha$71 (E20) Ala→Val]. Characterization using FAB- and electrospray-mass spectrometric techniques, *Biochem. Biophys. Acta* **1162:**203–208.
34. Covey, T. R., Conboy, J. J., and Henion, J. D., 1989, Collision-induced dissociation of multiply charged peptides, in *Proc. 37th Conf. Am. Soc. on Mass Spectrom. and Allied Topics,* Miami Beach, FL, May 21–26, 1989, pp. 905–906.
35. Covey, T. R., Huang, E. C., and Henion, J. D., 1991, Structural characterization of protein tryptic peptides via liquid chromatography/mass spectrometry and collision-induced dissociation of their doubly charged molecular ions, *Anal. Chem.* **63:**1193–1200.

36. Witkowska, H. E., Bitsch, F., and Shackleton, C. H. L., 1993, Expediting rare variant hemoglobin characterization by combined HPLC/electrospray mass spectrometry, *Hemoglobin,* **17:**227–242.
37. Falick, A. M., Shackleton, C. H. L., Green, B. N., and Witkowska, H. E., 1990, Tandem mass spectrometry in the clinical analysis of variant hemoglobins, *Rapid Commun. Mass Spectrom.* **4:**396–400.
38. Jensen, O. N., Höjrup, P., and Roepstorff, P., 1991, Plasma desorption mass spectrometry as a tool in characterization of abnormal proteins. Application to variant human hemoglobins, *Anal. Biochem.* **199:**175–183.
39. Jensen, O. N., Roepstorff, P., Rozynov, B., Horanyi, M., Szelenyi, J., Hollan, S. R., Aseeva, E. A., and Spivak, V. A., 1991, Plasma desorption mass spectrometry of haemoglobin tryptic peptides for the characterization of a Hungarian α-chain variant, *Biol. Mass Spectrom.* **20:** 579–584.
40. Jensen, O. N., and Roepstorff, P., 1991, Application of reversed phase high performance liquid chromatography and plasma desorption mass spectrometry for the characterization of a hemoglobin variant, *Hemoglobin* **15:**497–507.
41. Rubin, E. M., Witkowska, H. E., Spangler, E., Curtin, P., Lubin, B. H., Mohandas, N., and Clift, S. M., 1991, Hypoxia-induced *in vivo* sickling of transgenic mouse red cells, *J. Clin. Invest.* **87:**639–647.
42. Imai, K., Fushitani, K., Miyazaki, G., Ishimori, K., Kitagawa, T., Wada, Y., Morimoto, H., Morishima, I., Shih, D. T., and Tame, J., 1991, Site-directed mutagenesis in haemoglobin. Functional role of tyrosine-α42(C7) at the α1-β2 interface, *J. Mol. Biol.* **218:**769–778.
43. Ishimori, K., Imai, K., Miyazaki, G., Kitagawa, T., Wada, Y., Morimoto, H., and Morishima, I., 1992, Site-directed mutagenesis in hemoglobin: Functional and structural role of inter- and intrasubunit hydrogen bonds as studied with 37β and 145β mutations, *Biochemistry* **31:**3256–3264.
44. Coghlan, D., Jones, G., Denton, K. A., Wilson, M. T., Chan, B., Harris, R., Woodrow, J. R., and Ogden, J. E., 1992, Structural and functional characterization of recombinant human haemoglobin A expressed in *Saccharomyces cerevisiae, Eur. J. Biochem.* **207:**931–936.
45. Wada, Y., Matsuo, T., and Sakurai, T., 1989, Structural elucidation of hemoglobin variants and other proteins by digit-printing method, *Mass Spectrom. Rev.* **8:**379–434.
46. Oliver, R. W. A., and Green, B. N., 1991, The application of electrospray mass spectrometry to the characterization of abnormal or variant haemoglobins, *Trends Anal. Chem.* **10:**85–91.
47. Ferrige, A. G., Seddon, M. J., Green, B. N., Jarvis, S. A., and Skilling, J., 1992, Disentangling electrospray spectra with maximum entropy, *Rapid Commun. Mass Spectrom.* **6:**707–711.
48. Alter, B. P., Modell, C. B., Fairweather, D., Hobbins, J. C., Mahoney, M. J., Frigoletto, F. D., Sherman, A. S., and Nathan, D. G., 1976, Prenatal diagnosis of hemoglobinopathies. A review of 15 cases, *N. Engl. J. Med.* **295:**1437–1443.
49. Larsen, B. S., and McEwen, C. N., 1991, An electrospray ion source for magnetic sector mass spectrometers, *J. Am. Soc. Mass Spectrom.* **2:**205–211.
50. Wada, Y., Tamura, J., Musselman, B. D., Kassel, D. B., Sakurai, T., and Matsuo, T., 1992, Electrospray ionization mass spectra of hemoglobin and transferrin by a magnetic sector mass spectrometer. Comparison with theoretical isotopic distributions, *Rapid Commun. Mass Spectrom.* **6:**9–13.
51. Loo, J. A., Quinn, J. P., Ryu, S. I., Henry, K. D., Senko, M. W., and McLafferty, F. W., 1992, High-resolution tandem mass spectrometry of large biomolecules, *Proc. Natl. Acad. Sci. U.S.A.* **89:**286–289.
52. Huisman, T. H. J., and Wilson, J. B., 1984, The use of HPLC in the study of human hemoglobin variants, *Proteides Biol. Fluids* **32:**1029–1036.

53. Wilson, J. B., 1990, Separation of human hemoglobin variants by high performance liquid chromatography, in *HPLC of Biological Macromolecules, Methods and Applications* (R. Goodieng and F. Regnier, ed.), Marcel Dekker, Inc., New York, pp. 457–472.
54. Wada, Y., Fujita, T., Hayashi, A., Sakurai, T., and Matsuo, T., 1989, Structural analysis of protein variants by mass spectrometry: Characterization of haemoglobin Providence using a grand-scale mass spectrometer, *Biomed. Environ. Mass Spectrom.* **18:**563–565.
55. Abraham, E. C., Reese, A., Stallings, M., and Huisman, T. H. J., 1976/77, Separation of human hemoglobin by DEAE-cellulose chromatography using glycine–KCN–NaCl developers, *Hemoglobin* **1:**27–44.
56. McDonald, M. J., Shapiro, R., Bleichman, M., Solway, J., and Bunn, H. F., 1978, Glycosylated minor components of human adult hemoglobin. Purification, identification, and partial structural analysis, *J. Biol. Chem.* **253:**2327–2332.
57. Righetti, P. G., Gianazza, E., Bianchi-Bosisio, A., and Cossu, G., 1986, Conventional isoelectric focusing and immobilized pH gradients for hemoglobin separation and identification, in *The Hemoglobinopathies* (T. H. J. Huisman, ed.), Churchill Livingstone, Edinburgh, pp. 47–71.
58. Rahbar, S., Louis, J., Lee, T., and Asmerom, Y., 1985, Hemoglobin North Chicago (β36[C2]proline→serine): A new high affinity hemoglobin, *Hemoglobin* **9:**559–576.
59. Ek, K., Bjellqvist, B., and Righetti, R. G., 1983, Preparative isoelectric focusing in immobilized pH gradients. I. General principles and methodology, *J. Biochem. Biophys. Methods* **8:**135–155.
60. Witkowska, E., Bitsch, F., Kleman, K., Pinkoski, L., Kletke, C., Roitman, E., and Shackleton, C., 1992, Integration of ESI MS and LC/ESI MS into the repertoire of techniques used in a hemoglobinopathy laboratory, *Proc. Kyoto '92 Int. Conf. Biol. Mass Spectrom.*, (T. Matsuo, ed.), San-ei Publishing Co., Kyoto, Japan, p. 268–269.
61. Vettore, L., DeMatteis, M. C., Bassetto, M. A., and Pepe, G. M., 1978, Biosynthetic ratio of labelled globin chains in human reticulocytes, determined by electrophoresis on cellulose acetate, *Hemoglobin* **2:**129–141.
62. Tegos, C., and Beutler, E., 1980, A simplified method for studies of haemoglobin biosynthesis, *Clin. Lab Haemat.* **2:**191–197.
63. Righetti, P. G., Gianazza, E., Gianni, A. M., Comi, P., Giglioni, B., Ottolenghi, S., Secchi, C., and Rossi-Bernardi, L., 1979, Human globin chain separation by isoelectric focusing, *J. Biochem. Biophys. Methods* **1:**45–57.
64. Clegg, J. B., Naughton, M. A., and Weatherall, D. J., 1966, Abnormal human haemoglobins. Separation and characterization of the alpha and beta chains by chromatography and the determination of two new variants, Hb Chesapeake and Hb J (Bangkok), *J. Mol. Biol.* **19:**91–108.
65. Pucci, P., Ferranti, P., Marino, G., and Malorni, A., 1989, Characterization of abnormal human haemoglobins by fast atom bombardment mass spectrometry, *Biomed. Environ. Mass Spectrom.* **18:**20–26.
66. Huisman, T. H. J., Webber, B., Okonjo, K., Reese, A. L., and Wilson, J. B., 1981, The separation of human hemoglobin chains by high pressure liquid chromatography, in *Advances in hemoglobin analysis,* (S. M. Hanash and G. J. Brewer, eds.), Alan R. Liss, New York, pp. 23–38.
67. Congote, L. F., and Kendall, A. G., 1982, Rapid analysis of labeled globin chains without acetone precipitation or dialysis by high-pressure liquid chromatography and ion-exchange chromatography, *Anal. Biochem.* **123:**124–132.
68. Shelton, J. B., Shelton, J. R., and Schroeder, W. A., 1984, High performance liquid chromatographic separation of globin chains on a large-pore $C_4$ column, *J. Liquid Chromatog.* **7:**1969–1977.

69. Rahbar, S., and Asmerom, Y., 1989, Rapid HPLC techniques for globin chain synthesis studies, *Hemoglobin* **13**:475–487.
70. Katakuse, I., Ichihara, T., Nakabushi, H., Matsuo, T., Matsuda, H., Wada, Y., and Hayashi, A., 1984, Secondary ion mass spectra of tryptic peptides of human hemoglobin chains, *Biomed. Mass Spectrom.* **11**:386–391.
71. Schwartz, W. E., Smith, P. K., and Royer, G. P., 1980, N-(β-iodoethyl)trifluoroacetamide: a new reagent for the aminoethylation of thiol groups in proteins, *Anal. Biochem.* **106**:43–48.
72. Stone, K. L., LoPresti, M. B., Crawford, J. M., DeAngelis, R., and Williams, K., 1989, Enzymatic digestion of proteins and HPLC peptide isolation, in *A Practical Guide to Protein and Peptide Purification for Microsequencing,* (P. T. Matsudaira, ed.), Academic Press, San Diego, pp. 33–47.
73. Kostka, V., and Carpenter, F. H., 1964, Inhibition of chymotryptic activity in crystalline trypsin preparations, *J. Biol. Chem.* **239**:1799.
74. Stachowiak, K., and Dyckes, D. F., 1989, Peptide mapping using thermospray LC/MS detection: Rapid identification of hemoglobin variants, *Peptide Res.* **2**:267–274.
75. Craik, C. S., Largman, C., Fletcher, T., Roczniak, S., Barr, P. J., Fletterick, R., and Rutter, W. J., 1985, Redesigning trypsin: Alteration of substrate specificity, *Science* **228**:291–297.
76. Lehman, H., and Huntsman, R. G., 1974, *Man's Hemoglobin,* North Holland, Amsterdam.
77. Medzihradszky, K. F., Falick, A. M., Schindler, P. A., and Burlingame, A. L., 1993, Side reactions and non-specific cleavages observed during mass spectrometric analysis of proteins, *Anal. Biochem.*
78. Shackleton, C. H. L., Falick, A. M., Green, B. N., and Witkowska, H. E., 1991, Electrospray mass spectrometry in the clinical diagnosis of variant hemoglobins, *J. Chromatog.* **562**:175–190.
79. Jekel, P. A., Weijer, W. J., and Beintema, J. J., 1983, Use of endoproteinase Lys-C from *Lysobacter enzymogenes* in protein sequence analysis, *Anal. Biochem.* **134**:347–354.
80. Berezin, I. V., and Martinek, K., 1970, Specificity of α-chymotrypsin, *FEBS Lett.* **8**:261.
81. Wada, Y., Ikkala, E., Imai, K., Matsuo, T., Matsuda, H., Lehtinen, M., Hayashi, A., and Lehmann, H., 1987, Structure and function of a new hemoglobin variant, Hb Meilahti ($\alpha_2\beta_2$ 36(C2)Pro→Thr), characterized by mass spectrometry, *Acta Haemat.* **78**:109–113.
82. Huisman, T. H. J., and Jonxis, J. H. P., 1977, *The Hemoglobinopathies: Techniques of Identification,* Marcel Dekker, New York.
83. Houmard, J., and Drapeau, G. R., 1972, Staphylococcal protease: A proteolytic enzyme specific for glutamyl bonds, *Proc. Natl. Acad. Sci. U.S.A.* **69**:3506–3509.
84. Pucci, P., Ferranti, P., Malorni, A., and Marino, G., 1990, Fast atom bombardment mass spectrometric analysis of haemoglobin variants: Use of V-8 protease in the identification of Hb M Hyde Park and Hb San Jose, *Biomed. Environ. Mass Spectrom.* **19**:568–572.
85. Folk, J. E., 1970, Carboxypeptidase B (porcine pancreas), in *Proteolytic Enzymes* (G. E. Perlmann and L. Lerand, eds.) *Methods in Enzymology* (S. P. Colowick and N. O. Kaplan, eds.), Vol. XIX, Academic Press, New York, pp. 504–508.
86. De Biasi, R., Spiteri, D., Caldora, M., Iodice, R., Pucci, P., Malorni, A., Ferranti, P., and Marino, G., 1988, Identification by fast atom bombardment mass spectrometry of Hb Indianapolis [β112(G14)Cys→Arg] in a family from Naples, Italy, *Hemoglobin* **12**:323–336.
87. Hayashi, R., Moore, S., and Stein, W. H., 1973, Carboxypeptidase from yeast, *J. Biol. Chem.* **248**:2296–2302.
88. Kuhn, R. W., Walsh, K. A., and Neurath, H., 1974, Isolation and partial characterization of an acid carboxypeptidase from yeast, *Biochemistry* **13**:3871–3877.
89. Ryle, A. P., 1970, The porcine pepsin and pepsinogens, in *Proteolytic Enzymes* (G. E. Perlmann and L. Lerand, eds.) *Methods in Enzymology.* (S. P. Colowick and N. O. Kaplan, eds.), vol. XIX, Academic Press, New York, p. 316.

90. Casey, R., and Lang, A., 1975, Specific cyanylation and cleavage at cysteine-104 of human haemoglobin α-chain, *Biochem. J.* **145**:251–261.
91. Piot, J. -M., Guillochon, D., Zhao, Q., Ricart, G., Fournet, B., and Thomas, D., 1989, Identification of peptides, from a peptic haemoglobin hydrolysate produced by a pilot-plant scale, by high-performance liquid chromatography and mass spectrometry, *J. Chromatog.* **481**:221–231.
92. Piot, J. M., Zhao, Q., Guillochon, D., Dhulster, P., Ricart, G., and Thomas, D., 1990, Semi-preparative purification and characterization of peptides from complex haemoglobin hydrolysate by HPLC-mass spectrometry, *Chromatographia* **30**:205–210.
93. Witkowski, A., Naggert, J., Witkowska, H. E., Randhawa, Z. I., and Smith, S., 1992, Utilization of an active site serine 101→cysteine mutant to demonstrate the proximity of the catalytic serine 101 and histidine 237 residues in thioesterase II, *J. Biol. Chem.* **267**:18,488–18,492.
94. Jacobsen, G. R., Schaffer, M. H., Stark, G. R., and Vanaman, T. C., 1973, Specific chemical cleavage in high yield at the amino peptide bonds of cysteine and cystine residues, *J. Biol. Chem.* **248**:6583–6591.
95. Chait, B. T., and Kent, S. B. H., 1992, Weighing naked proteins: practical, high-accuracy mass measurement of peptides and proteins, *Science* **257**:1885–1894.
96. Juhasz, P., Papayannopoulos, I. A., Zeng, C., Papov, V., and Biemann, K., 1992, The utility of matrix-assisted laser desorption for the direct analys of enzymatic digests of proteins, *Proc. 40th ASMS Conf. Mass Spectrom. Allied Topics,* May 31–June 5, 1992, Washington, D.C., pp. 1913–1914.
97. Hunt, D. F., Alexander, J. E., McCormack, A. L., Martino, P. A., Michel, H., Shabanowitz, J., Sherman, N., Moseley, M. A., Jorgenson, J. W., and Tomer, K. B., 1991, Mass spectrometric methods for protein and peptide sequence analysis, in *Techniques in Protein Chemistry II,* (J. J. Villafranca, ed.), Academic Press, San Diego, pp. 441–454.
98. Katta, V., Chowdhury, S. K., and Chait, B. T., 1991, Use of a single-quadrupole mass spectrometer for collision-induced dissociation studies of multiply charged peptide ions produced by electrospray ionization, *Anal. Chem.* **63**:174–178.
99. Lee, T. D., and Vemuri, S., 1990, MacProMass: A computer program to correlate mass spectral data to peptide and protein structures, *Biomed. Environ. Mass Spectrom.* **19**:639–645.
100. Bitsch, F., Witkowska, H. E., Nugent, K., King, D., and Shackleton, C. H. L., 1992, Assessing structural alterations in natural and recombinant proteins by RP-LC/ESI MS, *Proc. 40th ASMS Conf. Mass Spectrom. Allied Topics,* May 31–June 5, 1992, Washington, D.C., pp. 428–429.
101. Henzel, W. J., Bourell, J. H., and Stults, J. T., 1990, Analysis of protein digests by capillary high-performance liquid chromatography and on-line fast atom bombardment mass spectrometry, *Anal. Biochem.* **187**:228–233.
102. Davies, M. T., and Lee, T. D., 1992, Analysis of peptide mixtures by capillary high performance liquid chromatography: A practical guide to small-scale separations, *Protein Sci.* **1**:935–944.
103. Udseth, H. R., Barinaga, C. J., and Smith, R. D., 1990, Developments in combined capillary electrophoresis-ESI-MS., *Proc. 38th ASMS Conf. Mass Spectrom. Allied Topics,* June 3–8, 1990, Tucson, AZ, pp. 1210–1211.
104. Ferranti, P., Malorni, A., Pucci, P., Fanali, S., Nardi, A., and Ossicini, L., 1991, Capillary zone electrophoresis and mass spectrometry for the characterization of genetic variants of human hemoglobin, *Anal. Biochem.* **194**:1–8.
105. Johansson, I. M., Huang, E. C., Henion, J. D., and Zweigenbaum, J., 1991, Capillary electrophoresis-atmospheric pressure ionization mass spectrometry for the characterization of

peptides. Instrumental considerations for mass spectrometric detection, *J. Chromatog.* **334:**311–327.

106. Wahl, J. H., Goodlett, D. R., Udseth, H. R., and Smith, R. D., 1992, Attomole level capillary electrophoresis-mass spectrometric protein analysis using 5-μm-i.d. capillaries, *Anal. Chem.* **64:**3194–3196.
107. Yates III, J. R., Griffin, P. R., Hood, L. E., and Zhou, J. X., 1991, Computer aided interpretation of low energy MS/MS mass spectra of peptides, in *Techniques in Protein Chemistry II* (J. J. Villafranca, ed.), Academic Press, San Diego, pp. 477–485.
108. Loo, J. A., Edmonds, C. G., and Smith, R. D., 1990, Primary sequence information from intact proteins by electrospray ionization tandem mass spectrometry, *Science* **248:**201–204.
109. Light-Wahl, K. J., Loo, J. A., Edmonds, C. G., Smith, R. D., Witkowska, H. E., Shackleton, C. H. L., and Wu, C. -S. C., 1993, Collisionally activated dissociation and tandem mass spectrometry of intact hemoglobin β-chain variant proteins with electrospray ionization, *Biol. Mass Spectrom.* **22:**112–120.
110. Molchanova, T. P., Mirgorodskaya, O. A., Abaturov, L. V., Podtelezhnikov, A. V., Tokarev, Y. N., and Grachev, S. A., 1989, Localization of amino acid substitutions in human hemoglobin. Mass-spectral approximate analysis of tryptic peptides, *Molekulyarnaya Biologiya (Russian Original)* **23:**169–180; English transl. August 1989, Consultants Bureau, New York.
111. Wada, Y., Matsuo, T., Papayannopoulos, I. A., Costello, C. E., and Biemann, K., 1992, Fast atom bombardment and tandem mass spectrometry for the characterization of hemoglobin variants including a new variant, *Int. J. Mass Spectrom. Ion. Processes* **122:**219–229.
112. Lacombe, C., Promé, D., Blouquit, Y., Bardakdjian, J., Arous, N., Mrad, A., Promé, J. -C., and Rosa, J., 1990, New results of hemoglobin variant structure determinations by fast atom bombardment mass spectrometry, *Hemoglobin* **14:**529–548.
113. Boissel, J., Kasper, T. J., Shah, S. C., Malone, J. I., and Bunn, H. F., 1985, Amino-terminal processing of proteins: Hemoglobin South Florida, a variant with retention of initiator methionine and $N^{\alpha}$-acetylation, *Proc. Natl. Acad. Sci. U.S.A.* **82:**8448–8452.
114. Foldi, J., Horanyi, M., Szelenyi, J. G., Hollan, S. R., Aseeva, E. A., Lutsenko, I. N., Spivak, V. A., Toth, O., Rozynov, B. V., 1989, Hemoglobin Siriraj found in the Hungarian population, *Hemoglobin* **13:**177–180.
115. Frigeri, F., Pandolfi, G., Camera, A., Rotoli, B., Ferranti, P., Malorni, A., and Pucci, P., 1990, Hb G-San José: Identification by mass spectrometry, *Clin. Chem. Enzym. Commun.* **3:**289–294.
116. De Angioletti, M., Maglione, G., Ferranti, P., de Bonis, C., Lacerra, G., Scarallo, A., Pagano, L., Fioretti, G., Cutolo, R., Malorni, A., Pucci, P., and Carestia, C., 1992, Hb City of Hope [β69 (E13) Gly→Ser] in Italy: Association of the gene with haplotype IX, *Hemoglobin* **16:**27–34.
117. Williamson, D., Nutkins, J., Rosthoj, S., Brennan, S. O., Williams, D. H., and Carrell, R. W., 1990, Characterization of Hb Aalborg, a new unstable hemoglobin variant, by fast atom bombardment mass spectrometry, *Hemoglobin* **14:**137–145.
118. Lane, P., Witkowska, H. E., Falick, A. M., Houston, M. L., and McKenna, J. D., 1993, Hemoglobin D-Ibadan–β°thalassemia: Detection by neonatal screening and confirmation by electrospray ionization mass spectrometry, *Am. J. Hematol.,* **44:**158–161.
119. Marsh, G., Marino, G., Pucci, P., Ferranti, P., Malorni, A., Kaeda, J., Marsh, J., and Luzzatto, L., 1991, A third instance of the high oxygen affinity variant, Hb Heathrow [β103 (G5) Phe→Leu]: Identification of the mutation by mass spectrometry and by DNA analysis, *Hemoglobin* **15:**43–51.
120. Wada, Y., Hayashi, A., Oka, Y., Matsuo, T., Sakurai, T., Matsuda, H., and Katakuse, I., 1989,

Mass spectrometric characterization of a haemoglobin variant, haemoglobin Riyadh, *Int. J. Mass Spectrom. Ion Processes* **91:**79–84.

121. Wada, Y., Fujita, T., Kidoguchi, K., and Hayashi, Y., 1986, Fetal hemoglobin variants in 80,000 Japanese neonates: high prevalence of Hb F Yamaguchi ($^{A}\gamma^{T}$80 Asp→Asn), *Hum. Genet.* **72:**196–202.
122. Fujita, S., Ohta, Y., Saito, S., Kobayashi, Y., Naritomi, Y., Kawaguchi, T., Imamura, T., Wada, Y., and Hayashi, A., 1985, Hemoglobin $A_2$ Honai ($\alpha_2\beta_2$90(F6)Glu→Val): A new delta chain variant, *Hemoglobin* **9:**597–607.
123. Johnson, R. S., Martin, S. A., and Biemann, K., 1988, Collision-induced fragmentation of $[M+H]^+$ ions of peptides: Side-chain-specific ions, *Int. J. Mass Spectrom. Ion Proc.* **86:**137–154.

# 5

# Analysis of Urinary Nucleosides

*Thomas D. McClure and Karl H. Schram*

## 1. INTRODUCTION

Conventional diagnosis, disease staging, and evaluation of therapy for cancer and acquired immune deficiency syndrome (AIDS) rely heavily on clinically observable measurements rather than biochemical tests. In the case of cancer, these measurements include tumor imaging and tissue biopsy and are accurate only after many cancer cells have been produced as a consequence of disease progression. A diagnosis of AIDS is established by detection of host-produced antibodies to the human immunodeficiency virus (HIV), whereas staging, measurement of response to therapy, and determining the type and severity of opportunistic infections (OI) suffered by AIDS patients is determined by measurement of the T- and CD-4 cell populations (1). For both cancer and AIDS, these methods can be unreliable; also, in some instances, they are painful and increase risk to the patient. For example, detection of mesothelioma (cancer of the pleural covering of the lung, most often the result of exposure to asbestos fiber) is an invasive procedure requiring rib resection to obtain tissue biopsy for cytological analysis. A dangerous level of unreliability in detecting HIV infection has been observed and results in a long lead time between HIV infection and production of antibodies. In a recent study (2), patients were shown to be infected with HIV for as long as 35 months before antibodies were detected. As a result of these deficiencies in clinical test methods, considerable research

*Thomas D. McClure and Karl H. Schram* • Department of Pharmacology/Toxicology, College of Pharmacy, University of Arizona, Tucson, Arizona 85721.

*Mass Spectrometry: Clinical and Biomedical Applications, Volume 2*, edited by Dominic M. Desiderio. Plenum Press, New York, 1994.

effort has been expended toward finding alternative indicators or "biomarkers" of cancer and AIDS.

The goal of AIDS and cancer biomarker research is to discover biochemical compounds that are more sensitive, specific, quantitative, and reproducible in detecting very early infection/occurrence and disease staging (3). An ideal biomarker of cancer or AIDS should possess the following characteristics (4):

1. The marker(s) should be present in body fluids, thus eliminating the need for invasive surgery.
2. The marker(s) should not only be specific for cancer, but should indicate the type of cancer.
3. Body-fluid levels of the marker(s) should correlate to tumor burden or disease stage.
4. The incidence of false positive and false negative results should be statistically acceptable.
5. The methods for detecting the marker(s) should be amenable to screening large populations.

A number of classes of biomedicals has been studied as biomarkers, including enzymes involved in DNA synthesis and tRNA (transfer RNA), methylating enzymes; serum polypeptides such as carcinoembryonic antigen, tissue polypeptide antigen, and placental alkaline phosphatase; the urinary polyamines putrescine, spermidine, and spermine; and urinary modified nucleosides (5–7). The last class of compounds, urinary modified nucleoside(s) (UMN), show considerable promise as biomarkers for both cancer and AIDS and are the subject of this review. Ideally, a UMN biomarker would be below detection limits in nondisease control samples but would be present in samples from cancer or AIDS patients at concentrations indicative of the stage of the disease. The term "modified nucleoside" has been used to describe a nucleoside that has been singly methylated in the aglycone portion of the molecule or on the 2′-hydroxyl within the ribose moiety. Modifications to nucleosides can be much more extensive and are often termed hypermodified nucleosides (8). For this discussion, the term "modified nucleoside" will be used to refer to any nucleoside that has a structure different from adenosine, cytidine, guanosine, thymine-riboside, uridine, and their deoxy- counterparts. Table 5.1 contains a list of modified nucleosides that have been found in urine, with the corresponding structures found in Fig. 5.1. (Numbers in bold type below provide reference to Table 5.1 and Fig. 5.1 for substances indicated.)

### *1.1. Urinary Modified Nucleosides as Biomarkers for Cancer*

The search for UMN as biomarkers for cancer involves two different experimental approaches. One is the structural elucidation of the new nucleo-

*Table 5.1*
*Urinary Nucleosides Discovered in the Urine of Control and Cancer Patients*

| # | Abbreviation | Compound | Urine | Reference |
|---|---|---|---|---|
| 1 | A | **adenosine** | 0.23±0.07[a] | 30 |
| | | | 1.04±0.42[a] | 75 |
| | | | 0.29±0.09[a] | 29 |
| | | | trace | 76 |
| 2 | Am | 2′-O-methyl- | * | 77 |
| 3 | dA | 2′-deoxy- | 0.76±1.13[a] | 75 |
| 4 | $M^1A$ | 1-methyl- | 4.32±1.58[a] | 78 |
| | | | 1.77±0.58[a] | 22 |
| | | | 5.91[b] | 25 |
| | | | 1.8[a] | 6 |
| | | | 2.02±0.35[a] | 30 |
| | | | 2.15±0.4[a] | 29 |
| 5 | $m^6A$ | $N^6$-methyl- | 2.1–5.2[b] | 76 |
| 6 | $i^6A$ | isopentenyl- | 0.050±0.019[b] | 39 |
| 7 | $Suc^6A$ | $N^6$-succinyl- | nq[c] | 78 |
| 8 | $m^{1,6}{}_2A$ | 1,$N^6$-dimethyl- | [d] | 9 |
| | | **N-[N-(9-β-D-ribofuranosyl** | | |
| 9 | $t^6A$ | -purine-6-yl)carbamoylthreonine | 0.3[b] | 76 |
| | | | 2.2±0.9[b] | 39 |
| 10 | MTA | 5′-deoxy-5′-methylthio- | [d] | 80 |
| 11 | MTA-SO | -5′-methylthio sulfoxide | 0.34±0.21[a] | 81 |
| | MTA-SO | | [d] | 10 |
| 12 | C | **Cytidine** | trace | 76 |
| 13 | Cm | 2′-O-methyl- | 0.2–0.5[b] | 76 |
| | | | [d] | 77 |
| 14 | dC | 2′-deoxy- | trace | 76 |
| 15 | $m^5C$ | 5-methyl- | 0.11±0.03[a] | 30 |
| 16 | $ac^4C$ | $N^4$-acetyl- | nq | 44 |
| 17 | G | **Guanosine** | trace | 76 |
| 18 | Gm | 2′-O-methyl- | [d] | 77 |
| | | | 0.17±0.09[b] | 39 |
| 19 | $m^2_2G$ | $N^2$,$N^2$-dimethyl- | 1.36±0.90[a] | 78 |
| | | | 1.2±0.30[a] | 82 |
| | | | 0.18–2.00[a] | 83 |
| | | | 2.27[a] | 25 |
| | | | 0.8[a] | 6 |
| | | | 105±0.12[a] | 30 |
| | | | 7.6±5.9[a] | 75 |
| | | | 3.27±0.40[a] | 84 |
| | | | 1.25±0.30 [a] | 29 |
| | | | 1.1–2.1[b] | 76 |
| | | | 0.05±0.022[b] | 21 |

(*continued*)

*Table 5.1 (Continued)*

| # | Abbreviation | Compound | Urine | Reference |
|---|---|---|---|---|
| 20 | $m^1G$ | 1-methyl- | 0.4–0.6[b] | 76 |
| | | | 0.96±0.44[a] | 78 |
| | | | 1.41[a] | 25 |
| | | | 0.6[a] | 6 |
| | | | 0.70±0.04[a] | 85 |
| | | | 1.83±0.29[a] | 84 |
| | | | 0.71±0.16[a] | 29 |
| | | | 0.68[a] | 27 |
| 21 | $m^2G$ | $N^2$-methyl- | 0.2–0.3[b] | 76 |
| | | | 1.29±1.08[a] | 78 |
| | | | 1.20±0.15[a] | 82 |
| | | | 0.35±0.14[a] | 22 |
| | | | 2.38[b] | 25 |
| | | | 0.3[a] | 6 |
| 22 | $m^7G$ | 7-methyl- | 1.46[a] | 25 |
| 23 | MTG | 5′-deoxy-5′-methylthio- | [d] | 62 |
| | | **7-β-D-ribofuranosyl** | | |
| 24 | 7-R-Hyp | -hypoxanthine | [d] | 11 |
| | | **imidazole** | | |
| 25 | 5,4-ACRI | 5-amino-4-carboxamide-riboside- | 0.18[b] | 76 |
| 26 | I | **inosine** | 0.18 | 86 |
| 27 | Im | 2′-O-methyl | [d] | 87 |
| 28 | $m^1I$ | 1-methyl- | 2.1–2.9 | 86 |
| 29 | 5′-dI | 5′-deoxy | [d] | 10 |
| 30 | Ψ | pseudouridine | 42–62 [b] | 76 |
| | | | 0.78±0.206[b] | 21 |
| | | | 0.07±0.078[b] | 88 |
| | | | 17.77±6.36[a] | 78 |
| | | | 22.4–2 6.7[a] | 22 |
| | | | 22.6±4.32[a] | 82 |
| | | | 8.57–27.64[a] | 83 |
| | | | 24.13[a] | 25 |
| | | | 0.30±0.08 | 28 |
| | | | 23.5[a] | 6 |
| | | | 24.80±4.8[a] | 30 |
| | | | 66.7±22.2[a] | 75 |
| | | | 33.19±3.9[a] | 84 |
| | | | 20.89±4.32[a] | 29 |
| | | | 20.33[a] | 27 |

*Table 5.1 (Continued)*

| # | Abbreviation | Compound | Urine | Reference |
|---|---|---|---|---|
| 31 | U | **uridine** | trace | 76 |
| 32 | Um | 2′-O-methyl- | 0.1[b] | 89 |
| | | | [d] | 77 |
| 33 | D | 5,6-dihydro- | nq | 46 |
| 34 | $m^3U$ | 3-methyl- | nq | 64 |
| | | | 0.32 | 63 |
| 35 | $ncm^5U$ | 5-carbamoylmethyl- | [d] | 90 |
| 36 | $ncm^5s^2U$ | 5-carbamoylmethyl-2-thio- | nq | 91 |
| 37 | $acp^3U$ | 3-(3-amino-3-carboxypropyl)- | [d] | 92 |
| | | **1-(β-D-ribofuranosyl)-pyridine** | | |
| 38 | PCNR | -5-carboxamide-2-one | 0.16[a] | 93 |
| 39 | 4,3-PCNR | -3-carboxamide-4-one | 1.8[b] | 52 |
| | | | 0.87±0.25[a] | 93 |
| | | **1-(α-D-ribofuranosyl)-pyridin** | | |
| 40 | α-4,3-PCNR | -3-carboxamide-4-one | 0.19[b] | 94 |
| 41 | X | **xanthosine** | | |
| 42 | 5′-dX | 5′-deoxy- | nq | 69 |

[a]Values in nmol/μmol creatinine.
[b]Values in mg/24h.
[c]nq = not quantitated.
[d]Identified in urine samples from cancer patients, thus quantities may vary.

sides usually present in low levels in body fluids of cancer patients but absent in control samples from control subjects. Whereas this approach is qualitative, at least during the identification stage, the potential for high selectivity toward the type and level of cancer is great because these markers *may* be unique.

The rationale for the appearance of these novel nucleosides is based on the premise that, upon transformation from the normal state to malignancy—a process known to involve nucleic acids—cellular biochemistry is altered. The appearance of novel nucleosides in urine (or other body fluids) may result from these nonspecific changes. Recent examples of UMN that *may* be specific for cancer include: $m^{1,6}_2$ A (**8**) (9), 5′-dI (**29**) (10), and 7-R-Hyp (**24**) (11) in the urine of cancer patients; further studies of these novel UMN may reveal a quantitative relationship between their concentration in urine and the type or level of cancer.

The second approach, which is quantitative, has focused on comparing the levels of previously identified UMN in control and cancer-patient samples. The

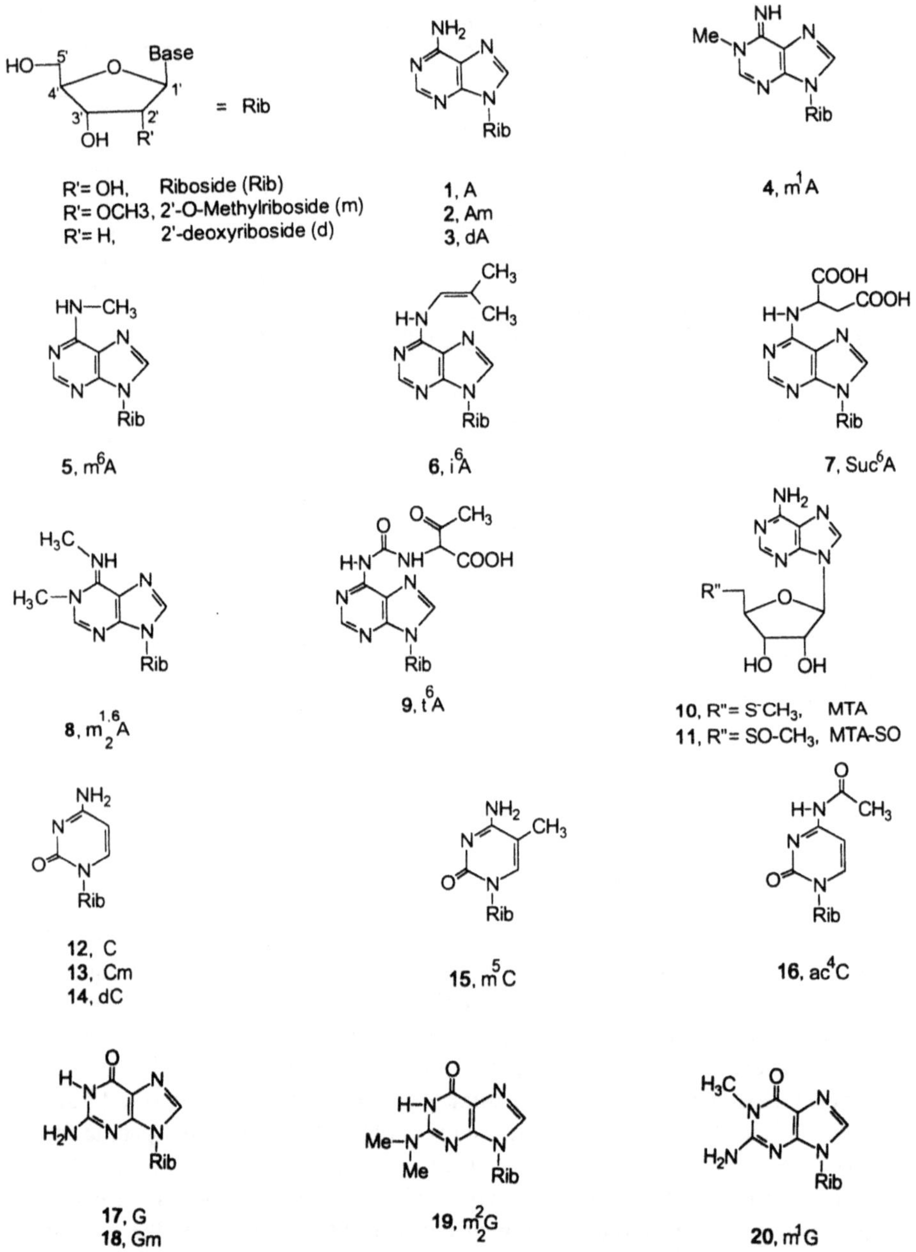


*Figure 5.1.* Structures of urinary nucleosides found in Table 5.1.

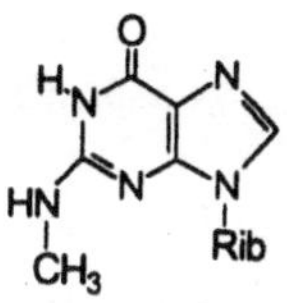

**21**, m$^2$G

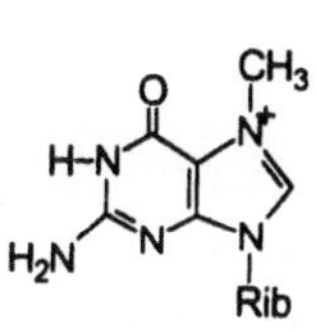

**22**, m$^7$G

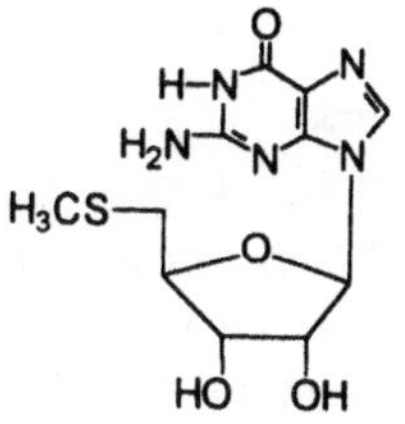

**23**, MTG

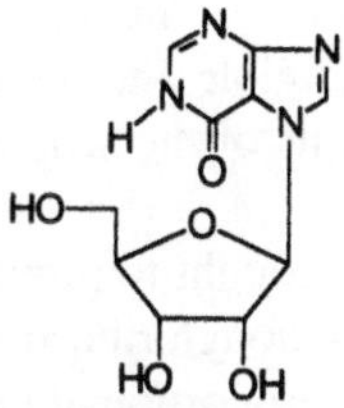

**24**, 7R-Hyp

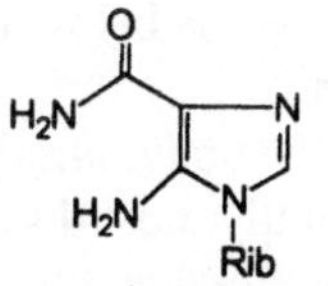

**25**, 5,4-AICR

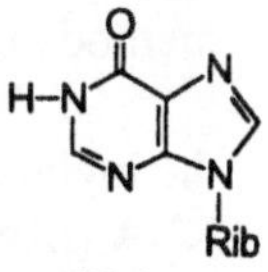

**26**, I
**27**, Im

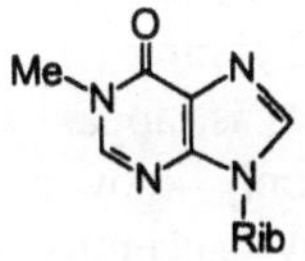

**28**, m$^1$I

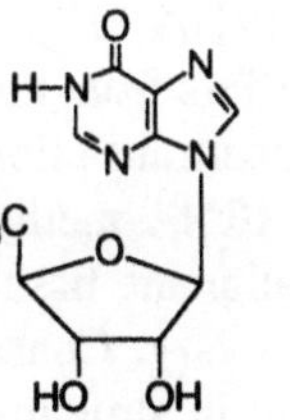

**29**, 5'-dI

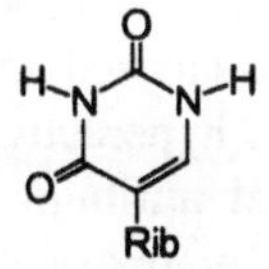

**30**, Ψ

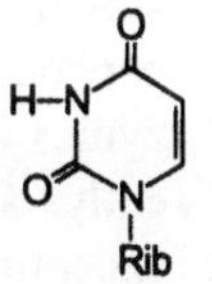

**31**, U
**32**, Um

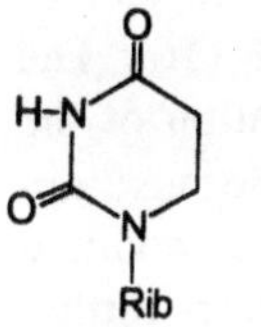

**33**, D

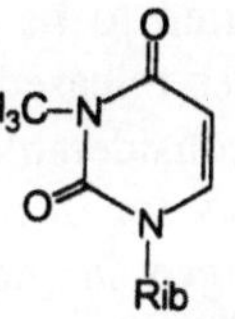

**34**, m$^3$U

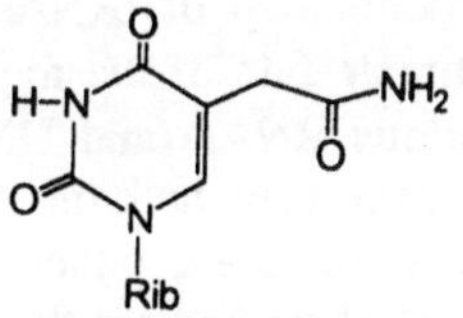

**35**, mcm$^5$U

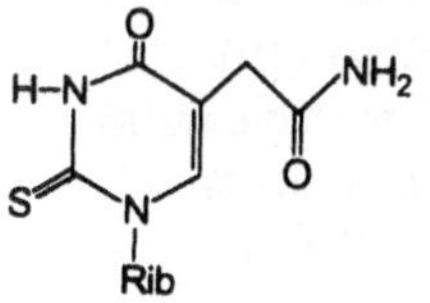

**36**, mcm$^5$s$^2$U

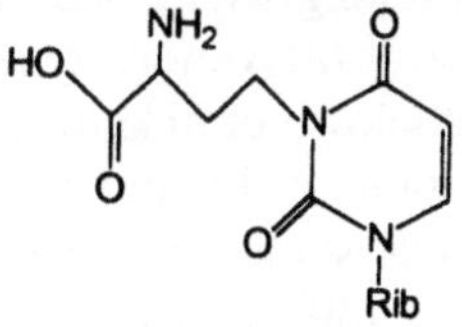

**37**, acp$^3$U

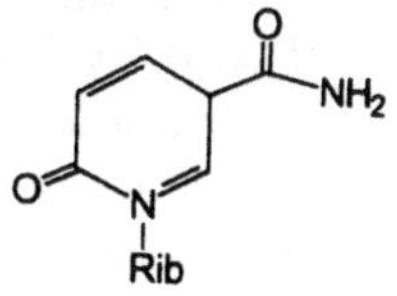

**38**, PCNR

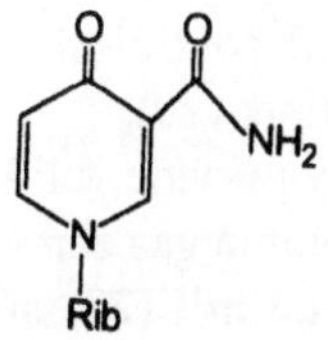

**39**, 4,3-PCNR

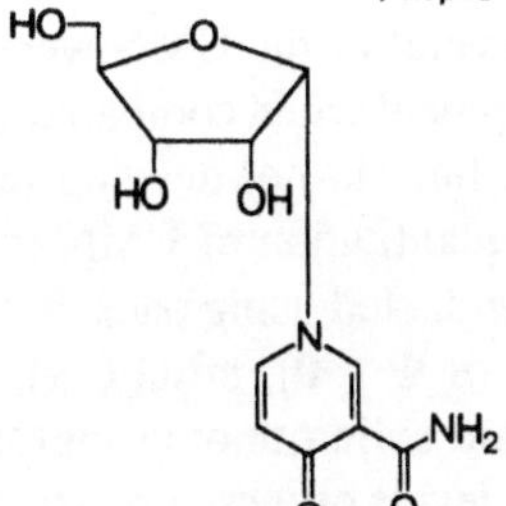

**40**, α-4,3-PCNR

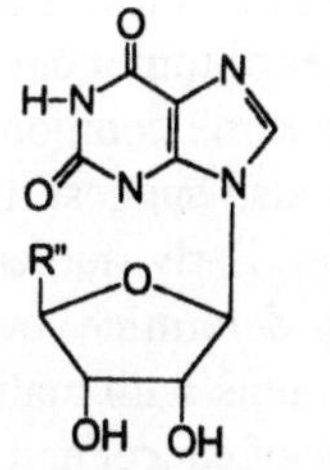

**41**, R'' = $CH_2OH$, X
**42**, R'' = $CH_3$ , 5'-dX

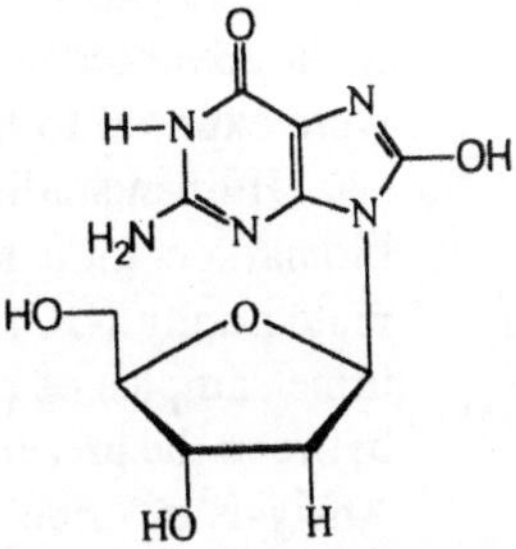

**43**

*Figure 5.1.* (*Continued*)

biochemical rationale for differences in the levels is the belief that UMN are derived almost exclusively from RNA. Even though UMN have been identified from other sources, e.g., 8-hydroxy-2′-deoxyguanosine, **43,** is formed as a result of oxidative damage to DNA by free radical species (12). The level of UMN formed by such "minor pathways" is considered negligible and, unless special techniques are employed, is below the detection limits of the standard methods used in quantitation.

Of the RNA present in the body, turnover of tRNA is thought to provide the highest levels and structurally most diverse UMN. Even though ribosomal RNA (rRNA) contains a larger total number of nucleosides, proportionally the number of modified nucleosides in rRNA is only about 3% compared to the 15–20% of modified nucleosides in tRNA (13). Furthermore, the primary structure of tRNA in neoplastic cells differs from that found in control cells, in part because of increased activity of posttranslational tRNA methylating enzymes (14). This hyperactivity of the tRNA methylating enzymes is, apparently, a universal attribute of all malignant tissues, with the only known exception being tumors of the human ovary. Consequently, every cancerous tissue contains tRNAs that are different in structure from their control tissue counterpart (15). Posttranslational modification of the nucleosides in the anticodon loop of tRNAs has been reported to be as high as 25% (16). The ultimate fate of the modified nucleosides released during degradation of the various RNAs (and DNA) is also of considerable importance. Because eucaryotic cells lack the enzymes necessary for their reutilization, the modified nucleosides are either excreted unchanged in the urine or undergo further metabolism and are then excreted (17). The levels of selected UMN have also been suggested as a means of determining whole-body RNA catabolism (18) and even for differentiating turnover of the various types of RNA (19).

Do the levels of these UMN, of known or unknown structure, have utility as biomarkers of cancer? Based on a relatively large body of data, the answer appears to be yes. For example, animal studies have shown that the excretion rate of the nucleosides $ac^4C$ (**16**), Ψ (**30**), and $m^1A$ (**4**), is increased upon exposure to cancer-inducing compounds even before the tumor is detectable by standard methods, and that levels of several of the UMN were highly elevated in the advanced stages of tumor development when compared with controls that were exposed to the same compounds but did not develop tumors (20).

Human studies also suggest that quantitation of UMN may provide valid biomarkers for cancer. Early studies conducted using packed-column gas chromatography (GC) to determine levels of Ψ (**30**), $m^2_2G$ (**19**), and $m^1I$ (**28**) in urine samples of patients with malignant solid tumors suggested a correlation between the presence of cancer and the levels of these urinary nucleosides (21). Analysis of urine from 27 patients representing 13 different malignancies for various UMN, including Ψ (**30**), $m^1A$ (**4**), $m^1I$ (**28**), $m^2G$ (**21**), and $m^2_2G$ (**19**) found that 26 of the 27 patients showed elevated levels of these potential biomarkers (22). The significance of these observations has been validated using multivariate statistical analysis methods to show that UMN can be used as markers for diagnosis and evaluation of follow-up therapy for various types of cancer (23).

Levels of known modified nucleosides in the urine of patients with specific types of cancer have also been examined in efforts to develop noninvasive methods for diagnosis or staging of the disease. Urinary concentrations of Ψ (**30**), $m^1A$ (**4**), $m^2_2G$ (**19**), $m^1I$ (**28**), $m^1G$ (**20**), and $m^2G$ (**21**) were used to evaluate 22 patients with breast carcinoma, 16 with and six without metastases. Fourteen of the 16 patients with metastatic cancer showed elevated levels of one or more nucleosides, whereas only one of the six without metastases had elevated levels of UMN (24). Similarly, lymphoma-bearing patients have been shown to excrete elevated levels of selected UMN compared to controls, but no correlation could be established as to the type of cancer and the level or structure of the modified nucleosides examined (25).

Mentioned earlier was the difficulty in diagnosing mesothelioma. In an effort to develop a noninvasive diagnostic procedure for this type of cancer, the levels of UMN in a subpopulation known to be at risk through occupational exposure to asbestos were measured and compared to levels observed in the urine of patients confirmed as having mesothelioma. Both groups showed elevated levels of UMN, with the mesothelioma group displaying the greatest variation in type and the highest levels of modified nucleosides (4). In a related double-blind study, nine out of nine cases of mesothelioma were diagnosed using UMN as biomarkers (15,26).

Other types of cancer in which levels of UMN have been used as potential biomarkers include chronic myelogenous leukemia (27,28), childhood acute lymphoblastic leukemia (29), and nasopharyngeal carcinoma (30).

In addition to cancer, levels of UMN have been suggested as being diagnostic of certain immunodeficiency diseases in children. One such disorder is adenosine deaminase deficiency, a condition that can lead to lymphocyte dysfunction and alters purine and pyrimidine metabolism, as is evidenced by a change in the urinary nucleoside profile. The specific nucleosides studied for this condition were $m^1A$ (**4**), $m^1I$ (**28**), A (**1**), MTA-SO (**11**,), $m^1G$ (**20**), C (**12**) and dC (**14**). The conclusion of this study was that the UMN mentioned can be used to diagnose adenosine deaminase deficiency (31,32).

### *1.2. Urinary Modified Nucleosides as Biomarkers for AIDS*

Although fewer data have been accumulated on UMN levels in AIDS patients, reports in the literature strongly suggest that AIDS may be diagnosed and possibly staged using UMN as biomarkers.

The biochemical rationale for higher levels of UMN associated with AIDS is unclear but may be in part a result of an increase in the turnover of tRNA in infected cells. This increase turnover is due to retrovirus–host cell interactions and not cellular transformation. Such a mechanism has been invoked to explain the increased levels of excreted Ψ (**30**) observed in chicken fibroblasts infected with Rous sarcoma virus, which, like HIV, is a retrovirus (33,34).

Whatever the reason, recent studies have reported that the levels of the modified nucleosides PCNR (**38**), $m^1A$ (**4**), $m^1I$ (**28**), $m^1G$ (**20**), $m^2G$ (**21**), $m^2_2G$ (**19**), $ac^4C$ (**16**), and $t^6A$ (**9**) are increased in the urine of AIDS patients. Figure 5.2 shows a comparison of the levels of these nucleosides in the urine of control subjects versus those in patients considered at high risk for HIV infection and those with AIDS (35,36). Of particular significance in these results is the finding that the urine of persons in the high-risk group who were asymptomatic and whose enzyme-linked immunosorbent assay (ELISA) test was negative for the presence of HIV showed elevated levels of some of these UMN. In addition, the level and identity of the UMN may allow differentiation of the stage of the disease (37). Consequently, modified nucleosides in urine may hold promise as biomarkers for the early detection of HIV infection as well as for staging of the disease.

## 1.3. Conventional Methods for the Analysis of Urinary Nucleosides

Several analytical techniques have been applied to the qualitative and quantitative analysis of UMN. Early quantitative methods relied on packed-

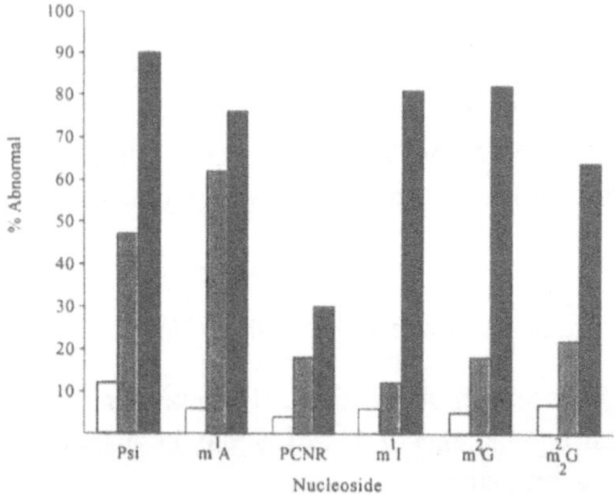


*Figure 5.2.* A comparison of nucleoside levels among urine samples from ☐ control, ▒ high-risk, and ■ AIDS patients. Reprinted with permission from Fischbein, *et al.*, 1985.

column GC for separation of the individual nucleoside components. Following their isolation from the urine matrix by adsorption onto activated carbon, the nucleosides were converted to their trimethylsilyl (TMS) derivatives and analyzed. This approach was used to determine the levels of Ψ (**30**), $m^1I$ (**28**), and $m^2_2G$ (**19**) in urine from a control subject and from a patient with colon cancer. The major problem with the earlier packed-column GC methods was a lack of chromatographic resolution, so that the procedure was incapable of separating components with similar retention times or artifacts, e.g., the 4-hydroxyacetanilide interference in the quantitation of $m^2_2G$ (**19**) (38).

More recently, immunoassay methods have been used to quantify UMN. Radioimmunoassay (RIA) was applied to the analysis of Gm (**18**), $i^6A$ (**6**), and $t^6A$ (**9**) in urine samples from control and cancer patients (39). Monoclonal antibodies were produced and used in an ELISA-based quantitation of $m^1A$ (**4**) in the urine of cancer patients (40). These immunoassay methods are very sensitive (picomole range) and highly specific, but suffer from the need for a specific antibody for each compound being evaluated. Consequently, these methods are not applicable to the identification and quantitation of previously unknown nucleosides.

The most frequently used method for the quantitation of UMN has been high-pressure liquid chromatography (HPLC). HPLC methods of analysis have the advantage over GC methods in that the compounds being studied need not be volatile, eliminating the need for derivatization. With modern instrumentation, HPLC with ultraviolet (UV) detection at 254 nm has the sensitivity necessary to detect urinary nucleosides (41). In one study, 13 nucleosides were separated using isocratic normal phase conditions with detection limits as low as 0.2 ng and linearity spanning three orders of magnitude (42). Using reverse phase (RP) conditions, modified nucleosides in urine from control and cancer patients can be detected at the 1-ng level (43,44). Multidimensional HPLC chromatography has been employed to increase the chromatographic separation and linearity of response for nucleosides in physiological fluids. This technique uses a boronate–gel affinity column coupled by switching techniques to a C-18 RP column (45).

The primary problems with HPLC-UV quantitation of nucleosides include the following:

1. At a fixed wavelength, UV absorption is a general property of many molecules and is not structurally specific. Thus, the potential of co-elution of unsuspected components, resulting in erroneous quantitative data, is a constant problem.
2. Little or no structural information is available for identification of unknown components. For example, Fig. 5.3 shows a reconstructed total-ion chromatogram (obtained using GC-MS) of an HPLC fraction

thought to contain only one nucleoside, but which turned out to contain a number of compounds including other nucleosides.

3. Detection requires the presence of a UV-absorbing chromphor; whereas that factor is generally not a problem, quantitation of non-UV-absorbing nucleosides such as D (**33**) is not possible (46). At present, D (**33**) is the only known non-UV-absorbing nucleoside, but the existence of other non-UV-absorbing nucleosides cannot be discounted.

### *1.4. Application of Mass Spectrometry to Analysis of Nucleosides*

Mass spectrometry has been used with considerable success in the structural elucidation of modified nucleosides from a variety of biological sources including tRNA (47), DNA (48), urine (49–52), and other sources (53).

Gas chromatography–mass spectrometry (GC-MS) is currently the only analytical technique with both the sensitivity and the specificity needed to establish unequivocally the identity of a modified nucleoside isolated from a complex biological source such as urine. Specificity is very high with a mass spectrometer because ions characteristic of the molecule or portions of the molecule are observed, rather than general physical properties like UV absorbance, as discussed for UV-HPLC. In addition to molecular weight and structure-elucidating fragmentation from a mass spectrum, sample introduction

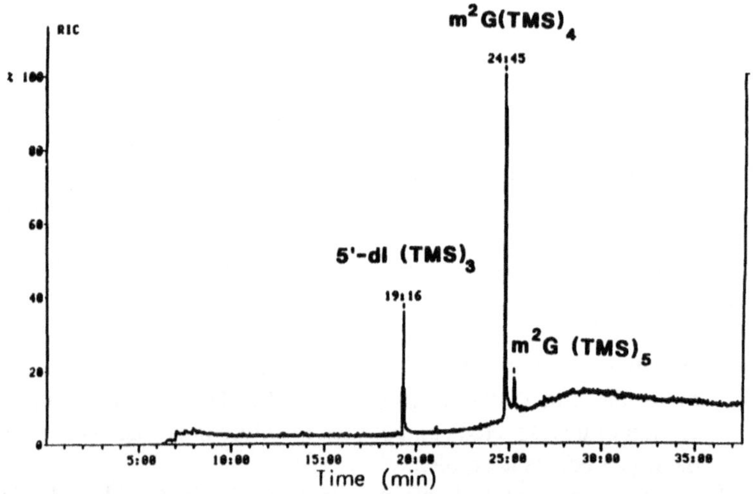


*Figure 5.3.* Reconstructed total ion current (RTIC) chromatogram (using GC-MS) of an HPLC fraction (using UV detection) formerly thought to contain only m$^2$G (**21**).

using a high-resolution capillary GC column provides an additional compound-specific parameter, retention time.

## 2. EXPERIMENTAL RESULTS

### 2.1. Qualitative Analysis of Urinary Modified Nucleosides

Two approaches have been developed for the detection and structural elucidation of previously unidentified modified nucleosides found in human urine. One method involves isolation of the nucleoside fraction using boronate–gel affinity chromatography followed by RP-HPLC separation and collection of the individual nucleoside fractions with subsequent GC-MS analysis of the TMS-derivatized component(s). This method is used in the authors' laboratory. The second approach involves a large-scale purification of nucleosides with activated carbon and ion-exchange chromatography columns (54). Discussed below are the advantages and limitations of each of these methods.

#### 2.1.1. Boronate–Gel Based Qualitative Analysis

*2.1.1a. Sample Preparation.* A summary of the boronate–gel method is shown in Fig. 5.4. A 24-hr urine sample collected from a subject is centrifuged to remove particular matter and stored at –20° until needed.

Boronate–gel affinity chromatography is used to isolate the nucleoside fraction from urine and is a modification of a previously published procedure (43). The stationary phase is Affi-gel 601 (Bio-Rad Labs, Richmond, CA), having a specific affinity for *cis*-hydroxyl groups, and is packed in a 60 × 9 mm I.D. column with a bed volume of 0.83 ml. Prior to sample introduction, the column is equilibrated with 20 ml of 0.25 M ammonium acetate (pH 8.8) and washed with 20 ml of 0.1 M formic acid. The sample is loaded onto the column as a 10-ml aliquot of urine, with the pH adjusted to 8.8 using 2.5 M ammonium acetate (pH 9.5). The column is washed with 8 ml of 0.25 M ammonium acetate (pH 8.8) and the nucleosides eluted with four ml of 0.2 M formic acid. This eluent is dried *in vacuo,* and the residue is dissolved in 1 ml of water for HPLC purification.

*2.1.1b. HPLC Purification.* The HPLC system used in these studies is a Shimadzu (Kyoto, Japan) system with an LC-6A control and dual pumps, a SPD-6AV UV-Vis detector, and a C-R3A Chromatopac integrating recorder. The stationary phase is a 5-μm Develosil ODS-5 (Nomura Chemicals, Nagoya, Japan) reverse-phase 250 × 4.6 mm I.D. column. This column is protected by a 15–30 μm Develosil ODS-5 precolumn and a 1-μm mesh size filter installed

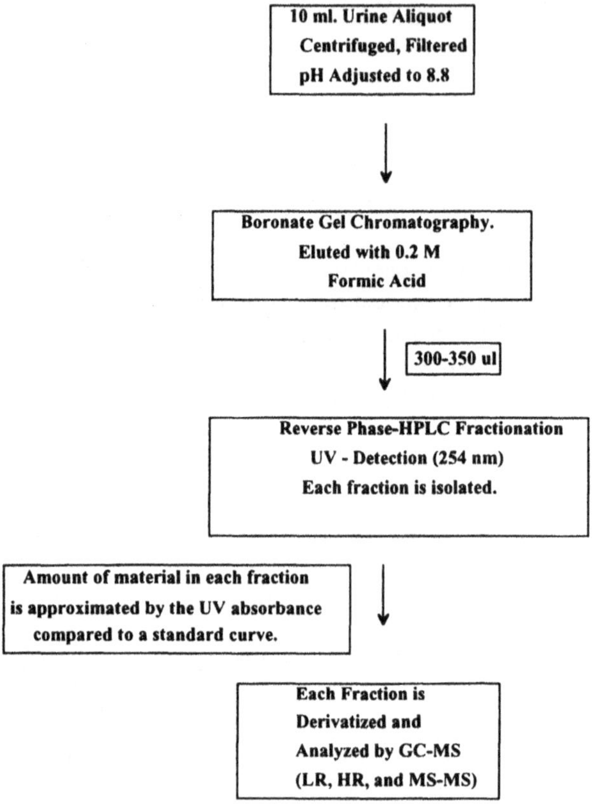


*Figure 5.4.* Boronate–gel sample preparation scheme for qualitative analysis of urinary modified nucleosides.

between the injector and the precolumn. The following solvent gradient program was used to separate the nucleoside components:

0 min—ultrapure water, pH 4.2, adjusted with 0.2 M formic acid;
25 min—13:87 methanol–water (v/v);
35 min—45:55 methanol–water (v/v);
40 min—the column was washed with water at pH 4.2.

The injection volume was 300 or 350 μl with a mobile phase flow of 1.1 ml per minute. Detection of the modified nucleosides was accomplished by absorption at 260 nm, with initial structural assignment of the major component in each fraction being based on retention times corresponding to known reference samples. Each individual nucleoside fraction was collected and lyophilized (55).

*2.1.1c. GC-MS Analysis.* Each of the lyophilized fractions from the HPLC purification step was reconstituted in an appropriate volume of water to afford a concentration of one µg/µl (based on UV absorbance). A 10-µl aliquot of the working solution was taken to dryness under a stream of nitrogen at room temperature and the TMS derivative prepared using previously reported methods (56–58); i.e., 40 µl of N,O-*bis*(trimethylsilyl)trifluoroacetamide (BSTFA) containing 1% trimethylchlorosilane (TMCS) and 10 µl of pyridine is combined with the dry sample, and the mixture is heated in a sealed vial for one hour at 100°.

The GC-MS analysis of the derivatized samples was performed using a MAT 90 double-focusing (BE, reverse geometry) mass spectrometer (Finnigan MAT, San Jose, CA) equipped with a Varian 3400 gas chromatograph (Varian Instrument Co., Palo Alto, CA). One µl of the derivatized sample was introduced by splitless injection onto a bonded-phase DB-5 capillary column (30 m × 0.25 mm; 0.25 µm film thickness; J&W Scientific, Folsom, CA) that was directly coupled to the ion source. The GC conditions used were as follows: injection temperature, 250°; column conditions, 10 PSIG head pressure; temperature program 150° for 3 min, 6° $min^{-1}$ to 300°, 300° for 10 min; interface temperature, 250°.

The mass spectrometer conditions used depend on the type of analysis being performed. Ionization was achieved by electron ionization (EI at 70 eV) and chemical ionization (CI using $CH_4$ or $NH_3$ as the reagent gas at approximately $10^{-5}$ torr source pressure). Mass analysis was performed using either the low (LR, $R = 1000$) or high resolution (HR, $R > 7000$) mode. First, fieldfree region product ion scans were obtained with the accelerating voltage held constant and the magnetic (**B**) and electric (**E**) fields scanned according to the expression **B/E**≐constant (59).

### *2.1.2. Activated Carbon–Based Qualitative Analysis*

*2.1.2a. Sample Preparation.* The activated carbon method for separation and isolation of UMN was developed by Chheda (54). A 24-hr urine sample is collected in a container with 50 ml of toluene. The toluene is removed, the pH is adjusted to 3, and the sample is allowed to stand for 2 hr before filtering. The entire sample is desalted using a column (3.8 × 50 cm) containing activated carbon and celite. After sample addition, the column is washed with approximately 3 l of water and the compounds of interest eluted with 3 l of 2N $NH_4OH$ in 50% aqueous $CH_3CH_2OH$. The eluate is concentrated to approximately 50 ml, and organic acids are removed using a 3.5 × 35 cm ion-exchange resin column (AG1-X8, BioRad Laboratories). The column is washed with water until UV absorbance (at 260 nm) falls below 0.2 AUFS, and the eluent is taken to near dryness. The residue is dissolved in 50 ml of 0.15 M boric acid and the *cis*-diol containing compounds separated from other components on a DEAE cellulose-borate containing column. Monitoring the UV absorbance (cutoff point is 0.3 $A_{260}$ units) of the column effluent, the non-*cis*-diol nucleosides (e.g., 2′-O-methyl nucleosides) are removed with 0.15 M boric acid. Most of the *cis*-diol nucleosides (and other compounds) are removed with a 0.7-M boric acid wash. The more acidic nucleosides, such as $t^6A$, are removed from the column using 1 N formic acid. Each of the collected fractions is taken to near dryness and further purified using RP HPLC.

*2.1.2b. HPLC Purification.* The conditions for collection of individual nucleosides present in the above mixtures were as follows: column, 21.4 mm × 25 cm Zorbax ODS $C_{18}$ (8-μm particle size); maximum sample size, 5 ml; mobile phase solvent program over the course of 60 min progressed linearly from 0 to 25% methanol in 0.1 M ammonium acetate buffer (pH 6.8); flow rate, 8 ml $min^{-1}$; detection, 254 nm. Figure 5.5 shows an HPLC chromatogram of the 0.7-M boric acid eluent from a control urine sample, indicating the elution order of nucleosides (54).

*2.1.2c. Instrumental Analysis of Fractions.* The large volume of urine used in the activated carbon method (relative to the boronate–gel method) allows several instrumental methods to be applied to structure elucidation. Thus, sufficient material is available for UV, NMR (both $^1H$ and $^{13}C$), and mass spectrometry. The $\lambda_{max}$ values obtained from the UV spectra of unknown nucleosides, when compared with model compounds at various pH values,

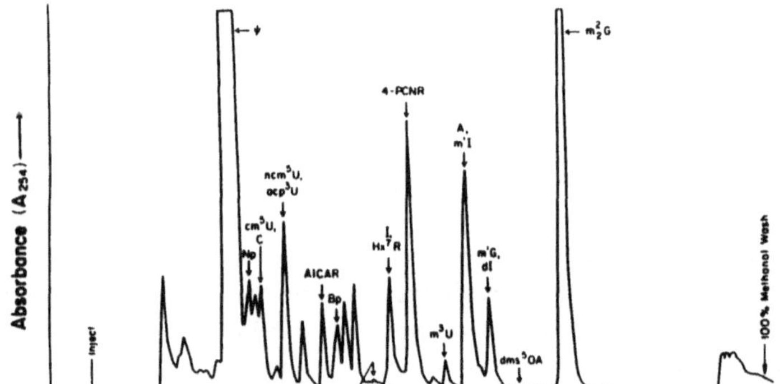


*Figure 5.5.* An HPLC-UV chromatogram showing the elution order of modified nucleosides found in a control urine sample. The sample preparation involved the activated-carbon technique. Reprinted with permission from Ref. 54.

provide information about the chromophore, and in some instances, the degree and position of substitution within the chromophore. These results can be obtained on levels as low as 100 ng (54).

The use of $^{1}H$ NMR spectrometry permits further structural information to be obtained, particularly as to the position of substitution within the aromatic base portion of the molecule. On the other hand, $^{13}C$ NMR provides information on the chemical environment of the carbon atoms in the base and the sugar (54).

The derivatization procedure and conditions for GC-MS were fundamentally the same as those described above in the boronate–gel-based isolation scheme.

### 2.1.3. *Analysis of Mass-Spectral Results*

As mentioned previously, nucleosides are thermally labile compounds, so MS analysis in the EI and CI modes requires preparation of TMS (or other) derivatives. In addition to increasing the volatility of the samples, forming this derivative increases the mass of the molecular ion to a region, typically between 400 and 800 amu, where instrument background is low. Using EI, these derivatized nucleosides produce numerous structurally significant fragment ions (60). These structurally significant ions are grouped into three ion series:

1. molecular-ion-related fragments (M series), resulting from loss of small neutral or radical species from the molecular ion;
2. base-related fragments (B series), containing the nucleobase plus portions of the sugar; and
3. sugar-related fragments (S series), formed after cleavage of the glycosidic bond between the base and C-1′.

Each of these ion series provide information important in determining the type and position of modification in the nucleoside.

*2.1.3a. M-Series Ions.* Typical ions present in this series are listed in Table 5.2. One of the most useful is the $[M-15]^{+}$ ion formed by loss of a methyl radical from the molecular ion $[M]^{+\cdot}$. This ion is useful for confirming the identity of the $[M]^{+\cdot}$ ion because a loss of 15 from other ions is not a major process in TMS-nucleoside spectra. Furthermore, the ratio $[M]^{+\cdot}/[M-15]^{+}$ provides some clues to structure. For most nucleosides, this ratio is less than one; however, for G (**17**), dA (**3**), dX, and a few other nucleosides, this ratio is greater than one (50,60,61). In some instances, other important ions are indicative of selected modifications in the structure of the modified nucleoside; for example, $[M-47]^{+}$ is representative of the loss of $[CH_3\text{-}S]^{\cdot}$ from $[M]^{+\cdot}$ and is indicative of a methylthionucleosides such as 5′-deoxy-5′-methylthioguanosine (MTG, **23**) and the corresponding adenosine analog (62).

*Table 5.2*
*Structural Significance of the Molecular Ion Series*[a]

| Ion | Structural significance and comments[b] |
|---|---|
| $M^{+\cdot}$ | Molecular weight of intact molecule. Use of $^2H$-labeled reagents will permit determination of the number of TMS groups added to the free nucleoside. High-resolution data important for determining the elemental composition of the intact molecule. |
| M − 15 | Permits ready identification of the molecular ion. If ratio $M/(M - 15) > 1$, guanosine *may* be present. |
| M − 90 | TMSOH is lost exclusively from O-2′. |
| M − 103 | Loss of the 5′-position as $CH_2OTMS$. Indicator of the 5′-position. |
| M − 105 | M − 15 − 90. No additional structural information. |
| M − 118 | M − 90 − CO. CO lost primarily from the O-4′ position. |
| M − 131 | Loss of c-1′, C-2′, O-2′, and O-4′. |
| M − 180 | M − 90 − 90. Useful for recognizing C-nucleosides. |

[a]Reprinted with permission from Ref. 50.
[b]Based on data from Ref. 60.

*2.1.3b. B-Series Ions.* Table 5.3 contains a list of the ions routinely found in this series. These ions provide structural information concerning the following:

1. The type of modification in the aglycone portion of the molecule;
2. The type and position of modifications to the sugar moiety, because different positions of the sugar are represented in this series; and
3. In some instances, the position of attachment of the sugar in the aglycone moiety for purine nucleosides [e.g., 7-R-Hyp, **24** (11)].

*2.1.3c. S-Series Ions.* The routinely observed ions of this series are collected in Table 5.4. The S-series ions confirm the position of sugar modification and provide some information as to aglycone structure. As a general rule, the S-series ions are more intense for pyrimidine than for purine nucleoside TMS derivatives (60). Anomeric configuration of modified nucleosides can be inferred by a comparison of the spectra of the α and β species. The relative abundance of the $[S\text{–}H]^{+\cdot}$ ion is typically greater for the α anomer than for the β anomer (60). As with the B series, the $[S\text{–}H]^{+\cdot}$ ion is also important in differentiating the site of attachment of ribose at the C-7 versus C-9 position in purine nucleosides (11).

All three of these ion series have been used to assign a tentative structure to previously unidentified modified nucleosides regardless of isolation methods or other instrumental techniques employed. GC-MS alone can in some cases

*Table 5.3*
*Structural Significance of the Base-Ion Series*[a]

| Ion | Structural significance and comments[b] |
|---|---|
| B + 204 | Same as B + 132 but with an additional TMS group. |
| B + 188 | Shifted from B + 116 by addition of TMS but not structurally related. |
| B + 132 | Of major structural importance in locating sites of modification in the sugar. Contains C-1′, C-2′, O-2′, O-4′, and a rearranged hydrogen. Formed by loss of the 3′ through 5′ positions of the sugar. Structurally related to the B + 60 ion of free nucleosides by addition of a TMS group to the hydroxyl group. Sugar modifications are evident. |
| B + 116 | Structurally related to the B + 44 ion of free nucleosides (contains C-1′, C-2′, and O-2′), but with a TMS group added to the hydroxyl group. Sugar modifications are evident. |
| B + 100 | Formed from B + 116 by loss of $CH_4$. |
| B + 74 | Related to the $BH^{+\cdot}$ ion, but with TMS group transferred from the sugar during glycosidic bond cleavage, instead of a hydrogen. Identification of base confirmed. |
| B + 58 | Formed from B + 74 by loss of $CH_4$. |
| B + 41 | Characteristically abundant in cytidine analogs. |
| B + 30 | Corresponds to the base plus C-1′, and the sugar-ring oxygen. Modifications to C-1′ are evident. Often a strong ion in purine nucleosides. B will be shifted by 72 daltons for each TMS present in the base. |
| B + 13 | Base plus CH observed in some but not all pyrimidine nucleosides. Absent in purine nucleosides. |
| B + 2H, B + H | Base ion. Establish nominal mass and elemental composition (with high resolution) of B, the aglycone. Shift of the B-series with $^2H$-labeled reagents dependent on number of TMS groups added to the base. |

[a]Reprinted with permission from Ref. 50.
[b]Based on data from Ref. 60.

provide sufficient information to assign a tentative structure to unknown UMN. Any information from other techniques (i.e., HPLC retention times and UV data, or, in the case of $m^{1,6}{}_2A$ (**8**), NMR data) is important in the process of structural assignment. Ultimate proof of structure lies in a comparison of the mass-spectral and chromatographic data between an authentic sample and the unknown. The strategy for using GC-MS for identification of unknown UMN will be demonstrated with examples from our laboratory.

### *2.1.4. Examples of Base-Modified Novel Nucleosides*

The first example, 3-methyluridine ($m^3U$, **34**), had been previously identified in control human urine on the basis of TLC retention factors and UV spectroscopy (63). We verified the structure using GC-MS and HPLC retention

*Table 5.4*
*Structural Significance of the Sugar-Ion Series*[a]

| Ion | Structural significance[b] Ribose | Deoxy | 2′-O-Me | Comments |
|---|---|---|---|---|
| S | 349 | 261 | 291 | Direct indicator of sugar modifictions. Isomers and epimers not distinguished with any confidence. Differentiation of α from β possible if both available. Also useful for establishing identity of base-related ions. |
| S–H | 348 | 260 | 290 | |
| S–32 | — | — | 259 | From O-methylated sugars only. |
| S–90 | 259 | 171 | 201 | Loss of TMSOH. |
| S–H–90 | 258 | 170 | 200 | Loss of TMSOH + H. |
| S–H–103 | 245 | 157 | 187 | Loss of $[CH_2OTMS]^{\cdot}$. If abundant, suggests unmodified 5′-position. |
| S–16–90 | 243 | 155 | 185 | |
| $C_4H_4O_2TMS_2$ | 230 | — | 230 | |
| $C_5H_7O_2TMS$ | — | — | 172 | |
| $C_3H_3O_2TMS_2$ | 217 | — | 217 | |
| $C_4H_6O_2TMS$ | — | — | 159 | |
| S–180 | 169 | 81 | — | |
| $CH_2OTMS$ | 103 | 103 | 103 | Suggestive of unmodified 5′-position, but not definitive. |

[a]Reprinted with permission from Ref. 50.
[b]Based on data from Ref. 60.

times in a comparison of the urinary component with a reference standard. Preliminary structural assignment was made after evaluating the EI-LR spectrum (see Fig. 5.6). The M-series ions ($[M]^{+\cdot}$ and $[M-15]^+$ at *m/z* 474 and 459, respectively) indicate the molecular weight, which is verified by CI-methane ($[MH]^+$, *m/z* 475, 100% relative intensity, RI). Correcting the molecular weight for the presence of three TMS groups (subtracting 216 Da, 72 for each TMS) gave a molecular weight for the free nucleoside of 258 Da. This molecular weight corresponds to a methylated uridine, and the spectrum is consistent with the presence of a pyrimidine nucleoside; i.e., the spectrum is dominated by S-series ions (60). The presence of the S-series ions ($[S]^+$, *m/z* 349; $[S-TMSOH]^+$, *m/z* 259; $[S-CH_4-TMSOH]^+$, *m/z* 243; $[C_3H_3O_2TMS_2]^+$, *m/z* 219; $[S-2TMSOH]^+$, *m/z* 169; and $[CH_2OTMS]^+$, *m/z* 103) is indicative of an unmodified pentafuranose, and thus the extra methyl group must be in the aglycone. Furthermore, a comparison of the B-series ions ($[B+H]^+$, *m/z* 126; $[B+58]^+$, *m/z* 183; $[B+74]^+$, *m/z* 199; $[B+102]^+$, *m/z* 227; $[B+188]^{+\cdot}$, *m/z* 313;

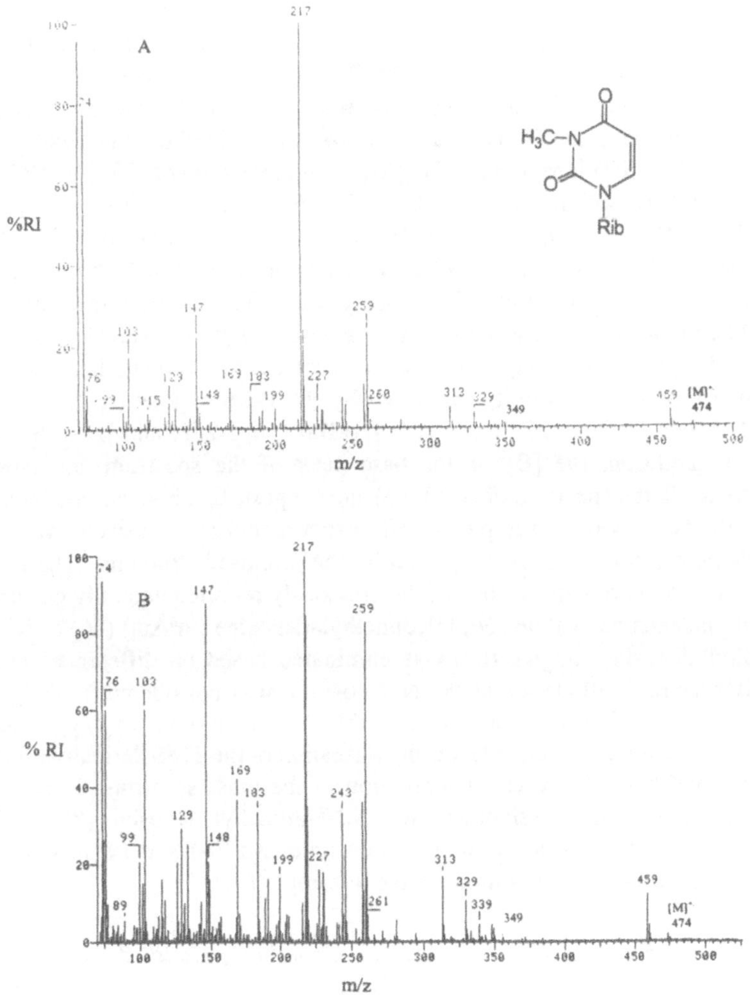


*Figure 5.6.* Low-resolution 70-eV EI mass spectra of (A) reference 3-methyluridine-$TMS_3$ and (B) the urinary component assigned as 3-methyluridine $TMS_3$. Reprinted with permission from Ref. 64.

and $[B+204]^+$, *m/z* 319) with those from the EI-LR spectrum of U-$TMS_3$showed a shift of 14 amu. High-resolution EI studies were consistent with the elemental composition of the ions mentioned above. A comparison of the HPLC retention times, GC retention times, and EI-LR spectra (Fig. 5.6) for the isolated UMN TMS derivative with the TMS from a reference sample of $m^3U$ (**34**) confirmed the structure as initially assigned (64).

The second example of a base-modified urinary nucleoside, isolated from the urine of a lung cancer patient, has considerable potential as a possible biomarker for this type of cancer. This UMN was discovered in an HPLC fraction co-eluting with three other nucleosides: adenosine (A, **1**), 5′-deoxy-5′-methylthoadenosine (MTA, **10**), and another unidentified nucleoside. The M-series (Fig. 5.7) ions at *m/z* 511 ($[M]^{+\cdot}$) and *m/z* 496 ($M-15]^+$) established the molecular weight as 511 Da. The S-series ions ($[S]^+$, *m/z* 349; $[S-TMSOH]^+$, *m/z* 149; $[S-2TMSOH]^+$, *m/z* 169; and $[CH_2OTMS]^+$, *m/z* 103) suggest that the modified nucleoside contains an unmodified ribose species. Subtracting the weight of the TMS-sugar portion (349 Da) from the molecular weight gives an aglycone weight of 162 Da. This weight is too low for inclusion of a TMA group in the aglycone, but corresponds to a dimethylated adenine minus a proton, a possibility supported by the B-series ions ($[B]^+$, *m/z* 162; $[B+H]^{+\cdot}$, *m/z* 163; $[B+2H]^+$, *m/z* 164; $[B+100]^+$, *m/z* 262; and $[B+116]^+$, *m/z* 278). In addition, the $[B]^+$ is the base peak of the spectrum, an unusual occurrence in the spectrum of A (**1**) (65) and suggests that N-6 was methylated. As with the previous example, EI-HR experiments are consistent with the ion-elemental compositions suggested by the proposed structures. The possibility that the unknown was one of the previously reported naturally occurring dimethyladenosine analogs, $N^6$, $O^{2'}$-dimethyladenosine ($m^6Am$) (66) or $N^6$,$N^6$-dimethyladenosine ($m^6_2A$), (67) was eliminated based on differences in the EI-LR spectra. Methylation at the N-7 position was considered unlikely because purine nucleosides methylated at this site incorporate an oxygen atom [from atmospheric oxygen (53)] during formation of the TMS derivative. There was no evidence of oxygen incorporation in the mass spectrum. Hence, the preliminary structural assignment was 1,$N^6$-dimethyladenosine (**8**), as was confirmed by comparison of GC retention time and EI-LR mass spectra (see Fig. 5.7) of the unknown with a synthetic sample (9).

### *2.1.5. Identification of Novel Nucleosides Containing Modified Sugars*

The first example of a sugar-modified nucleoside, discovered in control human urine, belongs to the novel class of 5′-deoxynucleosides that contains only two other members—5′-deoxyadenosine (5′-dA), a metabolite of cobolamine (vitamin $B_{12}$), and 5′-deoxyinosine (5′-d, **29**), identified in the urine of a chronic myelogenous leukemia patient (10). The unknown in the control urine sample was one of five nucleosides present in an HPLC fraction thought to contain only 2-methylthio-$N^6$-isopentenyladenosine ($ms^2i^6A$), based on HPLC retention time. The molecular weight was initially difficult to establish because no $[M]^{+\cdot}$ ion was present and only one M-series ion ($[M-15]^+$) was found in the molecular-ion region of the spectrum (Fig. 5.8). Subsequently, assignment of

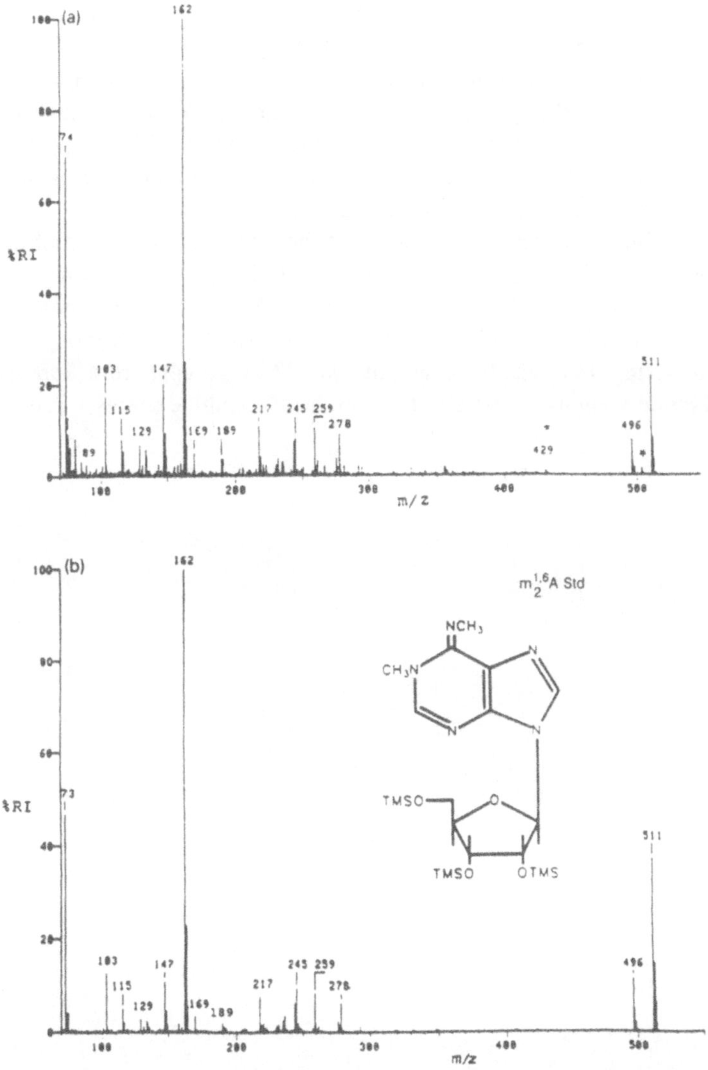


*Figure 5.7.* Low-resolution 70-eV EI mass spectra of (a) the urinary component assigned as I,N$^6$-dimethyladenosine-TMS$_3$ and (b) reference 1,N$^6$-dimethyladenosine-TMS$_3$. Background ions denoted by *. Reprinted with permission from Ref. 9.

the $[M-15]^+$ ion was confirmed using GC-CI to verify the molecular weight as 556 Da. The S-series ions characteristic of a ribose sugar were not observed. Instead, ions, indicating the presence of a deoxyribose sugar were present at *m/z* 260, $[S]^+$; *m/z* 219, $[C_3H_3O_2TMS_2]^+$; *m/z* 171, $[S-TMSOH]^+$; *m/z* 170, $[S-H-CH_2OTMS]^+$; *m/z* 129, $[C_3H_4OTMS]^+$; and *m/z* 103, $[CH_2OTMS]^+$. Although the possibility of a deoxyribose group was suggested by these results, the site of dehydroxylation (2′-, 3′-, or 5′- position) was not clear. Comparison of the S-series ions in the spectrum of the unknown with the spectra of 2′-, 3′-, and 5′-deoxyadenosine suggested most strongly the presence of a 5′-deoxyribose. However, these results were inconclusive.

Subtracting 260 Da, the mass of the TMS-deoxyribose moiety, from the molecular weight (556) gave a mass of 296 Da for the aglycone. At this point, preparation and EI-LR analysis of the TMS-$^2H_9$ derivative produced a mass shift equivalent to two TMS in the aglycone and two TMS groups in the sugar. Subtracting 144 Da, the mass of two TMS groups, from 296 gave a free aglycone weight of 152 Da, the weight of xanthine minus a proton. A com-

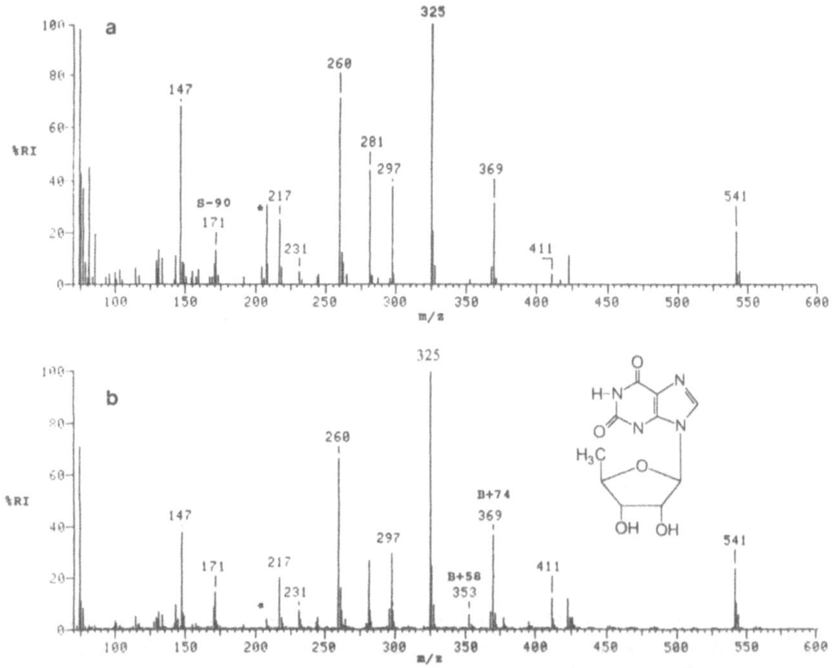


*Figure 5.8.* Low-resolution EI (70 eV) mass spectra of the TMS derivatives of (a) urinary component and (b) reference sample of 5′-deoxyxanthosine-$TMS_5$. Background ions indicated by *. Reprinted with permission from Ref. 68.

parison of the B-series ions (*m/z* 281, $[B-14]^+$; *m/z* 325, $[B+30]^+$; *m/z* 353, $[B+58]^+$; *m/z* 369, $[B+74]^+$; *m/z* 395, $[B+100]^+$; *m/z* 411, $[B+116]^+$; *m/z* 423, $[B+128]^+$) from the spectrum of the unknown with a spectrum of X–TMS indicated that the unknown aglycone was very probably xanthine.

The B-series ions, in addition to indicating that the aglycone was xanthine, provided ions supporting the presence of a 5′-deoxyribose moiety. For example, the $[B+116]^+$ ion contains the C-1′, C-2, and O-2′ (TMS attached) portions of the sugar ring and was observed in the spectrum of the unknown, verifying the presence of the 2′-hydroxy group. In like manner, the $[B+128]^+$ ion contains, in addition to other positions of the sugar, the 3′-OH group and was also found in the spectrum of the unknown, confirming the presence of a 3′-hydroxy group. High-resolution MS studies provided elemental composition information for the ions in the spectrum and were consistent with the assignments mentioned above. The tentative identification of this unknown urinary nucleoside as 5′-deoxyxanthosine (**42**) was verified by a comparison of the HPLC and GC-retention times, UV spectra, and EI-LR spectra (Fig. 5.8) of the unknown urinary component with a biochemically prepared reference sample (68). A subsequent study of the fragmentation of a series of 2′-, 3′-, and 5′-deoxynucleosides was generated as a result of the 5′-dX (**42**) project, and the results indicated that differentiation of the deoxyribose isomers based on differences in fragmentation patterns is possible (61).

Finally, an unknown nucleoside was isolated from the urine of a lung cancer patient. The HPLC fraction containing the unknown also contained two other known nucleosides, A (**1**) and MTA (**10**). Coincidentally, the fragmentation pattern in the spectrum of the unknown (Fig. 5.9) contained features similar to those in the spectrum of MTA (**10**). The molecular weight of the TMS derivative of this unknown was determined to be 601 Da using the M-series ions $[M]^{+\cdot}$ and $[M-15]^+$. As with MTA, the ion at *m/z* 554 ($[M-S-CH_3]^+$) was present in the spectrum and indicated the presence of a 5′-deoxy-5′-methylthionucleoside.

The S-series (*m/z* 175, $[S-C_4H_6S-OTMS]^+$; *m/z* 188, $[S-C_5H_6S-OTMS]^+$; *m/z* 217, $[S-OTMS]^+$; and *m/z* 245, $[S-H-CH_2OTMS]^+$) lacked ions typical of ribose, deoxyribose, and methyoxyoribose, but were analogous to the S-series ions in the spectrum of MTA (**10**, confirming the presence of a 5′-deoxy-5′-methylthioribose group.

Preliminary evidence for the assignment of guanine as the aglycone was found in the B-series ions (*m/z* 280, $[B-14]^+$; *m/z* 294, $[B]^+$; *m/z* 295, $[B+H]^+$; *m/z* 296, $[B+2H]^+$; *m/z* 324, $[B+30]^+$; *m/z* 352, $[B+58]^+$; *m/z* 368, $[B+74]^+$; *m/z* $[B+100]^+$; and *m/z* 410, $[B+116]^+$). Assignment of structure for the M, B, and S series of ions was supported by HR mass studies. Subtracting 294 Da (for the base) from the molecular weight of 601 gave a weight of 307 Da for the sugar portions. This mass is 42 Da less than a fully derivatized ribose minus

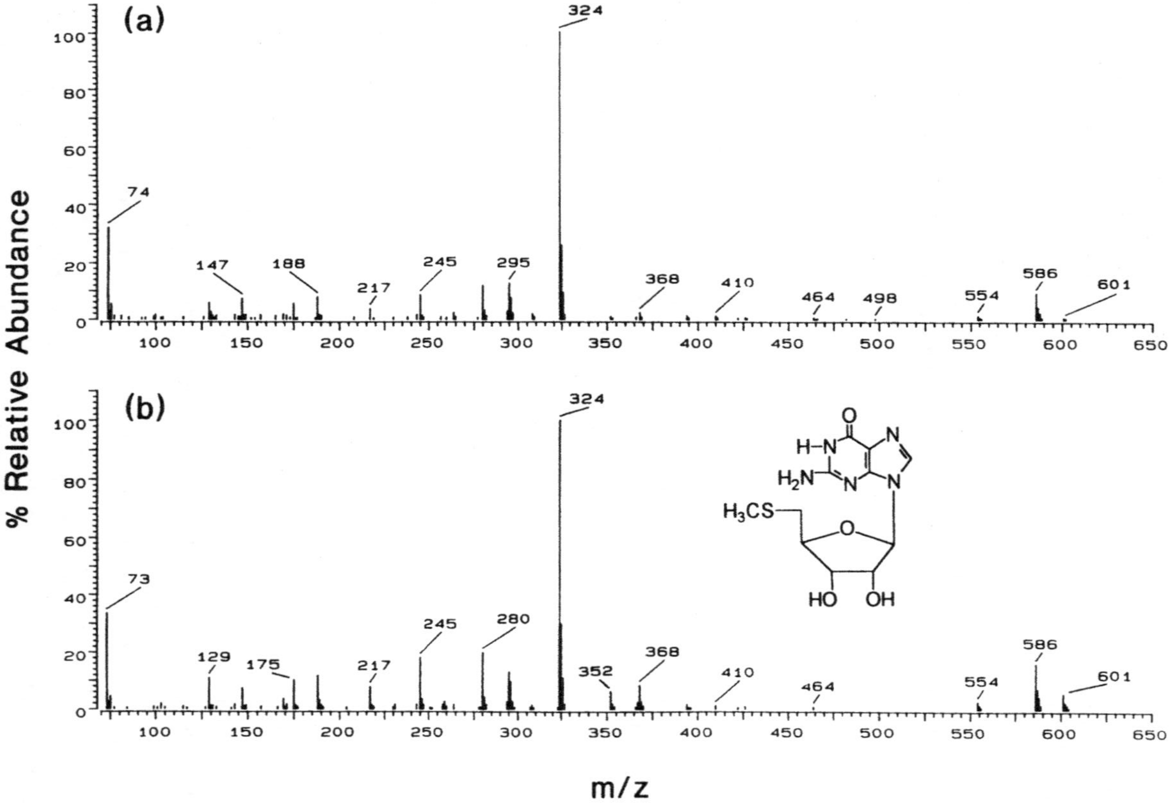


*Figure 5.9.* Low-resolution EI (70 eV) mass spectra of the TMS derivative of (a) the urinary component and (b) reference sample of 5′-deoxy-5′-methylthioguanosine-TMS$_4$. Reprinted with permission from Ref. 62.

a proton, a difference accounted for by the replacement of an —O—TMS group in ribose by —S—$CH_3$. Furthermore, subtracting 144 Da (the mass of two TMS groups) from 294 Da gave a weight of 150 Da for the free base; that mass equals the mass of guanine minus a proton. The preliminary structural assignment of the unknown as 5′-deoxy-5′-methylthio-guanosine (MTG, **23**) was verified by comparison of HPLC and GC retention times and EI-LR mass spectra (Fig. 5.9) of the urinary component with a synthetic reference standard of MTG (**23**) (62).

## 2.2. *Quantitative Analysis of Urinary Modified Nucleosides*

Virtually all of the quantitative analyses of UMN have been performed using HPLC with UV detection. However, as results of the work described in the previous section indicate, co-elution of known and unknown nucleosides (and other compounds) with the target nucleoside may interfere with quantitation when HPLC is used alone (this conclusion is highly dependent on the nucleoside being quantified and the levels of the biomarker in urine). We have used GC-MS to verify HPLC identification and as an alternative method of analysis capable of detecting and correcting for co-elution of unsuspected artifacts.

### 2.2.1. *HPLC–GC/MS Quantitation of UMN*

*2.2.1a. Sample Preparation and Analysis.* Samples are prepared using the boronate method described above. The UV absorbance of each fraction is recorded and compared to a standard curve for the nucleoside being quantified. The purity of each fraction is confirmed by GC-MS analysis of the TMS derivative. Purity of each UMN was measured as a percentage of the reconstructed total ion current (RTIC) for each fraction (69).

*2.2.1b. Results of Analysis.* Table 5.5 shows a comparison of levels of modified nucleosides isolated from urine of an advanced AIDS patient to "normal" values reported in the literature (54). Ten of the 12 modified nucleosides show differences that are statistically ($p$<0.01) significant from literature values for "normal" urine. This result confirms the hypothesis that levels of modified nucleosides are elevated in the urine of advanced AIDS patients (68). Unfortunately, the extensive sample handling necessary with using GC-MS to verify the purity of each HPLC fraction is a potential source of error, as well as being very time-consuming. Direct analysis of urine extracts by GC-MS requires much less sample manipulation and can be done in a much shorter time.

*Table 5.5*
*Nucleoside Levels in Urine from Control and an AIDS Patient*[a]

| Nucleoside | AIDS (mg/24hr) | PNU (mg/24hr) | % RIC (from GC-MS) | Value from Ref. 54 (mg/24hr) |
|---|---|---|---|---|
| Ψ | **142±8**[b] | 66±5 | 90 | 75±5 |
| $m^1A$ | 6.6±0.4 | 7.2 + 0.5 | 85 | 6.2±0.1 |
| 4,3-PCNR | **7.0±0.4** | 3.8±0.3 | 98 | 2.2±0.2 |
| X | **11.9±2.1** | 3.3±0.2 | 98 | 0.45 |
| $m^3U$ | 1.4±0.1 | 1.1±0.1 | 98 | 0.32 |
| $m^1I$ | **18.6±1.0** | 5.6±0.4 | 98 | 4.1±0.1 |
| $m^1G$ | **10.2±0.3** | 3.2±0.2 | 90 | 0.5±0.1 |
| $m^2G$ | **9.3±0.4** | 3.2±0.2 | 70 | 1.4±0.1 |
| A | **0.7±0.03** | 1.4±0.4 | 90 | N/A |
| $m^2_2G$ | **13.7±0.7** | 6.8±0.5 | 98 | 3.7±0.3 |
| $t^6A$ | **6.6±0.2** | 2.9±0.2 | N/A | 3.7±0.3 |
| MTA | **5.5±0.2** | 0.9±0.1 | 85 | N/A |

[a]Reprinted with permission from Ref. 69.
[b]Values in bold show statistically significant differences compared to PNU ($p<0.01$).

### *2.2.2. Solid-Phase Extraction and GC-MS Quantitation of UMN*

*2.2.2a. Sample Preparation.* Urine samples are collected as previously described for qualitative analysis. Figure 5.10 shows the general scheme for sample preparation prior to quantitation by GC-MS. Features of this sample preparation scheme include the following:

1. the use of disposable C-18 and boronate–gel columns, an important safety precaution in light of health concerns of workers handling potentially hazardous physiological liquids;
2. the small sample volume (one ml) needed for analysis; and
3. the short preparation time (up to six samples per hour have been prepared in our laboratory).

The internal standard used in these studies is 3-deazaadenosine, a modified nucleoside not known to occur naturally. The accuracy of quantitative results depends, in part, on selection of an internal standard that is not present in the sample matrix and behaves similarly to endogenous nucleosides (64).

*2.2.2b. GC-MS Analysis.* GC-MS conditions used are the same as presented for the qualitative method, with the exception that the mass analyzer is not scanned, but "jumps" to masses representing the $[M]^{+\cdot}$ and $[M-15]^+$ ions for each UMN being quantified. This method of mass analysis, selected ion

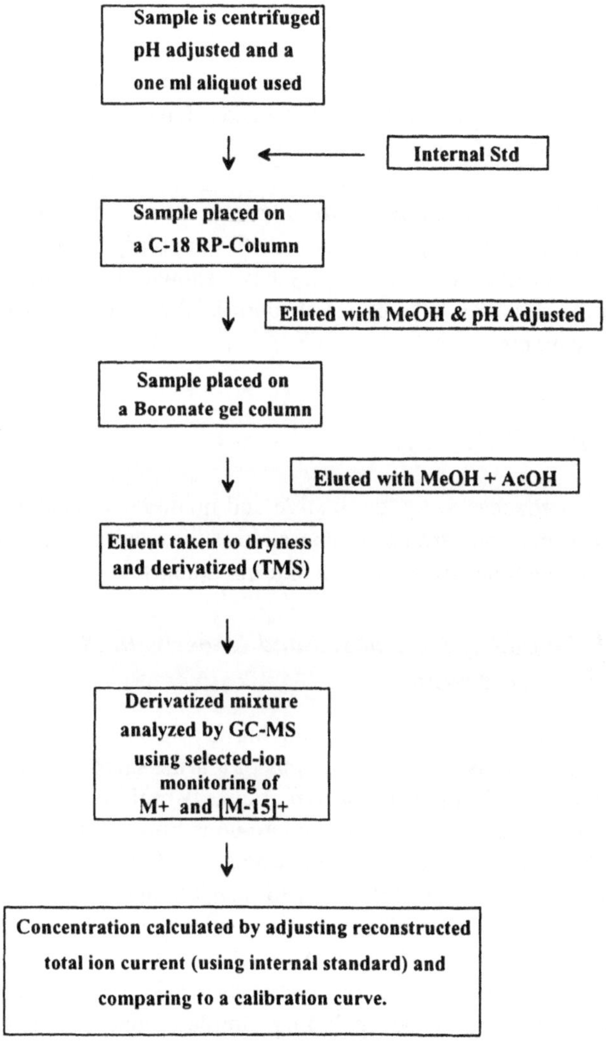


*Figure 5.10.* Solid-phase extraction sample preparation scheme used in the quantitation of urinary modified nucleoside–TMS derivatives by GC-MS.

monitoring (SIM), allows the mass spectrometer to dwell on only the ions of interest, thereby increasing the sample-related signal, decreasing the noise level, and leading to greater sensitivity. Though formal detection-limit experiments have not been completed, estimates as low as 50 pg of injected TMS nucleoside are reasonable. The RTIC for each of the ions studied is corrected as a function of the RTIC of the internal standard and compared to a standard curve (each concentration point in the standard curve has been taken through the same analytical scheme as the samples, thus correcting for recoveries) of less than 100 percent to produce a concentration value. Two ions for each nucleoside are used because not all nucleosides have large ion currents for both the $[M]^{+\cdot}$ and $[M-15]^{+}$ ions.

*2.2.2c. Results of Analysis.* Results from these experiments have not yet been published, and investigations are still ongoing. However, preliminary results from ten AIDS urine samples have shown significant differences in levels of $m^1G$ and $m^2G$ in the urines of control, HIV-infected but asymptomatic, and AIDS patients.

## 3. FUTURE PROSPECTS

New approaches to the qualitative and quantitative analysis of UMN as biomarker by mass spectrometry is possible as the result of continuing development of new methods of ionization and mass analysis.

### 3.1. FAB MS/MS of Derivatized and Underivatized Modified Nucleosides

The primary difficulty with studying UMN is the isolation step. One possible alternative is the direct analysis of a urine sample for the nucleosides of interest using MS/MS techniques (particularly multiple reaction monitoring, MRM, for quantitation). The use of continuous flow (CF) fast atom bombardment (FAB) with collision-induced dissociation (CID) of precursor ions on a tandem mass spectrometer (MS/MS) to detect product ions characteristic of both free (70) and TMS-derivatized nucleosides (71) have been reported. The time for separation and analysis of the nucleosides of interest from the matrix by this MS/MS method could be as short as 1 min per nucleoside of interest compared to almost 1 hr, not including sample preparation time, for GC-MS analysis (the same time as required for HPLC analysis). Furthermore, FAB ionization does not result in sample heating, so chemical derivatization is not necessary. However, FAB MS of TMS-derivatized nucleosides is more sensitive and provides more structurally useful information (comparable to the EI spectra) than FAB MS analysis of the free nucleosides (71–73).

Rapid advances in low-cost ion-trap mass analysis technology for automated accurate mass analysis and MS/MS experiments have been observed in the last five years. Coupling this type of mass analysis with CF-FAB represents a possible rapid method for in-clinic screening of nucleoside biomarkers in urine.

## *3.2. Application of Electrospray Mass Spectrometry*

Another approach with considerable promise in the identification and quantitation of nucleosides in urine is HPLC coupled to an electrospray ionization (ESP) source and a tandem mass analyzer. This approach has been successfully applied to the analysis of nucleotides in an HPLC effluent of 1:1 methanol/water (74). Although this ionization technique is usually used for higher-molecular-weight biological polymers, ESP also may be an important HPLC-MS interface. Modified nucleosides will be singly or at most doubly charged so that the MS-MS analysis of the precursor ion will provide significant structural information. Consequently, HPLC-ESP-MS/MS can also be used to search for UMN as a screening technique in the diagnosis, staging, and evaluation of chemotherapy for AIDS and cancer. Methods for quantitation will, however, have to be developed.

ACKNOWLEDGMENTS. Portions of the work reported in this review were supported by a National Institutes of Health grant (CA-43068) and a grant from the Arizona Disease Control Research Commission.

## *REFERENCES*

1. Royce, R. A., Luckmann, R. S., Fusaro, R. E., and Winkelstein, W. W., Jr., 1991, The natural history of HIV-1 infection: staging classifications of disease, *AIDS* **5:**355–364.
2. Imagawa, D. T., Lee, M. H., Wolinsky, S. M., Sano, K., Morales, F., Kwok, S., Sninsky, J. J., Nishanian, P. G., Giorgi, J., Fahey, J. L., Dudley, J., Visscher, B. R., and Detels, R., 1989, Human immunodeficiency virus type 1 infection in homosexual men who remain seronegative for prolonged periods, *NEJM* **320:**1458–1462.
3. Lippman, S. C., Lee, J. S., Lotan, R., Hittelman, W., Wargovich, M. J., Hong, W. K., 1990, Biomarkers as intermediate end points in chemoprevension trials, *J. Natl. Canc. Inst.* **82:**555–560.
4. Fischbein, A., Sharma, O. K., and Borek, E., 1985, Modified nucleosides and early detection of occupational cancer: A challenge for the future, *Mt. Sinai J. Med.* **52:**480–483.
5. Woo, K. B., Waalkes, T. P., Ahamann, D. L., Tormey, D. C., Gehrke, C. W., and Oliverio, V. T., 1978, A quantitative approach to determining disease response during therapy using multiple biologic markers, *Cancer* **41:**1685–1703.
6. Rasmuson, T., Bjork, G. R., Damber, L., Holm, S. E., Jacobsson, L., Jeppsson, A., Littbrand, B., Stigbrand, T., and Westman, G., 1983, Evaluation of carcinoembryonic antigen, tissue polypeptide antigen, placental alkaline phosphatase and modified nucleosides as biological markers in malignant lymphomas, *Recent Results in Cancer Res.* **86:**331–343.
7. Streeter, D. G., and Vold, B. S., 1984, Enzymes of nucleic acid metabolism, in *Markers of Chemically Induced Cancer* (G. Freeman and H. A. Milman, eds.), Noyes Publications, Park Ridge, NJ, pp. 19–36.
8. Dirheimer, G., 1983, Chemical nature, properties, location and physiological and pathological variations of modified nucleosides in tRNAs, in *Recent Results in Cancer Research,* Vol. 84, pp. 15–46.

9. Hammargren, W. M., Schram, K. H., Nakano, K., and Yasaka, T., 1991, Identification of a novel nucleoside, 1-,$N^6$-dimethyladenosine, in human cancer urine, *Anal. Chim. Acta* **247:** 201–209.
10. Chheda, G. B., Patrzyc, H. B., Bhargava, A. K., Crain, P. F., Sethi, S. K., McCloskey, J. A., and Dutta, S. P., 1987, Isolation and characterization of a novel nucleoside from the urines of chronic myelogenous leukemia patients, *Nucleosides Nucleotides* **6:**597–611.
11. Chheda, G. B., Dutta, S. P., Mittleman, A., Montgomery, J. A., Sethi, S. K., McCloskey, J. A., and Patrzyc, H. B., 1985, Isolation and characterization of a novel nucleoside, 7-β-D-ribofuranosylhypoxanthine, from the urine of a chronic myelogenous leukemia patient, *Cancer Res.* **45:**5958–5963.
12. Shigenaga, M. K., Gimeno, C. J., and Ames, B. N., 1989, Urinary 8-hydroxy-2′-deoxyguanosine as a biological marker of *in vivo* oxidative DNA damage, *Proc. Natl. Acad. Sci. U.S.A.* **86:**9697–9701.
13. Madden, B. E. H., 1990, Modified nucleotides in eukaryotic ribosomal RNA, *Nucleic Acid Methylation* **1990:**37–44.
14. Randerath, K., Agrawal, H. P., and Randerath, E., 1983, t-RNA alterations in cancer, *Recent Results in Cancer Res.* **84:**103–120.
15. Borek, E., Sharma, O. K., and Waalkes, T. P., 1983, New applications of urinary nucleoside markers, *Rec. Res. Cancer Res.* **84:**301–316.
16. Adamiak, R. W., and Gornicki, P., 1985, Hypermodified nucleosides of tRNA: synthesis, chemistry and structural features of biological interest, *Prog. Nuc. Acid Res.* 27–74.
17. Tworek, H. A., Bolanowska, W., Bhargava, A. K., Rachlin, E. M., and Chheda, G. B., 1986, Isolation and characterization of a novel nucleoside from human cancer urine, *Nucleosides Nucleotides* **5:**253–263.
18. Nakano, K., Nakao, T., Schram, K. H., Hammargren, W. M., McClure, T. D., Katz, M., and Petersen, E., Urinary Excretion of Modified Nucleosides as Biological Markers of RNA Catabolism in Patients with Cancer and AIDS, *Anal. Chem. Acta,* in press.
19. Schoch, G., Sander, G., Topp, H., and Heller-Schoch, G., 1990, Modified nucleosides and nucleobases in urine and serum as selective markers for the whole-body turnover of t-RNA, r-RNA, and m-RNA-Cap—future prospects and impact, in *Chromatography and Modification of Nucleosides, Part C* (C. W. Gehrke and K. C. T. Kuo, eds.), Elsevier, Amsterdam, pp. C389–C442.
20. Thomale, J., and Nass, G., 1982, Elevated urinary excretion of RNA catabolites as an early signal of tumor development in mice, *Cancer Lett.* **15:**149–159.
21. Waalkes, T. P., Gehrke, C. W., Zumwalt, R. W., Chang, S. Y., Lakings, D. B., Tormey, D. C., Ahmann, D. L., and Moertel, C. G., 1975, The urinary excretion of nucleosides of ribonucleic acid by patients with advanced cancer, *Cancer* **36:**390–398.
22. Speer, J., Gehrke, C. W., Kuo, K. C., Waalkes, T. P., and Borek, E., 1979, tRNA Breakdown products as markers for cancer, *Cancer* **44:**2120–2123.
23. Miller, J., Erb, N., Heller-Schoch, G., Lorenz, H., Winkler, K., and Schoch, G., 1983, Multivariate analysis of urinary RNA catabolites in malignancies: cross sectional and longitudinal studies, *Recent Results in Canc. Res.* **84:**317–330.
24. Vreken, P., and Tavenier, P., 1987, Urinary excretion of six modified nucleosides by patients with breast carcinoma, *Ann. Clin. Biochem.* **24:**598–603.
25. Salvatore, F., Colonna, A., Costanzo, F., Russo, T., Esposito, F., and Cimino, F., 1983, Modified nucleosides in body fluids of tumor-bearing patients, *Recent Results in Cancer Res.* **84:**360–377.
26. Sharma, O. K., Waalkes, T. P., Gehrke, C. W., and Borek, E., 1983, Applications of urinary nucleosides in cancer diagnosis and cancer management, *Cancer Detect. Prevent.* **6:**77–85.

27. Heldman, D. A., Grever, M. R., and Trewyn, R. W., 1983, Differential excretion of modified nucleosides in adult acute leukemia, *Blood* **61**:291–296.
28. Nielsen, H. R., and Killman, S. -A., 1983, Urinary excretion of γ-aminoisobutyrate and pseudouridine in acute and chronic myeloid leukemia, *J. Natl. Canc. Inst.* **71**:887–891.
29. Heldman, D. A., Grever, M. R., Miser, J. S., and Trewyn, R. W., 1983, Relationship of urinary excretion of modified nucleosides to disease status in childhood acute lymphoblastic leukemia, *J. Natl. Canc. Inst.* **71**:269–273.
30. Trewyn, R. W., Glaser, R., Kelley, D. R., Jackson, D. G., Graham, W. P., and Speicher, C. E., 1982, Elevated nucleoside excretion by patients with nasopharyngeal carcinoma, *Cancer* **49**:2513–2517.
31. Mills, G. C., Goldblum, R. M., Newkirk, K. E., and Schmalsteig, F. C., 1978, Urinary excretion of purines, purine nucleosides, and pseudouridine in adenosine deaminase deficiency, *Biochem. Med.* **20**:180–199.
32. Mills, G. C., Schmalstieg, F. C., and Goldblum, R. M., 1985, Urinary excretion of modified purines and nucleosides in immunodeficient children, *Biochem. Med.* **34**:37–51.
33. Esposito, F., Russo, T., Ammendola, R., Duilio, A., Salvatore, F., and Cimino, F., 1985, Pseuduoridine excretion and transfer RNA primers for reverse transcriptase in tumors of retroviral origin, *Cancer Res.* **45**:6260–6263.
34. Cimino, F., Russo, T., Colonna, A., Duilio, A., Ammendola, R., Costanzo, F., Oliva, A., Esposito, F., and Salvatore, F., 1987, Pseuduridine excretion in experimental neoplasias of retroviral origin, in *Human Tumor Markers* (F. Cimino, G. D. Birkmayer, J. V. Klavins, E. Pimentel, and F. Salvatore, eds.), Walter de Gruyter, Berlin, pp. 463–474.
35. Fischbein, A., Valciukas, J. A., Buschman, B. A., Kohn, M., Teirstein, A., Borek, E., Sharma, O. K., Bekesi, J. G., Apell, G., Bpesch, R. R., and Selikoff, I. J., 1985, Urinary syndrome (AIDS) and individuals at high risk of AIDS, *Canc. Detect. Prevent.* **8**:271–277.
36. Sharma, O. K., Buschman, F. L., Borek, E., Cohn, D. L., Penley, K. A., Judson, F. N., Dobozin, B. S., Zunich, K. M., and Kirkpatrick, C. H., 1987, Aberrant urinary excretion of modified nucleosides in patients with various manifestations of infection with HTLV-III/LAV, in *Human Tumor Markers* (F. Cimino, G. D. Birkmayer, J. V. Klavins, E. Pimentel, and F. Salvatore, eds.), de Gruyter, Berlin, pp. 545–558.
37. Borek, E., Sharma, O. K., Buschman, F. L., Cohn, D. L., Penley, K. A., Judson, F. N., Dobozin, B. S., Horsburgh, C. R., Jr., and Kirkpatrick, C. H., 1986, Altered excretion of modified nucleosides and γ-Aminoisobutyric acid in subjects with acquired immunodeficiency syndrome or at risk for acquired immunodeficiency syndrome, *Cancer Res.* **46**:2557–2561.
38. Leimer, K. R., Loeppky, R. N., and Gehrke, C. W., 1977, Identification of an interfering compound in the gas–liquid chromatographic determination of $N^2$,$N^2$-dimethylguanosine, *J. Chromatogr.* **143**:104–108.
39. Vold, B. S., Keigh, D. E., and Slavik, M., 1982, Urine levels of N-[9-(β-D-ribofuranosyl) purine-6-ylcarbamoyl]-*L*-threonine, $N^6$-(D2-isopentenyl)adenosine and 2′-O-methyl- guanosine as determined by radioimmunoassay for normal subjects and cancer patients, *Cancer Res.* **42**:5265–5269.
40. Itoh, K., Mizugaki, M., and Ishida, N., 1988, Preparation of a monoclonal antibody specific for 1-methyladenosine and its application for the detection of elevated levels of 1-methyladenosine in urines from cancer patients, *Jpn. J. Cancer Res.* **79**:1130–1138.
41. Karle, J. M., Anderson, L. W., Dietrick, D. D., and Clark, R. L., 1980, Determination of serum and plasma uridine levels in mice, rats, and humans by high-pressure liquid chromatography, *Anal. Biochem.* **109**:41–46.
42. Simpson, R. C., and Brown, P. R., 1986, High-performance liquid chromatographic profiling of nucleic acid components in physiological samples, *J. Chromatogr.* **379**:269–311.
43. Gehrke, C. W., Kuo, K. C., Davis, G. E., Suits, R. D., Waalkes, T. P., and Borek, E., 1978,

Quantitative high-performance liquid chromatography of nucleosides in biological materials, *J. Chromatogr.* **150**:455–476.
44. Gehrke, C. W., Zumwalt, R. W., McCune, R. A., and Kuo, K. C., 1983, Quantitative high-performance liquid chromatography analysis of modified nucleosides in physiological fluids, tRNA, and DNA, *Recent Results in Canc. Res.* **84**:344–359.
45. Schlimme, E., Boos, K. -S., Hagemeier, E., Kemper, K., Meyer, U., Hobler, H., Schnelle, T., and Weise, M., 1986, Direct clean-up and analysis of ribonucleosides in physiological fluids, *J. Chromatogr.* **378**:349–360.
46. Reimer, M. L. J., Schram, K. H., Nakano, K., and Yasaka, T., 1989, The identification of 5,6-dihydrouridine in normal human urine by combined gas chromatography/mass spectrometry, *Anal. Biochem.* **181**:302–308.
47. McCloskey, J. A., 1986, Experimental approaches to the characterization of nucleic acid constituents by mass spectrometry, in *Mass Spectrometry in Biochemical Research* (S. J. Gaskell, ed.), Wiley, New York, pp. 75–95.
48. McCloskey, J. A., 1990, Constituents of nucleic acids: Overview and strategy, in *Methods in Enzymology* Vol. 193 (J. A. McCloskey, ed.), Academic Press, San Diego, CA, pp. 771–781.
49. Schram, K. H., 1989, Purines and pyrimidines, in *Mass Spectrometry* (A. M. Lawson, ed.), deGruyter, Berlin, pp. 508–570.
50. Schram, K. H., 1990, Mass spectrometry of nucleic acid components, in *Biomedical Applications of Mass Spectrometry* (C. H. Suelter and J. T. Watson, eds.), Wiley, New York, pp. 203–287.
51. Schram, K. H., 1990, Mass spectrometry: applications in the analysis of nucleosides, nucleotides, and oligonucleotides, *Nucleosides Nucleotides* **9**:311–317.
52. Dutta, S. P., Crain, P. F., McCloskey, J. A., and Chheda, G. B., 1979, Isolation and characterization of 1-β-D-ribofuranosylpyridin-4-one-3-carboxamide from human urine, *Life Sci.* **24**:1381–1388.
53. McCloskey, J. A., 1991, Structural characterization of natural nucleosides by mass spectrometry, *Acc. Chem. Res.* **24**:81–88.
54. Chheda, G. B., Patrzyc, H. B., Yworek, H. A., and Dutta, S. P., 1990, Isolation and characterization of modified nucleosides from human urine, in *Chromatography and Modification of Nucleosides, Part C* (C. W. Gehrke and K. C. T. Kuo, eds.), Elsevier Amsterdam, pp. C185–C230.
55. Nakano, K., 1990, High-performance liquid chromatography analysis of nucleosides and bases in mucosa tissues and urine of gastorintestinal cancer patients, in *Chromatography and Modification of Nucleosides, Part C* (C. W. Gehrke and K. C. T. Kuo, eds.), Elsevier, Amsterdam, pp. 293–320.
56. Lawson, A. M., Stillwell, R. N., Tacker, M. M., Tsuboyama, K., and McCloskey, J. A., 1971, Mass spectrometry of nucleic acid components. Trimethylsilyl derivatives of nucleosides, *J. Amer. Chem. Soc.* **93**:1014–1023.
57. Schram, K. H., and McCloskey, J. A., 1979, Nucleosides and nucleotides, in *GLC and HPLC Determination of Therapeutic Agents* (K. Tsuji, ed.), Marcel Dekker, New York, pp. 1149–1190.
58. Schram, K. H., 1990, Preparation of trimethylsilyl derivatives of nucleic acid components for analysis by mass spectrometry, in *Methods in Enzymology, Volume 193* (J. A. McCloskey, ed.), Academic Press, San Diego, CA, pp. 791–796.
59. Bruins, A. P., Jennings, K. R., and Evans, S., 1978, The observation of metastable transitions in a double-focusing mass spectrometer using a linked-scan of the electric sector and magnetic sector fields, *Int. J. Mass Spectrom. Ion Phys.* **26**:395–404.
60. Pang, H., Schram, K. H., Smith, D. L., Gupta, S. P., Townsend, L. B., and McCloskey, J. A.,

1982, Mass spectrometry of nucleic acid constituents. Trimethylsilyl derivatives of nucleosides, *J. Org. Chem.* **47**:3923–3932.

61. Reimer, M. L. J., McClure, T. D., and Schram, K. H., 1989, Differentiation of isomeric 2′-, 3′-, and 5′-deoxynucleosides by electron ionization and chemical ionization linked-scanning mass spectrometry, *Biomed. Environ. Mass Spectrom.* **18**:533–542.
62. Hammargren, W. M., Luffer, D. R., Schram, K. H., Reimer, M. L. J., Nakano, K., Yasaka, T., and Moorman, A. R., 1992, The identificationof 5′-deoxy-5′-methylthioguanosine in human urine by gas chromatography/mass spectrometry, *Nucleosides Nucleotides* **11**:1275–1292.
63. Markiw, R. T., 1973, Isolation and identification of 3-methyluridine from normal human urine, *Biochem. Med.* **8**:182–187.
64. McClure, T. D., Schram, K. H., Nakano, K., and Yasaka, T., 1989, 3-Methyluridine in normal human urine, *Nucleosides Nucleotides* **8**:1417–1426.
65. Shaw, S. J., Desiderio, D. M., Tsuboyama, K., and McCloskey, J. A., 1970, Mass spectrometry of nucleic acid components. Analogs of adenosine, *J. Am. Chem. Soc.* **92**:2510–2522.
66. Crain, P. F., Choi, Y. C., Busch, H., and McCloskey, J. A., 1978, Characterization of $N^6,O^{2'}$-dimethyladenosine from nuclear RNA of Novikoff hepatoma, *Nucleic Acid Res.* **5**:771–776.
67. Gehrke, C. W., and Kuo, K. C., 1990, Ribonucleoside analysis of reversed-phase high performance liquid chromatography, in *Chromatography and Modification of Nucleosides, Part A,* (C. W. Gehrke and K. C. T. Kuo, eds.), Elsevier, Amsterdam, pp. A4–A71.
68. McClure, T. D., Schram, K. H., Nakano, K., and Yasaka, T., 1989, Identification and synthesis of 5′-deoxyxanthosine, a novel nucleoside in normal human urine, *Nucleosides Nucleotides* **8**:1399–1415.
69. Hammargren, W. M., Schram, K., Nakano, K., Yasaka, T., Katz, M., and Petersen, E., 1991, Analysis of urinary nucleosides from AIDS patients using GC/MS, *Nucleosides Nucleotides* **10**:659–661.
70. Sakurai, T., Matsuo, T., Kusai, A., and Nojima, K., 1989, Collisionally activated decomposition spectra of normal nucleosides and nucleotides using a four-sector tandem mass spectrometer, *Rapid Commun. Mass Spectrom.* **3**:212–216.
71. McClure, T. D., 1991, Biomedical applications of mass spectrometry, Ph. D. Thesis, University of Arizona, Tucson, AZ.
72. Schram, K. H., and Slowikowski, D., 1986, Fast atom bombardment of trimethylsilyl derivatives of nucleosides and nucleotides, *Biomed. Environ. Mass Spectrom.* **13**:263–264.
73. Slowikowski, D., and Schram, K. H., 1985, Fast atom bombardment mass spectrometry of nucleosides, nucleotides and oligonucleotides, *Nucleosides Nucleotides* **4**:309–345.
74. Whitehouse, C. M., Dreyer, R. N., Yamashita, M., and Fenn, J. B., 1985, Electrospray interface for liquid chromatographs and mass spectrometers, *Anal. Chem.* **57**:675–679.
75. Mills, G. C., Schmalsteig, F. C., Koolkin, R. J., and Goldblum, R. M., 1982, Urinary excretion of purines, purine nucleosides and pseudouridine in immunodeficient children, *Biochem. Med.* **27**:37–45.
76. Chheda, G. B., 1970, Purine and pyrimidine derivatives excreted in human urine, in *Handbook of Biochemistry,* 2nd edn. (H. A. Sober, ed.), The Chemical Rubber Company, Cleveland, Ohio, pp. G106–G113.
77. Dutta, S. P., Tworek, H. A., and Chheda, G. B., 1985, Detection and identification of 2′-O-methyl nucleosides in human cancer urine by capillary GC-MS, *Proc. 33rd ASMS Conf. Mass Spectrom. Allied Topics,* May 26–31, San Diego, CA, pp. 529–520.
78. Colonna, R., Russo, T., Cimino, F., and Salvatore, F., 1981, Modified nucleosides in biological fluids of cancer patients determined by high performance liquid chromatography, *J. Clin. Chem. Clin. Biochem.* **19**:640.
79. Chheda, G. B., 1977, Isolation and characterization of $N^6$-succinyladenosine from human urine, *Nuc. Acid Res.* **4**:739–746.

80. Kaneko, K., Fujimori, S., Kamatani, N., and Akaoka, I., 1984, 5′-Methylthioadenosine in urine from normal subjects and cancer patients, *Biochim. Biophys. Acta* **802**:687–695.
81. Mills, J. S., Mills, G. C., and McAdoo, D. J., 1983, Isolation and identification of 5′-methylthioadenosine sulfoxide from human urine, *Nucleosides Nucleotides* **2**:465–478.
82. Gehrke, C. W., Kuo, K. C., Waalkes, T. P., and Borek, E., 1979, Patterns of urinary excretion of modified nucleosides, *Cancer Res.* **39**:1150–1153.
83. Tormey, D. C., Waalkes, T. P., and Gehrke, C. W., 1980, Biological markers in breast carcinoma. Clinical correlations with pseudouridine, $N^2$,$N^2$-dimethylguanosine and 1-methylinosine, *J. Surg. Oncol.* **14**:267–273.
84. Heldman, D. A., Grever, M. R., Speicher, C. E., and Trewyn, R. W., 1983, Urinary excretion of modified nucleosides in chronic myelogenous leukemia, *J. Lab. Clin. Med.* **101**:783–792.
85. De, N. C., Mittelman, A., Dutta, S. P., Edmonds, C. G., Jenkins, E. E., McCloskey, J. A., Blakely, C. R., Vestal, M. L., and Chheda, G. B., 1981, Isolation and characterization of two new modified uracil derivatives from human urine, *J. Carbohydrates Nucleosides Nucleotides* **8**:363–389.
86. Borek, E., and Kerr, S. J., 1972, Atypical transfer RNAs and their origin in neoplastic cells, *Adv. Cancer Res.* **15**:163–190.
87. Dutta, S. P., Tworek, H. A., Bhargava, A. K., Patrzyc, H. B., and Chheda, G. B., 1988, Characterization of 2′-O-methylinosine and other 2′-O-methylnucleosides in human urine, *Proc. 36th ASMS Conf. Mass Spectrom. Allied Topics,* June 5–10, San Francisco, CA, pp. 494–495.
88. Lehtio, H., and Maenpaa, P. H., 1977, The urinary excretion of the nucleosides pseudouridine and 1-methylinosine during normal menstrual cycle, *Clin. Chim. Acta* **80**:181–185.
89. De, N. C., and Chheda, G. B., 1979, Isolation and characterization of 2′-O-methyluridine from human urine, *J. Carbohydrates Nucleosides Nucleotides* **6**:371–385.
90. Chheda, G. B., Antkowiak, J. G., Takita, H., Bhargava, A. K., and Tworek, H. A., 1990, Evaluation of 5-carbamoylmethyluridine as an indicator of tumor burden in lung cancer patients, *Nucleosides Nucleotides* **9**:441–442.
91. Chheda, G. B., Tworek, H. A., Bhargava, A. K., Antkowiak, J. G., Dutta, S. P., Patrzyc, H. B., 1988, Isolation and characterization of two new nucleosides from human urine, *Proc. 36th ASMS Conf. Mass Spectrom. Allied Topics,* June 5–10, San Francisco, CA, pp. 492–493.
92. Chheda, G. B., Tworek, H. A., Bhargava, A. K., Rachlin, E., Dutta, S. P., and Patrzyc, H. B., 1988, Isolation and characterization of 3-(3-amino-3-carboxypropyl)uridine from human urine, *Nucleosides Nucleotides* **7**:417–429.
93. Mills, G. C., Davis, N. J., and Lertratanangkoon, K., 1989, Isolation and identification of 1-ribosyl pyridone nucleosides from human urine, *Nucleosides Nucleotides* **8**:415–430.
94. Chheda, G. B., Dutta, S. P., Tworek, H. A., Bhargava, A. K., Takita, H., Antkowiak, J. G., and Patrzyc, H. B., 1989, Isolation and characterization of an unusual nucleoside from human urine, *Proc. 37th ASMS Conf. Mass Spectrom. Allied Topics,* May 21–26, Miami Beach, FL, pp. 84–85.

# 6

# Microdialysis/Mass Spectrometry

*Per E. Andrén, Shen-Nan Lin, and Richard M. Caprioli*

## 1. INTRODUCTION

Elucidation of the biochemistry of neurochemicals in the live animal has traditionally involved the sampling of blood, cerebrospinal fluid (CSF), and tissue from brain biopsy. Many other experimental studies of brain chemistry have been conducted with postmortem tissue, giving a poor representation of the dynamic chemistry that exists between cells and the interchange of compounds between cells and extracellular fluid. Total tissue levels may easily mask small but important neurochemical changes in specific brain regions. For these reasons, investigators have turned to techniques that directly sample the extracellular compartment of nervous tissue in living animals.

Some *in vivo* methods for estimation of the release of neurotransmitters have been known for many years, such as the push–pull perfusion (1) and cortical-cup (2) techniques. In the push–pull cannula method, two concentric tubes are stereotactically implanted in a localized brain area, and fluid is introduced by a perfusion pump through the inner cannula and removed by a second pump through the outer tube. The perfused fluid is in direct contact with brain tissue that surrounds the tip of the cannula, and endogenous substances are obtained in the perfusion fluid. Push–pull perfusion has been used to measure the release of endogenous compounds such as catecholamines and their metabolites, amino acids, peptides, and acetylcholine. The cortical-cup

*Per E. Andrén, Shen-Nan Lin, and Richard M. Caprioli* • Department of Biochemistry and Molecular Biology, and Analytical Chemistry Center, University of Texas Medical School, Houston, Texas 77225.

*Mass Spectrometry: Clinical and Biomedical Applications, Volume 2*, edited by Dominic M. Desiderio. Plenum Press, New York, 1994.

technique utilizes a small cup that is applied on a surface area, usually the cortex of the brain, and perfused with a physiological solution. Endogenous compounds diffuse from the brain tissue to the cup solution. Although both of these techniques have been shown to be useful under certain circumstances, they have severe limitations that have prevented them from being widely used.

## 2. *MICRODIALYSIS*

### 2.1. *Probe Design*

Tubes for *in vivo* brain dialysis were first reported by Ungerstedt (3,4), and this technique has been extensively used as a routine means for the measurement of many endogenous neurochemicals (5). Microdialysis sampling provides several advantages, detailed below, over the abovementioned techniques and provides a means of continuous sampling with no fluid loss. Samples of dialysate can often be analyzed directly by a suitable technique—e.g., liquid chromatography (LC)—without prior cleanup.

Conceptually, microdialysis imitates the function of a capillary blood vessel by perfusing a thin dialysis tube implanted in a tissue of interest. A number of different designs of microdialysis probes and membrane material with different molecular weight cutoff and dialysis properties have been used for *in vivo* sampling. The concentric design (illustrated in Fig. 6.1) commonly used consists of a small, hollow dialysis tube made of a polymer such as polycarbonate/polyether. One end of the dialysis tube is sealed by glue forming the tip of the probe, and the other end is usually glued onto a steel capillary tube, which becomes the shaft of the probe. A thin inner cannula extends through the shaft to the dialysis tip. The perfusion liquid enters the inner cannula and flows out through two small holes in the wall at the bottom. The liquid then sweeps the dialysate up and out of the device. The concentric probe can be made very thin, and the lengths of the membrane can be varied; typically, the outer diameter of the probe is about 0.25–0.5 mm and membrane length is 1–5 mm. A second microdialysis design, the loop-style (U-shape) probe, is composed of two parallel metal tubes connected by a dialysis tube forming a loop. The loop-type probe offers the advantage of a large membrane surface area, and because of the advantage of increased recovery it is beneficial in applications where its larger size can be tolerated.

### 2.2. *Perfusion Medium*

The perfusion medium, whose composition closely resembles that of the extracellular fluid of the brain or tissue of interest, is pumped by a microsyringe

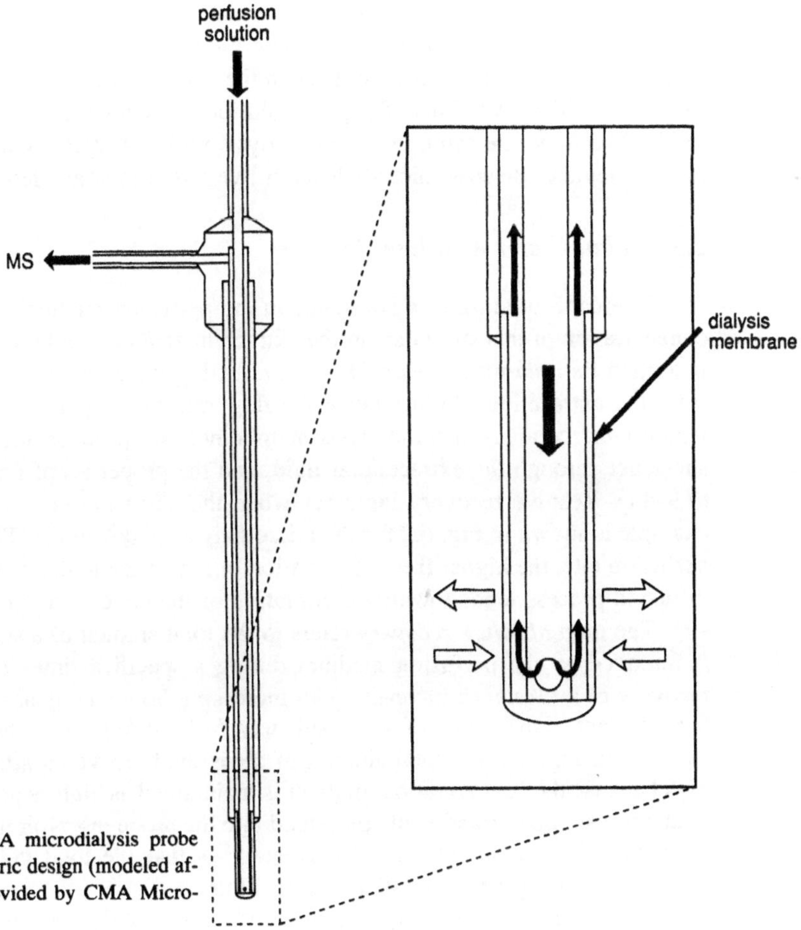


*Figure 6.1.* A microdialysis probe of the concentric design (modeled after probes provided by CMA Microdialysis AB).

pump through the microdialysis probe. Analytes outside the microdialysis probe diffuse into the probe reservoir and subsequently flow to the analytical device. For drug delivery, compounds may also be diffused into target tissue sites by diffusion from the perfusion medium into the extracellular fluid by the same means. A precisely controlled perfusion flow rate is important because it allows compounds to be reproducibly introduced or collected by the diffusion process across the microdialysis membrane wall. The diffusion over the membrane occurs along a logarithmic concentration gradient into or out of the probe, and the direction of this gradient depends upon the analyte concentrations in the fluids. Microdialysis can be performed locally in almost every organ or tissue in the body, including the blood. The perfusion medium passing through the microdialysis probe is usually an isotonic fluid, which contains physiological concentrations of salts similar to those in the extracellular fluid in the tissue of interest. Artificial CSF and Ringer's solution are commonly used perfusion fluids. The flow rate through a microdialysis probe is typically 0.5–5 μl/min. Several reviews discuss microdialysis in live tissue in some detail (3,6–8).

### 2.3. Relative and Absolute Recovery

The concentration of a substance in the perfusion medium relates to the concentration of this substance in the extracellular space, and this relationship is termed the *relative recovery.* Factors affecting the recovery of a substance from the extracellular fluid include the flow rate of the perfusion liquid, the length (surface area) of the dialysis membrane, the speed of diffusion of the substance through the extracellular fluid, and the properties of the membrane (7,9–11). Relative recovery increases when the perfusion rate decreases. An example is shown in Fig. 6.2 for the microdialysis of dopamine. The lower the perfusion rate, the higher the concentration of dopamine in the dialysate as the diffusion process tends toward equilibration of the concentration gradient.

The term *absolute recovery* refers to the total amount of a substance that is removed by the perfusion medium during a specified time. The absolute recovery of a substance increases with increasing flow rate, generally remaining constant above rates of approximately 5–10 μl/min. For the dopamine example in Fig. 6.2, the total amount of compound recovered after 10 min is high because the concentration gradient is maintained as high as possible. This gradient maximizes transfer of compound into the probe reservoir in the typical microdialysis experiment. The disadvantage is that the total volume of the sample is also high; i.e., the dialysate is dilute.

The relative recoveries of a number of neuropeptides are given in Table 6.1 for a 5-mm-long dialysis membrane using a 2.0-μl/min flow rate with CMA/10 (Carnegie Medicin) microdialysis probes (12). The percentage recoveries listed in the table are mean values obtained from three probes. Recovery variability between probes was <20% and within probe recoveries was <10%. The dialysis solution was Ringer's solution or 0.9% NaCl. It can be seen from these data that *in vitro* recoveries are somewhat low, especially for peptides such as β-endorphin, which tend to adhere to surfaces and subsequently are desorbed over long periods. Relative recovery data for other small compounds (13) of widely ranging structure indicate that 15–25% are typical values, although values range from 1–45%.

The properties of the microdialysis probe membrane can play an important role in the recovery of larger molecules such as neuropeptides, where the

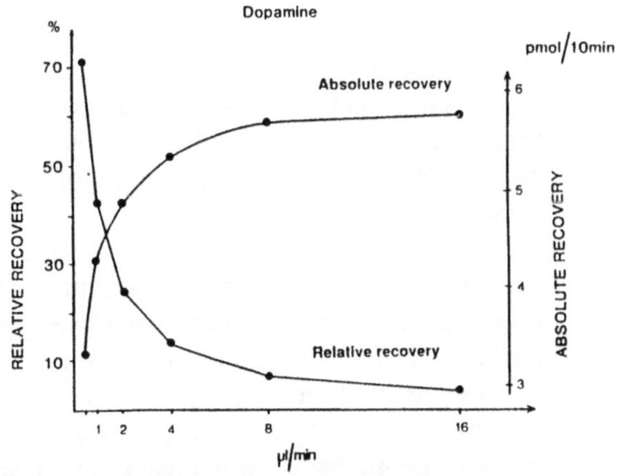


*Figure 6.2.* Relationship between perfusion flow rate and relative and absolute recovery for dopamine (courtesy CMA Microdialysis AB).

*Table 6.1*
In Vitro *Neuropeptide Recoveries of Microdialysis Probes*[a]

| Neuropeptide | Relative recovery (%) |
|---|---|
| Angiotensin | 19.0 |
| Arginine vasopressin | 18.3 |
| β-Endorphin | 3.0 |
| Bombesin | 16.6 |
| Cholecystokinin-8 | 12.7 |
| Corticotrophin releasing factor | 3.1 |
| Dynorphin 1-17 | 6.5 |
| Leutinizing hormone releasing hormone | 15.6 |
| [Leu]enkephalin | 20.9 |
| [Met]enkephalin | 24.8 |
| Neurotensin | 12.0 |
| Neuropeptide Y | 1.5 |
| Oxytocin | 16.4 |
| Substance K | 18.0 |
| Substance P | 15.5 |
| Thyrotropin releasing hormone | 19.4 |

[a]From Ref. 12.

limiting factor is usually not the speed of diffusion of the analyte through the extracellular fluid (9) but rather passage through the membrane. The physical and chemical properties of the membranes are important because the diffusion of solutes across the membrane is expected to vary depending on the interaction of the solute with the membrane. Several types of membrane material can be used, such as cellulose, cellulose acetate, cuprophan, polyacrylonitrile, and polycarbonate. Some of the properties of these materials are discussed in more detail below.

Microdialysis can be performed on anaesthetized animals or on awake animals with chronically implanted probes and guide cannulas. The anaesthetized animal setup is typically used when complicated experiments are performed, e.g., when more than one microdialysis probe or local injection cannulas are used. When using anaesthetized animals, a heating pad and subcutaneous injections of Ringer's solution during experiments of long duration are recommended to maintain the physiological integrity of the animal. Experiments on awake animals can be done using a guide cannula that has been implanted under anesthesia, the probe being introduced after the animal has regained consciousness. The obvious experimental advantage of making measurements in unanesthetized animals is counterbalanced by the fact that the animals can be susceptible to many influences, such as pain from the implanted probe, restraint by the tubing and wires, and stimuli from the environment. These considerations will not be covered in detail in this chapter, and the reader is referred to the reviews already cited for additional information and references.

## *3. GENERAL ASPECTS OF NEUROPEPTIDE ANALYSIS*

Microdialysis has become established as a useful sampling technique for *in vivo* assessment of a variety of low-molecular-weight compounds, although its use for studying neuropeptides and their release *in vivo* has not been extensive. This latter limit is in part because most techniques based on chemical detection currently available for the quantitative analysis of endogenous neuropeptides are generally not sufficiently sensitive for detection of the low femtomole levels of these compounds. Currently, radioimmunoassay techniques are being used for measurements of neuropeptide concentrations, but problems with cross-reactivity limit the usefulness of such methods.

Some peptides are known to bind easily to glass surfaces and membranes, and significant differences exist between the efficiencies of recoveries of such substances among the various membrane materials (12,14). In some cases, these investigators have shown, recovery may be improved by adding 0.025–0.5% bovine serum albumin (BSA) to the dialysate. In the study by Maidment and Evans (Ref. 14), several types of membrane materials were tested for their relative recovery of known concentrations of Leu-enkephalin. These membranes were made of cellulose, cellulose acetate, cuprophan, polyacrylonitrile, and polysulphone; also tested was the commercially available polycarbonate membrane probe (CMA Microdialysis AB, Stockholm, Sweden). Recovery experiments were carried out by continuous perfusion (2.7 μl/min) in unstirred artificial CSF with added synthetic Leu-enkephalin (see Table 6.2). After normalizing for surface area, the cellulose, cellulose acetate, and cuprophan membranes gave the poorest results (4–6% recovery), and the polyacrylonitrile and polycarbonate gave the best results (13–14% recovery).

The adsorption effect of peptides and their tendency to "stick" to the dialysis membrane have also been examined (14). The length of time required for the concentration of a peptide in the dialysate to reach maximal levels was measured after immersion in a sample solution of a peptide. Subsequently, the time-course of the decline following replacement in a peptide-free medium was also measured. The study included both labeled and unlabeled neurotensin and cholecystokinin-8 (CCK-8). The recovery of unlabeled neurotensin across polyacrylonitrile membranes reached maximal values more rapidly compared to the iodinated peptide, but some delay was seen when the peptide was put in the peptide-free solution. No adsorption was seen when the polycarbonate membrane was used. The unlabeled sulphated CCK-8 showed no tendency to bind to the polyacrylonitrile or to the polycarbonate membrane, whereas the iodinated unsulphated CCK-8 showed a significant amount of adsorption to the polycarbonate membrane but not to the polyacrylonitrile membrane. The study showed that changes in charge or hydrophobicity of a peptide can result in

*Table 6.2*
*Relative Recovery Values for Leu-Enkephalin with Several Types of Dialysis Membranes*

| Membrane | Mol. wt. (cutoff) | O.D. μm | Recovery (%)[a] [$^{125}$I]Leu-enk (*n*=4) | Recovery (%)[a] Leu-enk (*n*=4) | Normalized recovery (%)[b] |
|---|---|---|---|---|---|
| Cellulose | 5,000 | 250 | 3.2±0.1 | 2.4±0.1 | 6.4 |
| Cellulose acetate | 70,000 | 230 | 2.6±0.2 | 2.1±0.3 | 5.3 |
| Cuprophan (6.5 μm) | 12,000 | 215 | 5.6±0.6 | 5.1±0.4 | 13 |
| Cuprophan (19 μm) | 12,000 | 320 | 2.6±0.2 | 1.8±0.2 | 3.9 |
| Polyacrylonitrile | 40,000 | 300 | 8.0±0.7 | 8.8±0.9 | 13.5 |
| Polycarbonate | 20,000 | 500 | 12.6±0.8 | 13.3±0.7 | 13 |

[a]Experimental conditions: Membrane lengths of 5 mm, perfusion rate 2.7 μl/min, sample was 1 nM containing 1000 cpm ($^{125}$I)Leu-enkephalin. After 30 min equilibration, duplicate 30-min samples were collected.
[b]Normalized to surface area.

marked changes in recovery across different membranes. It is therefore prudent to validate each peptide for different membrane materials, if maximum recovery is vital for a given experiment.

## *4. MICRODIALYSIS AND LC/MS: APPLICATIONS*

Microdialysis has been used with many analytical techniques, and, in particular, has been shown to be compatible with liquid chromatography. The combination of microdialysis and mass spectrometry (MD/MS) has great potential because it provides the analysis with mass specificity and the capability for an automated direct interface with mass spectrometers. In a typical experiment, the microdialysis probe delivers a flow of about 1 μl/min, which can be combined on line with MS interfaces such as continuous-flow fast atom bombardment (CF-FAB) and electrospray ionization (ESI). The combined techniques of on-line microdialysis and CF-FAB were first reported by Caprioli and Lin (Ref. 15), who analyzed antibiotic drug levels in the blood of rats and rabbits. In one experiment, a microdialysis probe was inserted into the jugular vein of a rat and the antibiotic penicillin G was administered intramuscularly. The dialysis probe was perfused with sterile water containing 5% glycerol, and the perfusate was passed through a micro-C-18 column to concentrate the dialysate. The column was eluted with 60% acetonitrile, 35% water, and 5% glycerol, and the eluate was introduced into the mass spectrometer through a CF-FAB interface (Fig. 6.3). Samples were taken at 10-min intervals. The pharmacokinetic curve from this experiment is shown in Fig. 6.4, labeled "rat," along with curves, reported in the literature, for human and dog; the latter two were measured by non-mass-spectrometric techniques. The time-dependent rise and fall in the concentration of free (non-protein-bound) drug is clearly seen for the experiment (rat) and is similar to those for human and dog, although, of course, these curves need not be coincident, because of species differences. Drug recovery for the rat experiment was measured to be about 16%.

Several other investigators have reported the use of MD/MS in live-animal experiments, either on or off line. Lin *et al.* (16) have used multiple probes with mass spectrometric detection to monitor the pharmacokinetics of valproic acid (VPA), an antiepileptic drug, in brain and blood. CF-FAB/MS was used with switching valves to monitor the dialysate by flow injection methods. An IV dose of 50 mg/kg was given to a 250-g rat. Figure 6.5 shows the pharmacokinetic curves produced with peak brain levels of VPA approximately ten times lower than that in blood. In another study, the half-life of an organophosphate, *tris* (2-chloroethyl) phosphate (TRCP) in rat was measured by microdialysis with CF-FAB and tandem mass spectrometry (17). With the

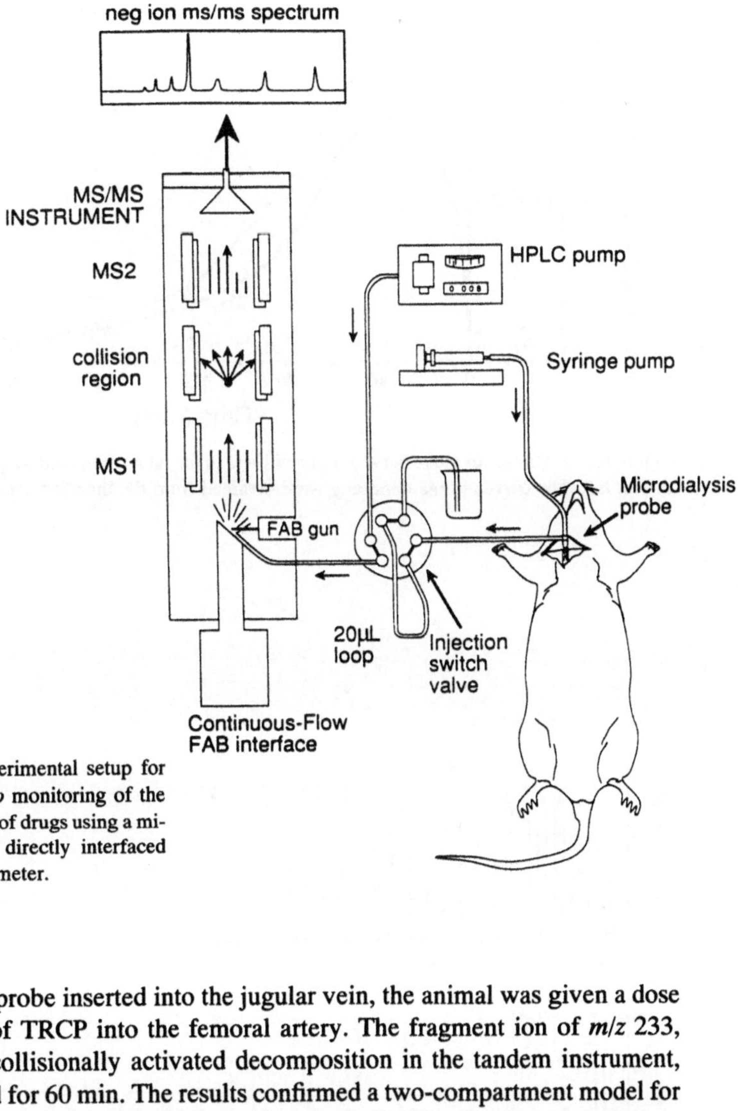


*Figure 6.3.* Experimental setup for the on-line *in vivo* monitoring of the pharmacokinetics of drugs using a microdialysis probe directly interfaced to a mass spectrometer.

microdialysis probe inserted into the jugular vein, the animal was given a dose of 20 mg/kg of TRCP into the femoral artery. The fragment ion of *m/z* 233, produced by collisionally activated decomposition in the tandem instrument, was monitored for 60 min. The results confirmed a two-compartment model for the compound in the blood of rats, with α half-life (distribution and elimination) of 2.9 min and β half-life (elimination) of 23.1 min. Eicosanoids have been successfully measured in the trachea of anesthetized guinea pigs using off-line microdialysis and GC/MS techniques (18). Mean levels of $PGE_2$, $PGD_2$, $PGF_{2\alpha}$, and $TXB_2$ were in the range of 20–60 pg/hr. Another study measured *in vivo* the release and metabolism of brain N-acetylaspartate (NAA) and N-acetylaspartylglutamate (NAAG) in the brain of a rat (19). Additional details of this work are given later in this chapter. Also, the time domain for the

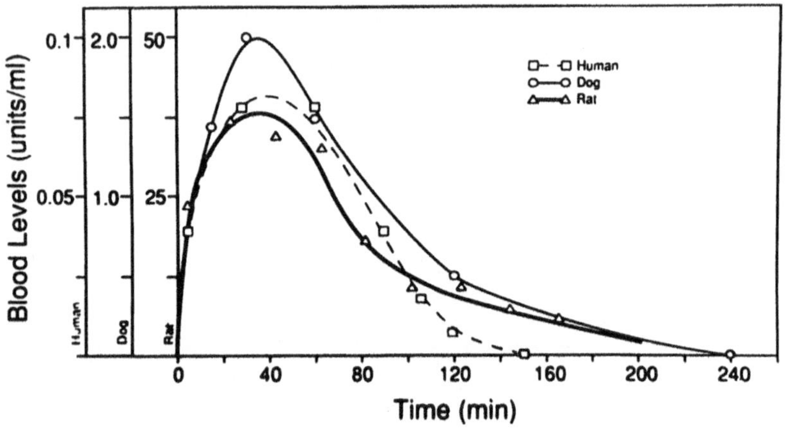


*Figure 6.4.* Pharmacokinetic curve for penicillin G in the rat as measured by the apparatus shown in Fig. 6.3. The curves in man and dog were obtained from the literature (see Ref. 15).

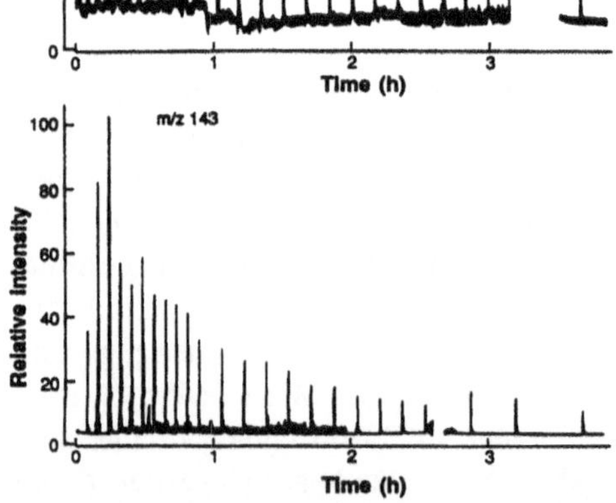


*Figure 6.5.* Pharmacokinetics of valproic acid in blood (bottom) and brain (top) of the rat as measured by microdialysis/mass spectrometry (see Ref. 16).

digestion of substance P by carboxypeptidase Y *in vitro* has been shown using microdialysis with CF-FAB (20) and electrospray (21) ionization.

In summary, the coupling of microdialysis and mass spectrometry is advantageous in that microdialysis provides a highly filtered, low volume, aqueous solution of polar analytes. The low volumes of solvent flowing into the mass spectrometer (0.5–5 μl/min in a typical experiment) obviate many of the problems in the interface between them. However, several aspects of microdialysis may cause difficulties in some cases; low analyte recovery, time dependence of sample aliquots, accurate calibration of an implanted probe, and the high concentrations of inorganic ions present in the perfusion media containing physiological salt solutions.

## 5. IN VIVO *METABOLISM OF SUBSTANCE P*

The *in vivo* metabolism of the neuropeptide substance P (SP) has been studied in brain in an anaesthetized rat using a variety of analytical techniques. This undecapeptide (Arg-Pro-Lys-Pro-Gln-Gln-Phe-Phe-Gly-Leu-Met-$NH_2$) is considered to be a neurotransmitter or neuromodulator with a wide CNS distribution (22) and has been proposed to mediate various physiological responses (23). The putative role of SP as a neurotransmitter has focused interest on the mechanisms of its inactivation. Both cytosolic and membrane-bound peptidases capable of degrading SP have been reported to exist in the brain (24). Membrane-bound peptidases may be of greater importance because termination of SP activity probably occurs by proteolytic breakdown of the peptide rather than by an uptake mechanism (25). Many purified enzymes have been shown to hydrolyse SP *in vitro* at a variety of peptide bonds, but it is uncertain how these enzymes perform *in vivo*.

Angiotensin-converting enzyme has been shown to hydrolyse SP between positions 8–9 and 9–10 (26,27), and endopeptidase 24.11 (enkephalinase 24.11) cleaves SP bonds between positions 6–7, 7–8, and 9–10 (28,29). Further, the SP-degrading enzyme hydrolysed the 6–7, 7–8, and 8–9 bonds (25), an SP-endopeptidase cleaved the 7–8 and 8–9 bonds (30), and dipeptidyl-aminopeptidase IV hydrolyses the 2–3 and 4–5 bonds (31). A bacitracin-sensitive enzyme (32) may also take part in the metabolism of SP in the CNS.

In initial experiments using microdialysis and electrospray mass spectrometry (33), the *in vivo* metabolism of SP was studied by monitoring fragment peptides produced over time after injection of small amounts of SP into specific brain regions. For example, rats were anaesthetized and the microdialysis probe stereotaxically implanted into the striatum of the brain. The schematic setup of the system is given in Fig. 6.6. The microdialysis probe was perfused at a flow rate of 0.3 μl/min with an artificial extracellular fluid (5 mM

KCl, 120 mM NaCl, 1.2 mM $MgCl_2$, 1.8 mM $CaCl_2$, 0.15% phosphate buffered saline; pH = 7.4). In one experiment, 2 hr after probe implantation, a 10-μM solution of SP in artificial extracellular fluid was introduced to the microdialysis probe and infused for two hours. The recovered dialysis samples were loaded into a 10-μl loop on a ten-port injection valve and injected onto a micro-reversed-phase C-18 column by a syringe pump at 10 μl/min. Electrospray ionization (ESI) (Vestec Corp., Houston, TX) was used to analyze the LC eluent flowing at 2 μl/min. The ESI process produces multiply charged peptide species through proton addition, i.e., $(M + H)^{+}$, $(M + 2H)^{+2}$, $(M + 3H)^{+3}$, etc. For substance P and its fragments, singly or doubly charged species are primarily observed, as shown in Table 6.3. Figure 6.7 shows the ESI mass spectrum for the *in vivo* infusion of SP into the rat brain striatum at 0.3 μl/min. For comparison, the spectrum from the control experiment is also shown. This control used the same conditions and apparatus as for the *in vivo* experiment, except that the rat was replaced by a small vial of physiological saline solution held at 37 °C. The mass spectrum for the *in vivo* SP experiment showed the presence

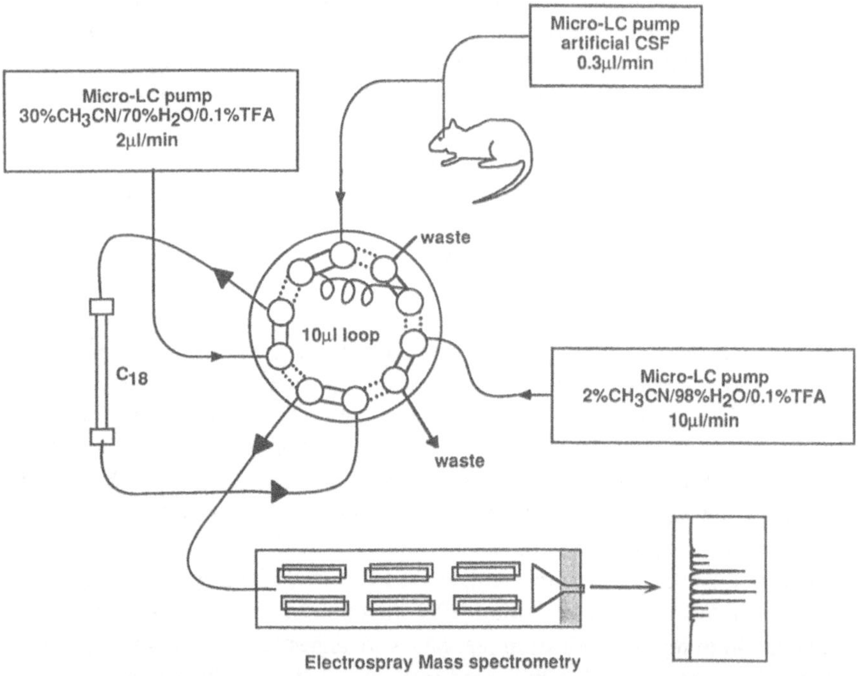


*Figure 6.6.* Experimental setup for the analysis of brain neuropeptide and neuropeptide metabolites in perfusate containing physiological levels of salt solutions.

*Table 6.3*
*m/z of Ions from SP Produced on Electrospray Ionization*

| SP fragment | Ions observed | |
|---|---|---|
| | $(M+H)^+$ | $(M+2H)^{+2}$ |
| 1–11 | — | 674.6 |
| 1–9 | — | 552.8 |
| 1–8 | — | 524.8 |
| 1–7 | — | 450.5 |
| 5–11 | 870.3 | — |
| 6–11 | 742.2 | — |
| 7–11 | 613.9 | — |
| 8–11 | 466.3 | — |

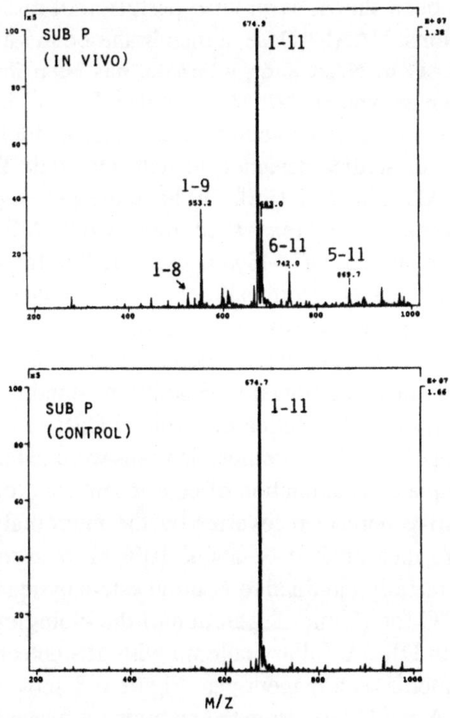


*Figure 6.7.* Electrospray mass spectrum of (top) *in vivo* metabolism of substance P in the anesthetized rat from infusion of the neuropeptide into the brain and (bottom) control spectrum obtained from apparatus identical except that the microdialysis probe was placed in Ringer's solution.

of fragments, 1–7, 1–8, 1–9, 8–11, 7–11, 6–11, and 5–11. Fragment 1–9 gave the most intense peak, suggesting that this fragment and not fragment 1–7 or 1–8 is the major metabolic species for the initial hydrolysis of SP in the striatum. Subsequent experiments showed that fragment 1–8 is not rapidly degraded in the brain; thus, its low concentration was probably not the result of rapid degradation. Although much is still to be learned about the *in vivo* metabolism of SP and other neuropeptides, it is clear that the combination of microdialysis and mass spectrometry will play a major role in unraveling the chemical intricacies of such systems.

## 6. IN VIVO *STUDIES OF NAAG AND NAA*

N-acetylaspartate (NAA) and N-acetylaspartylglutamate (NAAG) are neurosubstances of the central nervous system, and recent reports have suggested that NAAG may be a neurotransmitter (34). Application of exogenous NAAG was shown to be capable of inducing depolarization of various neurocells, including lateral septal neurons, hippocampal pyramidal cells, and pyriform cortex pyramidal cells. Stimulation-evoked release of endogenous NAAG has been shown by *in vitro* perfusion of rat brain slices obtained from several regions. NAALADase, a membrane-bound hydrolytic enzyme that hydrolyzes NAAG to NAA and glutamate, has been implicated in the metabolic breakdown of synaptic NAAG upon its release (35,36). However, although NAA has been shown to be abundant in brain, its biological function remains unclear.

In studies designed to help elucidate the biological roles of NAA and NAAG, Lin *et al.* (Ref. 19) have used microdialysis and capillary column gas chromatography/mass spectrometry (GC-MS) to analyze samples directly from a live animal. GC-MS was used off-line for sensitive and molecularly specific simultaneous determinations of brain concentrations of NAA and NAAG. A 3-mm microdialysis probe (CMA/10, Carnegie Medicin) was stereotaxically inserted into the hypothalamus of anesthetized adult male rats. The perfusion medium was artificial CSF solution containing 5-mM KCl set at a flow rate of 1.2 μl/min. A solution of artificial CSF containing 400 mM KCl was used as a depolarization stimulus. Samples were collected at 30-min intervals to ensure adequate accumulation of neurochemicals for quantitative measurements. The relative percent recoveries by the microdialysis probe for NAA and NAAG were measured to be about 10% in *in vitro* studies. Both compounds were chemically modified to N-butyl esters by reaction with 3N HCl in *n*-butanol at 65 °C for 15 min. Separation of the biological sample was carried out using a 15-m DB-5 capillary column with MS detection by chemical ionization using ammonia as a reagent gas. Figure 6.8 shows selective ion chromatograms of NAA and NAAG from the rat brain perfusate from this experiment. Overall, the detection sensitivity for NAA was 10 times greater than that for NAAG, and the concentration ranges of the two components in the 30-min fractions ranged from 8–60 ng for NAA and 0.3–5 ng for NAAG. NAA-$d_6$ (50 ng) and NAAG-$d_3$ (200 ng) were added to the microdialysate as internal standards. Figure 6.9 shows the release of NAA and NAAG *in vivo.* Rats were infused with artificial CSF for 2 hr to establish baseline concentrations of NAA and NAAG. Then, infusion was continued with 400 mM KCl in artificial CSF for 24 h. Consistent stimulation-induced increase of NAA was obtained for all of five animals. On the other hand, over the group of animals tested, only two animals showed elevation of NAAG upon depolarization. These data agree with the reported slight increase in brain NAAG observed by *in vivo* methods (37) but are in contrast with the several-fold NAAG release observed by *in vitro* methods

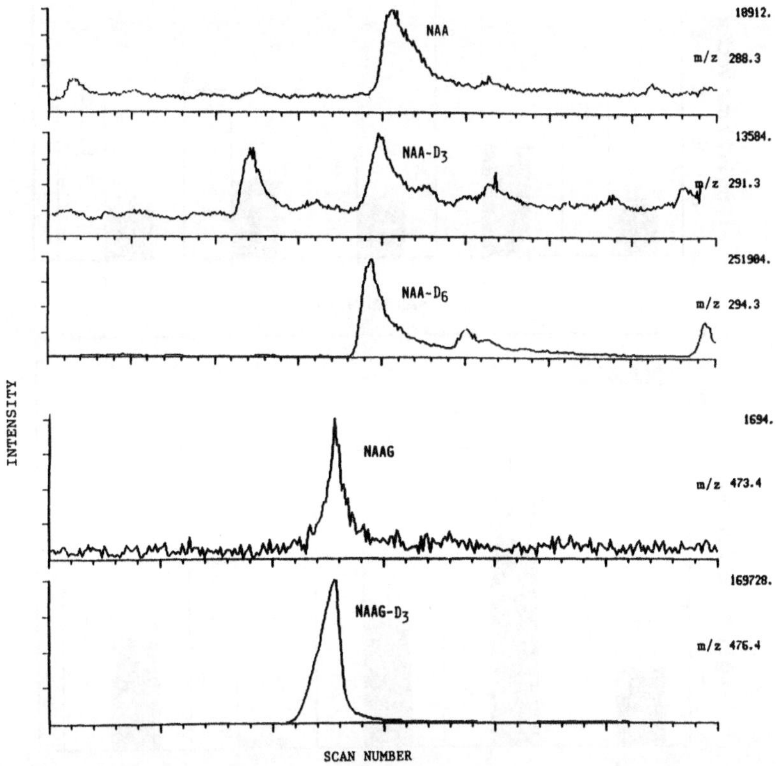


*Figure 6.8.* Analysis of brain levels of NAA and NAAG in the anesthetized rat by off-line microdialysis/GC-MS methods. The *m/z* for each of the selected ion chromatograms is shown to the right together with the relative ion intensity. The deuterated analogs were used as standards for quantitative purposes. For more details, see text and Ref. 19.

using rat brain slices (38). The assessment of the biochemical interrelationship between NAAG and NAA is a subject of great interest and importance in brain chemistry. It is likely that the distinctive and consistent increase in NAA levels upon depolarization observed in these studies may be the result of evoked release of a neurotransmitter or the product of the rapid cleavage of the stimulated-synaptic-released NAAG. Additional experiments are under way to further elucidate the biological roles of these compounds.

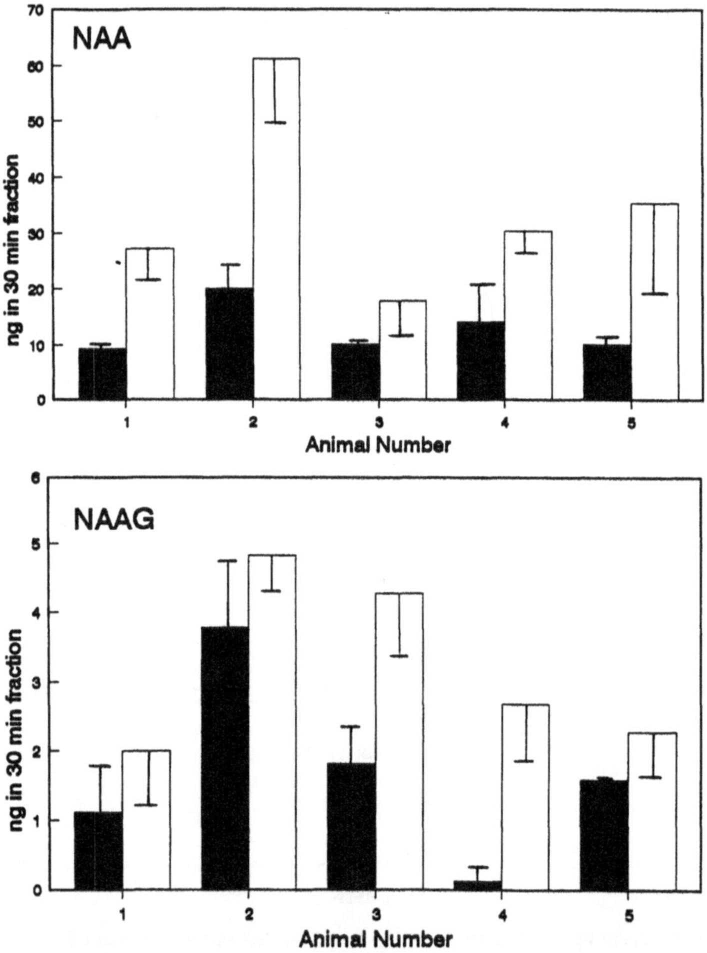


*Figure 6.9.* Release of NAA (top) and NAAG (bottom) in rat brain (hypothalamus) as measured by microdialysis/GC-MS methods. ■, Baseline values; □, 400 mM KCl. Depolarization was achieved using 400 mM KCl.

ACKNOWLEDGMENTS. The authors thank the National Institute of Health, the Swedish MRC, the Swedish Society of Medicine, and the Sweden–America Foundation for their support of some of the work presented here.

## *REFERENCES*

1. Gaddum, J. J., 1961, Push–pull cannulae, *J. Physiol.* **155:**1–2.
2. Pepeu, G., 1977, The release of acetylcholine from the brain: an approach to the study of central cholinergic mechanisms, *Prog. Neurobiol.* **2:**257–288.
3. Ungerstedt, U., 1984, Measurement of neurotransmitter release by intracranial dialysis, in *Measurement of Neurotransmitter Release* In Vivo (C. A. Marsden, ed.), Wiley, London, pp. 81–105.
4. Ungerstedt, U., and Pycock, C., 1974, Functional correlates of dopamine neurotransmission, *Bull. Schweiz. Akad. Med. Wiss.* **1278:**1–13.
5. Robinson, T. E., and Justice, J. B., Jr., eds., 1991, *Microdialysis in the Neurosciences,* Elsevier, Amsterdam.
6. Ungerstedt, U., 1986, Microdialysis—A new bioanalytical sampling technique, *Curr. Sep.* **7:**43–46.
7. Beneviste, H., 1989, Brain microdialysis, *J. Neurochem.* **52:**1667–1679.
8. Kendrick, K. M., 1989, Use of microdialysis in neuroendocrinology, in *Methods in Enzymology, Vol. 168* (P. M. Conn, ed.), Academic Press, San Diego, CA, pp. 182–205.
9. Amberg, G., and Lindefors, N., 1989, Intracerebral microdialysis: II. Mathematical studies of diffusion kinetics, *J. Pharmacol. Meth.* **22:**157–183.
10. Lindefors, N., Amberg, G., Ungerstedt, U., 1989, Intracerebral microdialysis: I. Experimental studies of diffusion kinetics, *J. Pharmacol. Meth.* **22:**141–156.
11. Bungay, P. M., Morrison, P. F., and Dedrick, R. L., 1990, Steady-state theory for quantitative microdialysis of solutes and water *in vivo* and *in vitro, Life Sci.* **46:**105–119.
12. Kendrick, K. M., 1990, Microdialysis measurement of *in vivo* neuropeptide release, *J. Neurosci. Meth.* **34:**35–46.
13. Collin, A. K., and Ungerstedt, U., 1988, *Microdialysis* Carnegie Medicin, p. 21.
14. Maidment, N. T., and Evans, C. J., 1991, Measurement of extracellular neuropeptides in the brain: microdialysis linked to solid-phase radioimmunoassays with sub-femtomole limits of detection, in *Microdialysis in the Neurosciences* (T. E. Robinson and J. B. Justice, Jr., eds.), Elsevier, Amsterdam, pp. 275–303.
15. Caprioli, R. M., and Lin, S. N., 1990, On-line analysis of penicillin blood levels in the live rat by combined microdialysis/fast-atom bombardment mass spectrometry, *Proc. Natl. Acad. Sci. U.S.A.* **87:**240–243.
16. Lin, S. N., Slopis, J. M., Andrén, P., Chang, S., Butler, I., and Caprioli, R. M., 1991, Microdialysis/MS in studies of valproic acid induced brain neurochemical changes, *Proc. 39th ASMS Conf. on Mass Spectrom. Allied Topics,* (R. M. Caprioli, ed.), Nashville, TN, May 1991, p. 583.
17. Deterding, L. J., Dix, K., Burka, L. T., and Tomer, K. B., 1992, On-line coupling of *in vivo* microdialysis with tandem mass spectrometry, *Anal. Chem.* **64:**2636–2641.
18. Yergy, J. A., Callaghan, D., Rousseau, P., and Masson, P., 1992, Direct measurement of eicosanoids in the respiratory system using microdialysis sampling and GC/MC detection, *Proc. 39th ASMS Conf. Mass Spectrom. Allied Topics* (A. Marshall, ed.), Nashville, May, 1991, p. 1011.
19. Lin, S. N., Slopis, J. M., Chang, S., Butler, I. J., and Caprioli, R. M., Microdialysis/mass spectrometry for *in vivo* studies on the release and metabolism of brain NAAG and NAA,

*Proc. 40th ASMS Conf. Mass Spectrom. Allied Topics* (A. Marshall, ed.), Washington, D.C., May, 1992, p. 1853.

20. Caprioli, R. M., 1990, Direct analysis of biological processes, in *Continuous-flow Fast Atom Bombardment Mass Spectrometry* (R. Caprioli, ed.), Wiley, New York, p. 91.
21. Kerns, E. H., Klohr, S. E., Volk, K. J., Lee, M. S., and Rosenberg, I. A., 1992, Microdialysis sampling combined with ionspray mass spectrometry for on-line reaction monitoring. *Proc. 40th ASMS Conf. Mass Spectrom. Allied Topics,* (A. Marshall, ed.), Washington, D.C., May 1992, p. 610.
22. Nicoll, R. A., Schenker, C., and Leeman, S. E., 1980, Substance P as a neurotransmitter candidate, *Ann. Rev. Neurosci.* **3**:227–268.
23. Pernow, B., 1983, Substance P, *Pharmacol. Rev.* **35**:85–141.
24. Krause, J. E., 1985, On the physiological metabolism of substance P, in *Substance P: Metabolism and Biological Actions* (C. C. Jordan and P. Oehme, eds.), Taylor and Francis, London, pp. 13–31.
25. Lee, C. M., Sandberg, B. E. B., Hanley, M. R., and Iversen, L. L., 1981, Purification and characterizationof a membrane-bound substance P degrading enzyme from human brain, *Eur. J. Biochem.* **114**:315–327.
26. Yokosawa, H., Endo, S., Oguro, Y., and Ishii, S. I., 1983, A new feature of angiotensin-converting enzyme in the brain: hydrolysis of substance P, *Biochem. Biophys. Res. Commun.* **116**:735–742.
27. Thiele, E. A., Strittmatter, S. M., and Snyder, S. H., 1985, Substance K and substance P as possible endogenous substrates of angiotensin converting enzyme in the brain, *Biochem. Biophys. Res. Comm.* **128**:317–324.
28. Matsas, R., Fulcher, I. S., Kenny, J., and Turner, A. J., 1983, Substance P and (Leu)enkephalin are hydrolyzed by an enzyme in pig caudate synaptic membranes that is identical with the endopeptidase of kidney microvilli, *Proc. Natl. Acad. Sci. U.S.A.* **80**:3111–3115.
29. Matsas, R., Kenny, A. J., and Turner, A. J., 1984, The metabolism of neuropeptides. The hydrolysis of peptides, including enkephalins, tachykinins and their analogues, by endopeptidase-24.11. *Biochem. J.* **223**:433–440.
30. Nyberg, F., LeGreves, P., Sundquist, C., and Terenius, L., 1984, Characterization of substance P (1–7) and (1–8) generating enzyme in human cerebrospinal fluid, *Biochem. Biophys. Res. Commun.* **125**:244–250.
31. Heymann, E., and Mentlein, R., 1978, Dipeptidyl peptidase IV hydrolyses of substance P, *FEBS Lett.* **91**:360–364.
32. Horsthemke, B., Schulz, M., and Bauer, K., 1984, Degradation of substance P by neurones and glial cells, *Biochem. Biophys. Res. Comm.* **125**:728–733.
33. Andrén, P. E., Emmett, M. R., DaGue, B. B., and Caprioli, R. M., 1993, Metabolism of substance P by *in vivo* microdialysis and LC/ESI/MS, *Proc. 41st ASMS Conference Mass Spectom. Allied Topics*, San Francisco, June 1993, p. 41a–41b.
34. Blakely, R. D., and Coyle, J. T., 1988, The neurobiology of N-acetylaspartylglutamate, *Int. Rev. Neurobiol.* **30**:39–100.
35. Slusher, B. S., Robinson, M. B., Tsai, G., Simmons, M. L., Richards, S. S., and Coyle, J. T., 1990, Rat brain N-acetylated α-linked acidic dipeptidase activity, *J. Biol. Chem.* **265**:21,297–21,301.
36. Slusher, B. S., Tsai, G., Yoo, G., and Coyle, J. J., 1992, Immunocytochemical localization of the N-acetyl-aspartyl-glutamate (NAAG) hydrolyzing enzyme N-acetylated α-linked acidic dipeptidase (NAALADase), *J. Comp. Neurol.* **315**:217–219.
37. Tsai, G., Stauch, B. L., Vornov, J. J., Deshpande, J. K., and Coyle, J. T., 1990, Selective release of NAAG from rat optic nerve terminals *in vivo, Brain Research* **518**:313–316.
38. Zollinger, M., Amsler, U., Do, K. Q., Streit, P., and Cuenod, M., 1988, Release of NAA on depolarization of rat brain slices, *J. Neurochem.* **51**:1919.

# Index

GPSR Compliance

*The European Union's (EU) General Product Safety Regulation (GPSR) is a set of rules that requires consumer products to be safe and our obligations to ensure this.*

*If you have any concerns about our products, you can contact us on ProductSafety@springernature.com*

In case Publisher is established outside the EU, the EU authorized representative is:

Springer Nature Customer Service Center GmbH
Europaplatz 3
69115 Heidelberg, Germany

**Batch number: 10761599**

Printed by Printforce, the Netherlands